Sustainability Solutions

An Interdisciplinary and Transdisciplinary Nexus

Sustainability Solutions addresses a wide range of global challenges with applied solutions in the interdisciplinary and transdisciplinary sustainability nexus. Titles in the series present a strong sustainable development focus, and topics which connect environment, economic and social applications are encouraged. Topics could include environmental impacts, vulnerability, risk assessment and reduction, human-environmental systems (e.g. food, water, health) and innovative solutions to climate change. Volumes in the series present novel solutions to societal challenges using innovative application, for example demonstrating a solutions-based approach using machine learning, artificial intelligence, decision support systems, advance geospatial technology. Case studies and shorter report volumes which present knowledge transfer for policymaking and governance are also encouraged.

Keywords:

Interdisciplinary environmental solutions, transdisciplinary knowledge transfer, geospatial technologies, natural resource management, decision support system, artificial intelligence, deep learning, smart technology, environmental science, economic, social and spatial, human-environment interactions, sustainable development

Md. Anwarul Abedin · Gulsan Ara Parvin · Nina Takashino · Rajib Shaw
Editors

Climate Crisis in South and East Asia

Analyzing Cross Cutting Issues and Sustainable Solutions from Gender Perspective

Springer

Editors
Md. Anwarul Abedin
Department of Soil Science
Bangladesh Agricultural University
Mymensingh, Bangladesh

Gulsan Ara Parvin
College of Policy Science, Osaka Ibaraki Campus
Ritsumeikan University
Osaka, Japan

Nina Takashino
College of Policy Science, Osaka Ibaraki Campus
Ritsumeikan University
Osaka, Japan

Rajib Shaw
Graduate School of Media and Governance
Keio University
Fujisawa, Japan

ISSN 3059-278X ISSN 3059-2798 (electronic)
Sustainability Solutions
ISBN 978-3-032-01011-7 ISBN 978-3-032-01012-4 (eBook)
https://doi.org/10.1007/978-3-032-01012-4

This Springer imprint is published by the registered company Springer Nature Switzerland AG
The registered company address is: Gewerbestrasse 11, 6330 Cham, Switzerland

Preface

The climate crisis is among the most pressing problems of the world today, and South and East Asia bear a disproportionate share of its impacts. It is no longer a threat for the future, but a fact of the present, it is rephrasing the social, economic, and ecological landscape of the world. Nowhere is this more visibly in evidence than in South and East Asia, countries famous for their ecological riches, socio-cultural diversity and developmental dynamism, yet equally famous for being afflicted with deeply ingrained vulnerabilities.

Within the above-mentioned context, this book, *Climate Crisis in South and East Asia: Analyzing Cross-Cutting Issues and Sustainable Solutions from Gender Perspective*, attempts to unravel the complexity of the interface between environmental degradation, social inequality, and gender disparity. It combines cross-disciplinary analysis and empirical findings to highlight the manner in which climate change differently affects women, men, and gender-diverse persons and also the imperative for equitable, gender-sensitive responses to sustainable adaptation and mitigation. The book comprises of around 18 chapters focusing on climate crisis, disaster issues and gender studies including gender dynamics, gender inclusion, gender mainstreaming, etc. in global south. This book is divided into four Parts: *Introduction, climate crisis and gender in south-Asia, climate crisis and gender in East Asia, and concluding remarks*. Part I comprises one chapter and highlights the intersection of climate crisis and gender, highlighting how environmental challenges disproportionately affect women and marginalized gender groups by examining case studies and global frameworks to promote equitable and sustainable climate solutions. Part II (eight chapters) discusses the gender responsive drought resilience, gender dynamics in climate induced disaster, multi-sectoral gender perspectives, gender inclusive strategies, gender focused social capital, and women leaded watershed management strategies in South-Asia. Part III (eight chapters) highlights the gender adaptive risk management, disaster management through gender-sensitive approaches, and gender transformative approach in East Asia. Part IV comprises only one concluding chapter which highlights the lessons, recommendations, and future action policies.

One of the central premises of this book is that sustainable climate solutions cannot be achieved without addressing gender inequality. Women and marginalized genders are not only victims of climate change—they are also powerful agents of change. Across South and East Asia, women are leading grassroots environmental movements, innovating climate-resilient farming techniques, managing community-based adaptation projects, and advocating for equitable climate policies. Recognizing and supporting this leadership is essential for building resilient and inclusive societies. This book is written for the students, researchers, practitioners, and policy makers in the fields of climate change, disaster risk reduction, and gender studies. We sincerely hope and believe that this book will help to connect all the stakeholders to enhance gender adaptive climate resilience in South and South East Asia. We also hope that this book will contribute meaningfully to the growing body of knowledge on gender and climate justice in Asia. We will be happy if the readers find the book useful and relevant to their work. Finally, we are thankful to several of our mentors, and we would like to dedicate this book to our parents who helped us to shaping our life.

Mymensingh, Bangladesh — Md. Anwarul Abedin
Osaka Ibaraki Campus, Japan — Gulsan Ara Parvin
Osaka Ibaraki Campus, Japan — Nina Takashino
Fujisawa, Japan — Rajib Shaw

About This Book

The intensity of climate change impacts has been observed all over the world because of extreme weather events and temperature fluctuations. It is predicted that climate change may push up to 158 million more women and girls into poverty and 236 million face more food insecurity by 2050. The climate crisis is far from "gender neutral." Women and men are affected differently by weather and climate, and therefore need gender-sensitive information and services. Furthermore, the climate crisis has a greater impact on women and often constrained by their gender roles. This is due to the unequal sharing of power between women and men, the gender gap in access to education and employment opportunities, the unpaid care burden, prevalence of gender-based violence, and all other forms of deep-rooted gender-based discrimination. Among them, lack of resources, less representation in climate related decision making, traditional gender roles; particularly the roles of caregiver and provider of food/water/ fuel, etc. boost gender more vulnerable to climate change. Women exhibit resilience and resourcefulness during response to recovery phases. They play a crucial but often unnoticed role in household resilience during disasters, employing various mitigation and adaptation strategies.

South and East Asia are the most affected countries that apprehend climate change crisis. With this contrast, this book attempts to insight of the genders' role and how they cope with these disasters happened through climate change. Moreover, a number of case studies from various countries in South and East Asia have been carried out/ illustrated here to understand the risks and challenges that climate change poses for women in South and East Asia. This book concludes with policy recommendations to mainstreaming gender in climate policies and strategies in the region, paving the way for a more inclusive and sustainable way where no one is left behind.

Contents

Editors and Contributors

About the Editors

Md. Anwarul Abedin is a professor and leader of Laboratory of Environment and Sustainable Development in the Department of Soil Science, Bangladesh Agricultural University, Bangladesh. He has got his Ph.D. from Kyoto University, Japan, and also did his JSPS-postdoc in Kyoto University and Commonwealth postdoc in Northumbria University, UK. Dr. Abedin was awarded the esteemed "Pan Asia Risk Reduction Fellowship" offered by START, USA, and carried out collaborative research among the BAU, START, and Kyoto University, Japan. He also worked as JSPS Invitational Fellow at College of Policy Science, Ritsumeikan University, Osaka-baraki Campus, Japan. He has 20 years teaching and research experiences in the field of agriculture, food safety and security and environmental issues viz. water resource management, solid waste management, environmental impact assessment, community-based adaptations and resilience mapping, climate change and disaster risk reduction, land use change, community participation. He has worked as consultant with FAO, JICA, IUCN, Oxfam, and GCRF. Dr. Abedin is collaborating in a number of research projects funded by DFID, NERC, British Council, USAID, JICA-JST, CARRIA, ESRC-UKRI, World Bank, Kurita Foundation, International Science Council, and China Government. Currently, he has research collaboration with UCDavis, USA; Salford University, UK; Sichuan University, China; Ritsumeikan University, Japan; IIT, Roorkee, India; AFU, Nepal; and many more. He has served as Editorial Board, reviewers in good number of national and international journal. He has published more than 120 research articles, books, and book chapters in reputed journals and publishers.

Gulsan Ara Parvin is a researcher and instructor of Collage of Policy Science, Ritsumeikan University, Japan. She served different graduate schools and research centers of Kyoto University, Japan, for about 12 years. She was an associate professor of Urban and Regional Planning Department of Bangladesh University of Engineering and Technology (BUET). She has conducted her Ph.D. from Urban Engineering Department of the University of Tokyo. Her expertise is in disaster risk reduction and community resilience, social capital, women empowerment, and microfinance in Bangladesh. She has published about 50 research papers in different journals and books. She has research and teaching experiences in Bangladesh, Japan, Malaysia, Thailand, and USA.

Nina Takashino is an associate professor in the College of Policy Science at Ritsumeikan University, Japan. She received her Ph.D. in Economics from Kobe University in 2007. Before joining Ritsumeikan University, she worked at Hokkaido University, the University of Tokyo, and Tohoku University. In 2018 and 2019, she spent her sabbatical as a visiting fellow at the International Gender Studies Centre at Oxford University and the Amsterdam Research Centre for Gender and Sexuality, funded by the Japanese government. Her research interests include gender and human security issues, development policy, poverty alleviation, and food policy. Her research activity is mainly devoted to analyzing people's behavior based on the field survey using questionnaires and deriving policy implications. She has research experience in Indonesia, India, Bangladesh, Nepal, Cambodia, the Philippines, Japan, and Cameroon.

Rajib Shaw a Japanese national of Indian origin, Rajib Shaw is the professor in Graduate School of Media and Governance in Keio University's Shonan Fujisawa Campus (SFC). He is also co-founder of a Delhi (India)-based social entrepreneur startup Resilience Innovation Knowledge Academy (RIKA) and RIKA Institute, a not-for-profit entity of RIKA. Professor Shaw is the recipient of prestigious "Pravasi Bharatiya Samman Award (PBSA)" in 2021 for his contribution in education sector. PBSA is the highest honor conferred on overseas Indians and persons of Indian origin from the President of India. Professor Shaw is also the recipient of United Nations Sasakawa Award for Disaster Risk Reduction in 2022 for his lifetime contributions in the field of disaster risk reduction. Sasakawa Award is the highest UN award for disaster risk reduction and was established by 1986 by the founding Chairman of the Nippon Foundation, Mr. Ryoichi Sasakawa. He is also the senior fellow of Institute of Global Environmental Strategies (IGES) Japan and the chairperson of SEEDS Asia and CWS Japan, two Japanese NGOs, and a distinguished professor of Institute of Disaster Management and Reconstruction (IDMR) of Sichuan University, China. Earlier, he was the executive director of the Integrated Research on Disaster Risk (IRDR), a decade-long research program co-sponsored by the International Council for Science (ICSU), the International Social Science Council (ISSC), and the United Nations International Strategy for Disaster Reduction (UNISDR). Previously, he served as a professor in the Graduate School of Global Environmental Studies of Kyoto University. His expertise includes community-based disaster risk management, climate change adaptation, urban risk management, and disaster and environmental education. Professor Shaw was the Chair of the United Nations Science Technology Advisory Group (STAG) for disaster risk reduction and is the co-chair of the Asia Pacific Science Technology Advisory Group (AP-STAG). He was also the CLA (coordinating lead author) for the IPCC 6th Assessment Report on Climate Change (Working Group II, Asia Chapter). He is the editor of a book series on disaster risk reduction, published by Springer, editor-in-chief of Progress in Disaster Science, a journal published by Elsevier. Professor Shaw has published more than 61 books and over 400 academic papers and book chapters.

Contributors

A. K. M. Abdul Ahad Biswas Department of Disaster Risk Management, Faculty of Environmental Science and Disaster Management, Patuakhali Science and Technology University, DumkiPatuakhali, Bangladesh

Md. Anwarul Abedin Department of Soil Science, Faculty of Agriculture, Bangladesh Agricultural University, Mymensingh, Bangladesh

Basanta Raj Adhikari Department of Civil Engineering, Pulchowk Campus, Institute of Engineering, Tribhuvan University, Lalitpur, Nepal

Md. Mazharul Anwar Department of Emergency Management, Faculty of Environmental Science and Disaster Management, Patuakhali Science and Technology University, Patuakhali, Bangladesh

Ahmad Shekib Arab National Agriculture and Food Research Organization, Tsukuba, Ibaraki, Japan

Sara Tokhi Arab Division of Intelligent Agriculture Machinery Research, Institute of Agriculture Machinery, National Agriculture and Food Research Organization, Kannondai, Tsukuba, Ibaraki, Japan

Shazmin Shareena Ab. Azis Faculty of Built Environment & Surveying, Real Estate, Universiti Teknologi Malaysia, Skudai, Johor, Malaysia

Nursyuhaida Aziz Faculty of Built Environment & Surveying, Real Estate, Universiti Teknologi Malaysia, Skudai, Johor, Malaysia

Shastitharran Baskaran Faculty of Built Environment & Surveying, Real Estate, Universiti Teknologi Malaysia, Skudai, Johor, Malaysia

Ranit Chatterjee Resilience Innovation Knowledge Academy, Delhi, India

Chealy Chet Royal University of Phnom Penh (RUPP), Toul Kork, Phnom Penh, Cambodia

Ambika Dabral Resilience Innovation Knowledge Academy, Delhi, India

Thuy Dung Dang School of Languages and Tourism, Hanoi University of Industry, Hanoi, Vietnam

Jessica Dator-Bercilla Ateneo de Manila University, Manila, Philippines

Sanjaya Devkota Institute of Himalayan Risk Reduction (IHRR), Patan, Nepal

Ranjan Dhungel Department of Civil Engineering, Pulchowk Campus, Institute of Engineering, Tribhuvan University, Lalitpur, Nepal

Md. Esraz-Ul-Zannat Department of Urban and Regional Planning, Faculty of Civil Engineering, Khulna University of Engineering & Technology, Khulna, Bangladesh

Aprillia Findayani Geography Department, Universitas Negeri Semarang, Semarang, Indonesia;
U-Inspire Indonesia, Jakarta, Indonesia

Suraj Gautam Institute of Himalayan Risk Reduction (IHRR), Patan, Nepal

Wittachchi Koralalage Athula Chammika Gnanapala Faculty of Management Studies, Sabaragamuwa University of Sri Lanka, Belihuloya, Sri Lanka

Shehla Gul Department of Geography and Geomatics, University of Peshawar, Peshawar, Pakistan

Huy Ngoc Ha Institute of Vietnam and World Economic, Vietnam Academy of Social Sciences, Hanoi, Vietnam

Md. Fuad Hasan Department of Emergency Management, Faculty of Environmental Science and Disaster Management, Patuakhali Science and Technology University, Patuakhali, Bangladesh;
Department of Environmental Science and Technology, German University Bangladesh, Gazipur, Bangladesh

Md. Hosenuzzaman Department of Soil Science, Bangladesh Agricultural University, Mymensingh, Bangladesh;
UWA School of Agriculture and Environment, The University of Western Australia, Perth, WA, Australia

Kamrunnahar Ishana Department of Soil Science, Government Brojomohun College, National University, Barishal, Bangladesh

Mohammad Kabirul Islam Department of Soil Science, Faculty of Agriculture, Patuakhali Science and Technology University, DumkiPatuakhali, Bangladesh

Md. Jashim Uddin Department of Social Work, Government Brojomohun College, National University, Barishal, Bangladesh

Sudeshika Jayasekara Faculty of Agricultural Sciences, Sabaragamuwa University of Sri Lanka, Belihuloya, Sri Lanka

Juhadi Geography Department, Universitas Negeri Semarang, Semarang, Indonesia

Sadiya Arefin Juthee Department of Crop Botany, Bangladesh Agricultural University, Mymensingh, Bangladesh

Swarna Bintay Kadir Institute of Disaster Management, Khulna University of Engineering & Technology, Khulna, Bangladesh

Suvendrini Kakuchi Resilience Innovation Knowledge Academy, Delhi, India

Malavika Kanathoor Indian Institute of Technology Roorkee, Roorkee, India

Minakshi Keeni Graduate School of Agricultural Science, Tohoku University, Sendai, Japan

Chansopheak Keng Royal University of Phnom Penh (RUPP), Toul Kork, Phnom Penh, Cambodia

A. B. M. Mahbub Morshed Khan Department of Agricultural Botany Faculty of Agriculture, Faculty of Agriculture, Patuakhali Science and Technology University, Dumki, Patuakhali, Bangladesh

Bonjun Koo The Great Hanshin-Awaji Earthquake Memorial Disaster Reduction and Human Renovation Institution, Kobe, Japan

Monu Kumari Resilience Innovation Knowledge Academy, Delhi, India;
Indian Institute of Technology Roorkee, Roorkee, India

Harshit Sosan Lakra Indian Institute of Technology Roorkee, Roorkee, India

Sainan Li Institute for Disaster Management and Reconstruction, Sichuan University—The Hong Kong Polytechnic University, Chengdu, China

Puja Maharjan Department of Civil Engineering, Pulchowk Campus, Institute of Engineering, Tribhuvan University, Lalitpur, Nepal

Md. Sagirul Islam Majumder Department of Soil Science, Faculty of Agriculture, Patuakhali Science and Technology University, Dumki, Patuakhali, Bangladesh

Surabhi Mehrotra Maulana Azad National Institute of Technology, Bhopal, India

Sonajharia Minz Jawaharlal Nehru University, New Delhi, India

Rina Suryani Oktari Family Medicine Department, Medical Faculty Syiah Kuala University, Banda Aceh, Indonesia;
Tsunami and Disaster Mitigation Research Center (TDRMC), Banda Aceh, Indonesia

Gulsan Ara Parvin College of Policy Science, Ritsumeikan University, Osaka, Japan

Li Peng College of Geography and Resource Science, Sichuan Normal University, Chengdu, China

Emma Porio Ateneo de Manila University, Manila, Philippines

Shobha Poudel Science Hub, Kathmandu, Nepal

Atta-ur Rahman Department of Geography and Geomatics, University of Peshawar, Peshawar, Pakistan

Nur Hannani Ab Rahman Faculty of Built Environment & Surveying, Real Estate, Universiti Teknologi Malaysia, Skudai, Johor, Malaysia

Ranathunga Arachchillage Danushka Chandimal Ranathunga Faculty of Management Studies, Sabaragamuwa University of Sri Lanka, Belihuloya, Sri Lanka

Md. Rasheduzzaman Department of Emergency Management, Faculty of Environmental Science and Disaster Management, Patuakhali Science and Technology University, Patuakhali, Bangladesh

Md. Saiful Islam Department of Soil Science, Faculty of Agriculture, Patuakhali Science and Technology University, DumkiPatuakhali, Bangladesh

Debashish Sarker Geodetic Detachment, Survey of Bangladesh (SoB), Tejgaon, Dhaka, Bangladesh

Sanjida Hossain Setu Department of Emergency Management, Faculty of Environmental Science and Disaster Management, Patuakhali Science and Technology University, Patuakhali, Bangladesh

Md. Shamsuzzoha Department of Emergency Management, Faculty of Environmental Science and Disaster Management, Patuakhali Science and Technology University, Patuakhali, Bangladesh

Rajib Shaw Graduate School of Media and Governance, Keio University, Endo, Fujisawa, Japan;
Resilience Innovation Knowledge Academy, Delhi, India;
Keio University, Tokyo, Japan

Kazi Farzana Shumi Bangladesh University of Professional (BUP), Dhaka, Bangladesh

Shweta Sinha Pridi Banomyong International College, Thammasat University, Bangkok, Thailand

Serey Sok Royal University of Phnom Penh (RUPP), Toul Kork, Phnom Penh, Cambodia

Nina Takashino College of Policy Science, Ritsumeikan University, Osaka, Japan

Leakhena Thoeun Royal University of Phnom Penh (RUPP), Toul Kork, Phnom Penh, Cambodia

Bingwei Tian Institute for Disaster Management and Reconstruction, Sichuan University—The Hong Kong Polytechnic University, Chengdu, China

Thi My Thi Tong Institute of Vietnam and World Economic, Vietnam Academy of Social Sciences, Hanoi, Vietnam

Thi Tuyet Tran Institute of Human Geography and Sustainable Development, Vietnam Academy of Social Sciences, Hanoi, Vietnam

Noralene Uy Ateneo de Manila University, Manila, Philippines

Deepthi Wickramasinghe University of Colombo, Colombo, Sri Lanka

Part I
Introduction

Chapter 1
Introduction to Climate Crisis and Gender

Md Anwarul Abedin, Sadiya Arefin Juthee, Md Hosenuzzaman, Md Shamsuzzoha, Gulsan Ara Parvin, Nina Takashino, and Rajib Shaw

Abstract The climate crisis, driven by global warming, environmental degradation, and resource depletion, exacerbates social inequalities, particularly in communities already vulnerable to climate change. Despite the growing recognition of the experiences and skills women contribute to development and environmental sustainability efforts, women continue to exhibit lower capabilities and heightened exposure to climate vulnerabilities due to their diminished influence in politics, the economy, and the law. Women in South and East Asia encounter distinctive challenges, including constrained access to resources, displacement, and elevated susceptibility to climate-related disasters. The chapter analyzes how gendered power dynamics influence the impacts of and responses to climate change and highlights the importance of integrating gender-sensitive approaches into climate policy. This chapter explores the intersection of the climate crisis and gender, highlighting how environmental challenges disproportionately affect women and marginalized gender groups, and examines case studies and global frameworks for promoting equitable and sustainable climate solutions.

Keywords Climate · Gender · Gender-sensitive approach · South and East Asia

M. A. Abedin (✉) · M. Hosenuzzaman
Department of Soil Science, Bangladesh Agricultural University, Mymensingh, Bangladesh
e-mail: m.a.abedin@bau.edu.bd

S. A. Juthee
Department of Crop Botany, Bangladesh Agricultural University, Mymensingh, Bangladesh

M. Hosenuzzaman
UWA School of Agriculture and Environment, The University of Western Australia, Perth, WA, Australia

M. Shamsuzzoha
Department of Emergency Management, Patuakhali Science and Technology University, DumkiPatuakhali, Bangladesh

G. A. Parvin · N. Takashino
College of Policy Science, Ritsumeikan University, Osaka Ibaraki Campus, Osaka, Japan

R. Shaw
Graduate School of Media and Governance, Keio University, Fujisawa, Japan

Md. A. Abedin et al. (eds.), *Climate Crisis in South and East Asia*, Sustainability Solutions, https://doi.org/10.1007/978-3-032-01012-4_1

1.1 Introduction

Climate change presents humanity with an unprecedented and pressing problem (IPCC, 2023). Global temperatures have risen by about 1.5 °C during the last 150 years due to a significant increase in greenhouse gas concentrations in the atmosphere (Hirt et al., 2023). Global food security and agricultural output are seriously threatened by the ensuing changes in climate dynamics, which show up as altered rainfall patterns, an increase in extreme weather events, and increased seasonal unpredictability. When taken as a whole, these developments threaten the stability of human societies and the institutions that support them (FAO, 2024a).

One of the biggest problems of our day is the growing climate crisis, which is marked by increasing global temperatures, accelerating environmental deterioration, and the relentless loss of natural resources (Dankelman & Naidu, 2020; Rao et al., 2019). The consequences of this phenomenon are widespread but not evenly distributed (Dankelman & Naidu, 2020). Rather, by adding to already existing social and economic disparities, the crisis unfairly and disproportionately burdens marginalized communities. Women and gender-diverse people are disproportionately affected by climate change in places like South and East Asia, especially where heightened sensitivity to climate change coexists with long-standing gender inequities (Castelo et al., 2024).

Even though the importance of women for environmental preservation and sustainable development is becoming more widely acknowledged, structural obstacles like their lack of political representation, economic marginalization, and legal exclusion still make it difficult for them to respond to and recover from climate-related shocks (Fernandez, 2023). These differences are not only coincidental; rather, they result from gendered power dynamics that influence decision-making, adaptation options, and resource access (Brady et al., 2022).

The complex relationship between gender and climate change is examined in this chapter, which also explains how environmental stresses can exacerbate preexisting gender-based vulnerabilities. The systemic basis of these disparities and the urgent need to incorporate gender-sensitive practices into climate governance are further explored in the text (Fig. 1.1). The conversation highlights the significance of inclusive climate action that considers the diverse experiences of women and marginalized gender groups, drawing on global frameworks, policy assessments, and regional case studies (FAO, 2024b). The causes and effects of gendered climate impacts are examined in this chapter. It aims to support a more resilient and fair future course for dealing with climate emergencies.

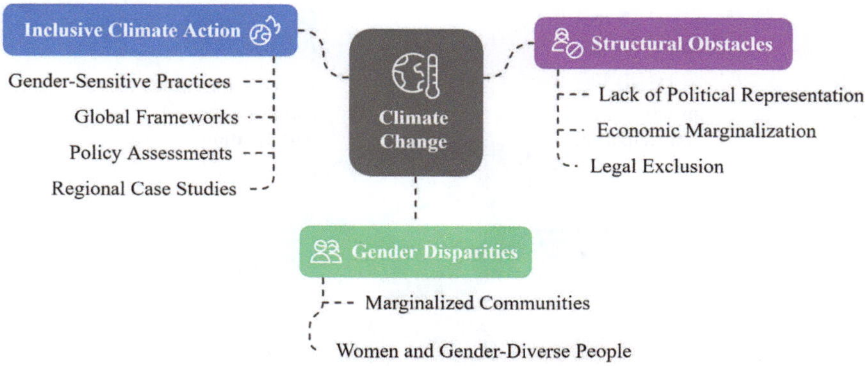

Fig. 1.1 Climate change and gender disparities

1.2 Climate Crisis and Gender Nexus

The complex relationship between gender and climate change is examined in this chapter, which also explains how environmental stresses can exacerbate preexisting gender-based vulnerabilities (UN Women, 2021). The systemic basis of these disparities and the urgent need to incorporate gender-sensitive practices into climate governance are further explored in the text. The conversation highlights the significance of inclusive climate action that considers the diverse experiences of women and marginalized gender groups, drawing on global frameworks, policy assessments, and regional case studies. The causes and effects of gendered climate impacts are examined in this chapter. It aims to support a more resilient and fair future course for dealing with climate emergencies.

Women are often the principal administrators of household energy, food, and water resources in different countries of the world (Naz & Doneys, 2022; Tewari et al., 2023; WB, 2021). As a result, they are among the first to feel the effects of changing climatic patterns and environmental deterioration. But even though they have been on the front lines and have contributed to sustainable practices, women often do not have access to the financial resources, land rights, and decision-making platforms that are essential for successful climate adaptation (Bryan et al., 2024a; FAO, 2020). Their ability to adapt to climate shocks is hampered by this structural exclusion, making them more susceptible to livelihood loss, food poverty, and displacement (Bradshaw & Fordham, 2015; Deininger et al., 2023; Erman et al., 2021).

Women's independence and fortitude are further limited in South and East Asia by institutional impediments and patriarchal customs. Women and girls are particularly vulnerable to health hazards, disruptions in their schooling, and even gender-based violence during climate-related disasters. These gender-based vulnerabilities are the result of long-standing power disparities in the political, economic, and legal institutions rather than occurring on their own (Ravera et al., 2016).

Fig. 1.2 Empowering women in climate action

Therefore, while developing inclusive and successful climate initiatives for climate action, it is essential to take the gender-climate relationship into account (Bryan et al., 2024b; FAO, 2022; UN, 2024b). The recognition of these distinct effects and the empowering of women as climate change agents in environmental governance are the foundations of gender-sensitive methods (Fig. 1.2). It is not only a question of justice when gender equity is incorporated into climate policy; it is a necessary precondition for the creation of resilient, adaptive, and sustainable communities (Aziz & Anjum, 2024; Mirza, 2003). To highlight the need to place gender at the center of climate discourse and action, this chapter examines these links using data from case studies and international frameworks (Wright et al., 2019).

1.3 Common Climatic Events

Frequent and more severe weather occurrences are signs of the growing climate crisis, which poses serious risks to ecosystems and human society (Fig. 1.3). Extreme weather events, such as heat waves, floods, droughts, cyclones, and sea level rise, are among the most common of these (Shamsuzzoha et al., 2021, 2022). Vulnerable groups may be severely and frequently irrevocably impacted by these occurrences. Due to environmental degradation brought on by irresponsible human activity and global warming, these occurrences are not only growing more often but also more severe (Kempa & Tyagi, 2025; Smith & Smith, 2013).

It has been demonstrated that droughts and floods have detrimental effects on food security, water availability, and agricultural livelihoods (Beer, 2018; Binte Mostafiz et al., 2021). Women, particularly those living in rural areas, are heavily involved in these fields. Monsoon fluctuations and extended dry spells directly affect women's ability to get water, grow food, and provide for their families in South and East Asia. Like this, tropical cyclones and coastal surges, which are becoming more severe due to rising sea temperatures, displace millions of people, interfere with necessary services, and exacerbate gender inequality by denying women access to resources for emergency relief, medical attention, and rehabilitation (Ahamed et al., 2021; Ginige et al., 2014).

Wildfires and heat waves are significant climate-related issues. Inadequate shelter, restricted mobility, and limited access to cooling facilities increase heat-related health concerns that are more common in women, especially pregnant and elderly women (Rahman et al., 2024). Additionally, social and health services are often disrupted by climate-related disasters, making women more susceptible to gender-based abuse and exploitation in makeshift shelters and camps for displaced people (Naim et al., 2024; Sadia et al., 2016).

These frequent climatic catastrophes highlight how urgent it is to understand and treat the distinct effects they have on people of different genders. Social inequality will continue to grow if inclusive climate resilience planning that is based on the

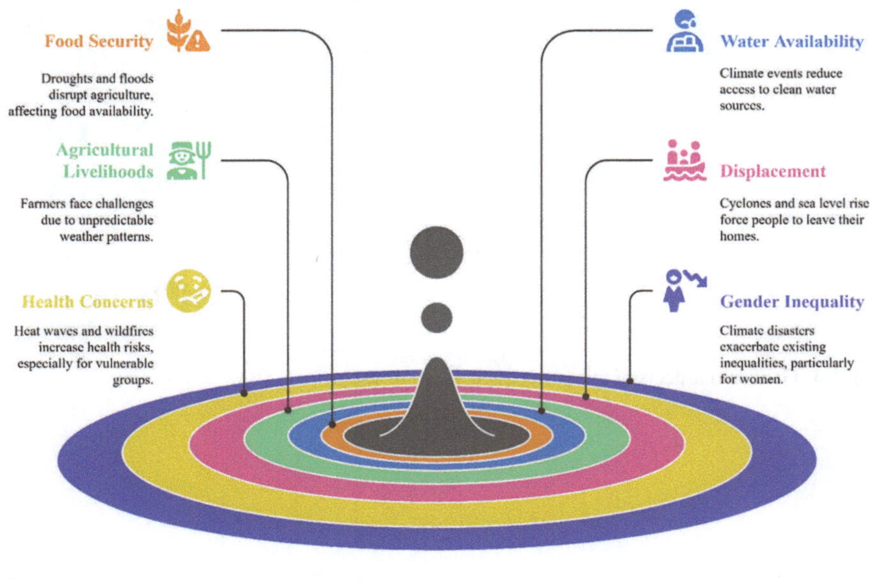

Fig. 1.3 Climate crisis impacts on gender

needs and experiences of women and other excluded groups is not implemented (Fig. 1.3). To build just, equitable, and sustainable adaptive capacities, it is essential to recognize the gendered aspects of climatic crisis induced disasters (Fig. 1.4).

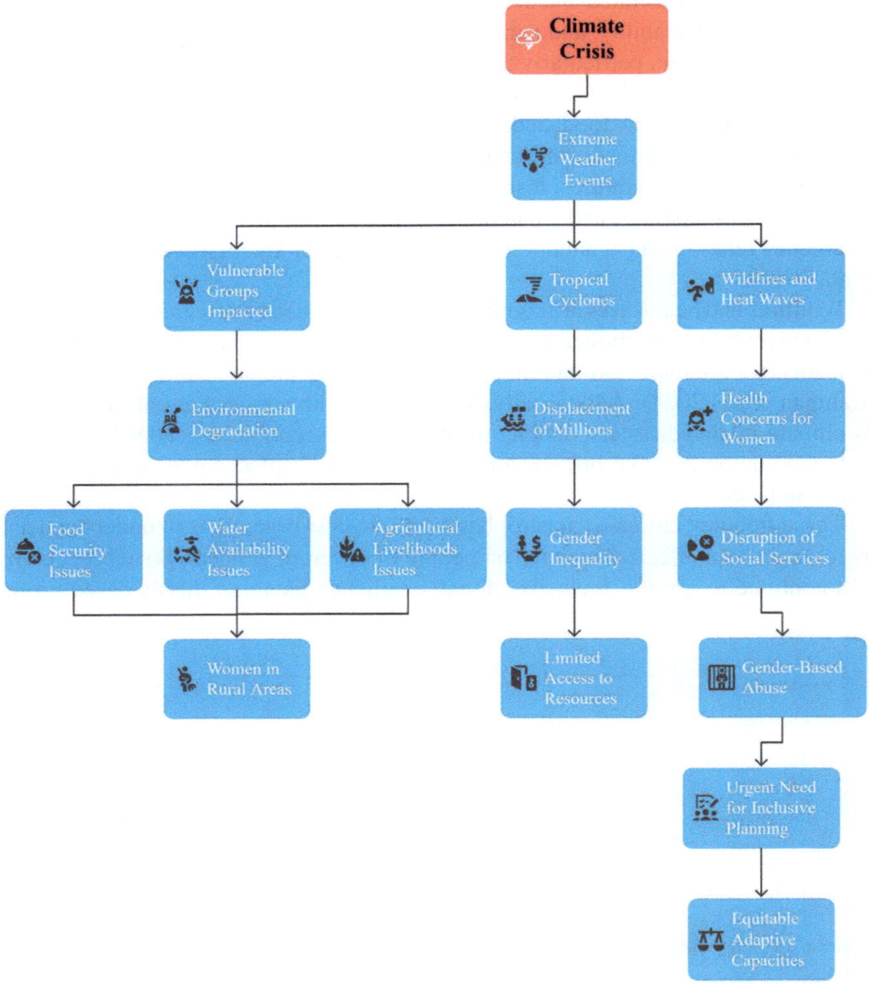

Fig. 1.4 Gendered aspects of climate crisis induced disasters

1.4 Gender and Environmental Indicator

For assessing the varying effects of climate change on various population segments, especially women and marginalized gender groups, gender and environmental indicators are essential tools. These indicators make it easier to quantify how social factors like gender, wealth, health, and education interact with environmental stresses, including resource shortages, ecosystem degradation, and climate-induced disasters (Naz et al., 2018; Sadia et al., 2016). We can learn more about how climate change exacerbates existing disparities and restricts the ability of disadvantaged groups to adapt effectively by looking at these occurrences (Fig. 1.5).

The gender gap in access to natural resources, such as clean energy, water, and land, is a prominent sign of this inequality. Women are heavily dependent on these resources for household and agricultural needs in the world, especially in South and East Asia (Kadetz & Mock, 2018; Yumarni et al., 2014). Nevertheless, their ownership rights and decision-making power over their use are restricted. The representation of women in environmental governance and policymaking is another important metric. Women often hold a disproportionate number of leadership roles in decision-making organizations that make important environmental legislation, despite their significant participation in community-based sustainability projects (Ikeda, 1995).

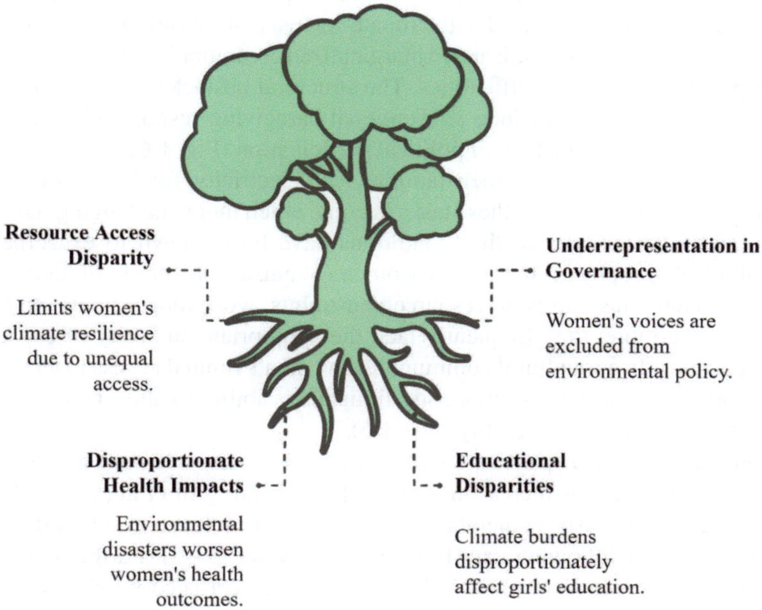

Fig. 1.5 Inadequate integration of gender-specific environmental indicators hinders effective climate change adaptation for vulnerable populations

Another important indicator, health outcomes, shows that after environmental disasters, women are more likely to experience respiratory illnesses, mental health problems, and hunger. The effect of environmental deterioration on girls' education is also shown in educational indicators, especially in rural areas where they are frequently disproportionately responsible for climate-related tasks like fetching water or taking care of displaced family members (Shanta et al., 2025).

Policymakers and academics can gain a deeper understanding of the complex vulnerabilities that impact people's experiences of climate change by incorporating gender-specific environmental indicators into climate assessments. These understandings are essential for creating focused interventions that advance gender parity, strengthen resilience, and guarantee inclusive and successful climate solutions (Graveline & Germain, 2022; Haque et al., 2021).

1.5 Gendered Vulnerabilities to Climate Change

Not every population is equally impacted by climate change (Cutter et al., 2010; MoEFCC, 2022b). Existing social, political, and economic frameworks frequently make these issues worse, which increases systemic inequality. Gender is an especially principal factor in determining vulnerability among these factors (Suresh et al., 2025; Wyles et al., 2014). When faced with climate-related disruptions, women, especially those living in the Global South and in marginalized communities, often face a unique and disproportionate set of difficulties. The structural obstacles that underline these gendered vulnerabilities include conventional caregiving responsibilities, unequal access to resources, and a lack of political participation (Fig. 1.6).

Women oversee providing their families with electricity, food, and water. When there is environmental stress, these tasks become much more challenging. Droughts, floods, and other severe weather conditions have been shown to exacerbate the psychological and physical strain on women (Shamsuzzoha et al., 2025a). This is because women must adjust to resource constraints, work longer hours, and travel farther. Furthermore, they frequently lack the appropriate authority or capacity to make decisions. In agricultural communities, women's limited access to technology, credit, and land ownership significantly limits their ability to alter their livelihoods and respond to climatic variability (Fig. 1.6).

Significant gender disparities are also revealed by displacement brought on by catastrophic disasters or increasing sea levels (Ariyaningsih et al., 2025). Women and girls are particularly vulnerable to gender-based violence, food hardship, and inadequate shelter following catastrophes (Pothiawala, 2015). Furthermore, it has been shown that preexisting health disparities are exacerbated by climate change. For example, pregnant women are more vulnerable to vector-borne and heat-related illnesses, which are increasing in frequency because of changing climatic patterns (Anwarul Islam et al., 2019).

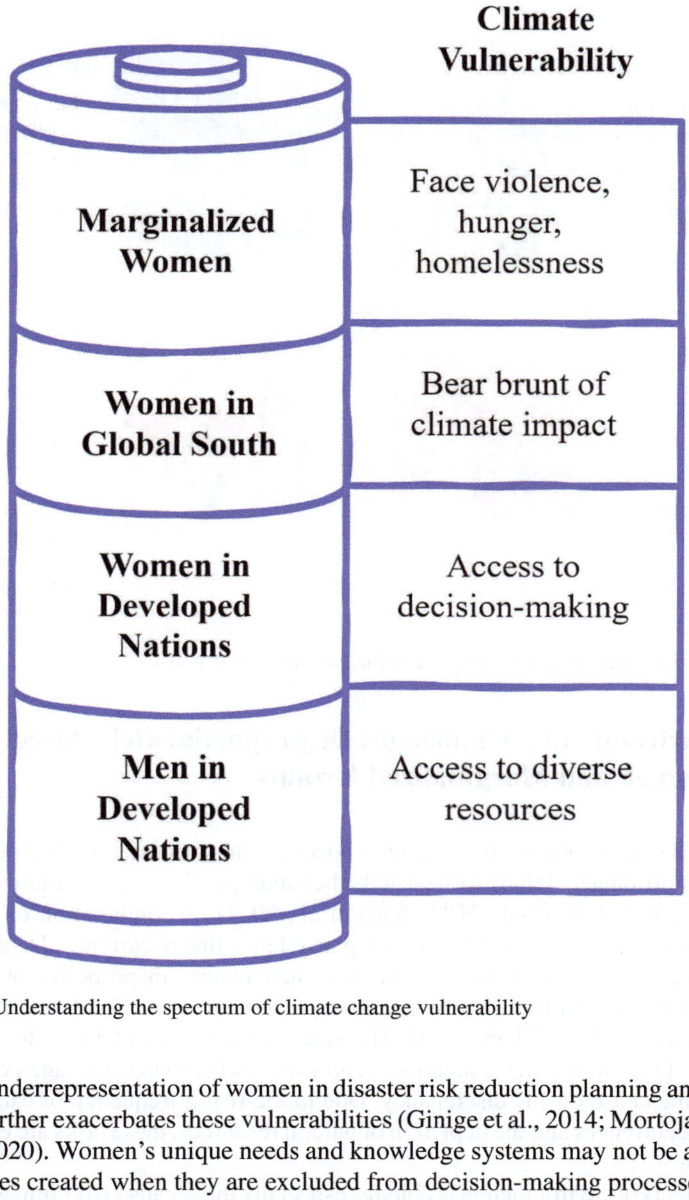

Fig. 1.6 Understanding the spectrum of climate change vulnerability

The underrepresentation of women in disaster risk reduction planning and climate policy further exacerbates these vulnerabilities (Ginige et al., 2014; Mortoja & Yigitcanlar, 2020). Women's unique needs and knowledge systems may not be addressed by policies created when they are excluded from decision-making processes, which could result in ineffective or harmful interventions (Fig. 1.7). These gendered aspects must be recognized and addressed since they are crucial to the creation of resilient communities and are not just a social justice issue (Aldrich, 2012; Faas & Jones, 2017). To effectively address the climate problem and make sure that no one is left behind in the pursuit of sustainability, inclusive, gender-sensitive measures are essential.

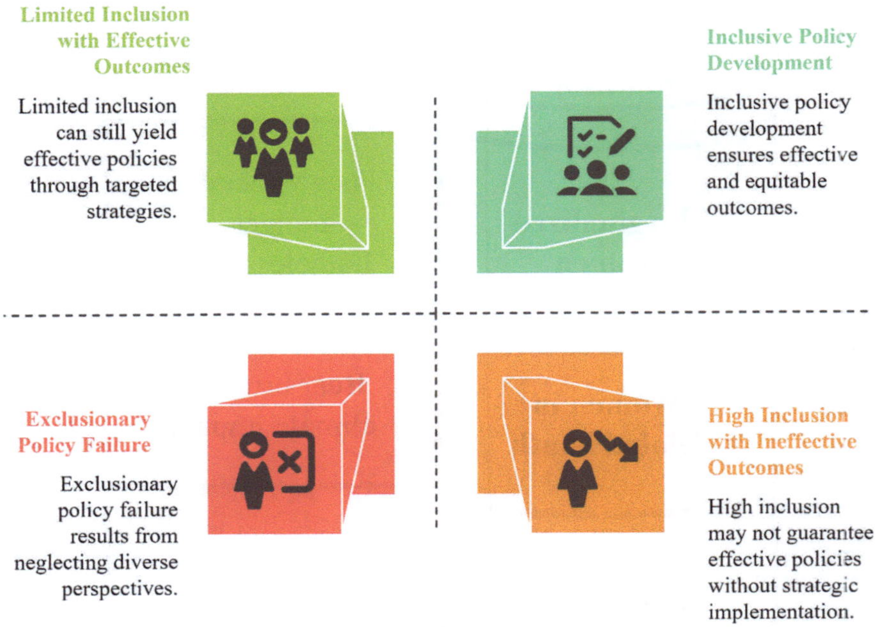

Fig. 1.7 Gender inclusion in disaster risk reduction and climate policy

1.6 Environmental Challenges Disproportionately Affect Women and Marginalized Groups

The effects of environmental problems, especially those caused by climate change, are disproportionately felt by women and other underprivileged communities (Barenstein et al., n.d.; Huq et al., 2015; Saha et al., 2024). A combination of political, cultural, and socioeconomic forces is responsible for this occurrence. Disasters and environmental degradation brought on by climate change disproportionately affect women, especially in rural and resource-dependent areas of South and East Asia (Fernando et al., 2025; Khan, 2024). The gendered division of labor, which assigns women to tasks like water collection, crop care, and resource management in the home, is the cause of this discrepancy. Due to the rising frequency of catastrophic weather occurrences and the depletion of natural resources, these tasks are especially vulnerable.

The effects of environmental problems, especially those caused by climate change, are disproportionately felt by women and other underprivileged communities. A combination of political, cultural, and socioeconomic forces is responsible for this occurrence. Disasters and environmental degradation brought on by climate change disproportionately affect women, especially in rural and resource-dependent areas of South and East Asia. The gendered division of labor, which assigns women to tasks like water collection, crop care, and resource management in the home, is the cause of this discrepancy (FAO, 2020; Ravera et al., 2016; UNW, 2024). Due to the rising frequency of catastrophic weather occurrences and the depletion of natural resources, these tasks are especially vulnerable.

Furthermore, displacement is a common outcome of climate-related disasters, including floods, droughts, and storms, which increases the vulnerability of women and vulnerable groups. Women are more likely to experience gender-based violence when they are displaced, according to research. Additionally, because women in refugee camps or temporary shelters have extremely limited access to healthcare, education, and career possibilities, displacement has been shown to increase their vulnerability to economic and health risks (Sphere Association, 2004, 2018). In crises, marginalized groups—such as low-income groups, persons with disabilities, and Indigenous communities—face even more obstacles when trying to obtain resources and relief in catastrophic situations. These groups lack the social and political capital to effectively advocate for their demands and are frequently left out of decision-making processes (Adger, 2003, 2010; Paul et al., 2016).

In addition to environmental variables, gendered power dynamics and systemic inequalities also contribute to the disproportionate impact of climate change on women and vulnerable groups. These populations have less access to financial support systems and climate adaptation technology, and they are more likely to be shut out of leadership roles in climate governance (Nasreen, 2021; Pal & Bhatia, 2018; Shamsuzzoha et al., 2024). As a result, their abilities to address environmental issues are usually hindered, and their experiences are commonly disregarded in international debates on climate policy.

Improving climate resilience requires addressing these gendered inequities. To develop more equitable and successful responses to the climate catastrophe, it is essential to acknowledge and include the unique needs, experiences, and capacities of women and marginalized groups into climate policy and adaptation plans (GoB, 2020; MoEFCC, 2022b). In addition to advancing gender equality, this strategy improves the general viability and effectiveness of initiatives to mitigate and adapt to climate change (Fig. 1.8).

Fig. 1.8 Disproportionate effect on women and marginalized groups due to climate change

1.7 Women's Empowerment is Key to Success of Climate Actions

In addition to international cooperation, tackling the climate problem calls for an inclusive strategy that recognizes and capitalizes on the critical roles that women play in sustainability and environmental management (Brouwer et al., 2007; Das et al., 2021). Despite their disproportionate susceptibility to the effects of climate change, women have special knowledge and abilities that can enhance climate resilience, especially in rural and vulnerable communities (Mitu Akter et al., 2024). Nonetheless, women's underrepresentation in local, national, and international decision-making processes has the impact of limiting their ability to fully contribute to climate action.

To achieve long-term, sustainable climate solutions, women's empowerment is essential. Often, women take on the major responsibility for managing natural resources, including family and community-level farming techniques, water management, and energy use (Kim et al., 2024). Their participation in environmental decision-making, especially in areas like agriculture, water management, and disaster preparedness, can lead to more effective, context-specific responses to climate change (Ersel, 2015). Furthermore, because women's interests tend to prioritize sustainability, community well-being, and equitable resource allocation, female leadership in climate adaptation and mitigation projects often produces better results.

Improving women's empowerment can significantly increase climate resilience in areas like South and East Asia, where women face barriers including limited access to land, education, and financial resources (Ahamed et al., 2021; UN Women, 2021). It has been shown that giving women access to climate-related information, technology, and financial tools enables them to take the lead in projects that lower emissions, support renewable energy, and strengthen their ability to adapt to climate challenges. Therefore, removing the gendered obstacles that stand in the way of women's potential and advocating for gender-sensitive policies that encourage their active involvement in climate governance are essential.

Incorporating women's perspectives into climate policies can also help improve global plans to make them more equal and inclusive (Fig. 1.9). A more comprehensive, varied, and cooperative strategy for tackling the climate catastrophe can be created by giving gender equality top priority in frameworks for climate action (CPHA, 2019; MoST, 2010). In addition to increasing women's ability to withstand the effects of climate change, this empowerment speeds up the world's shift to a more just and sustainable future for all (Tozier de la Poterie et al., 2023).

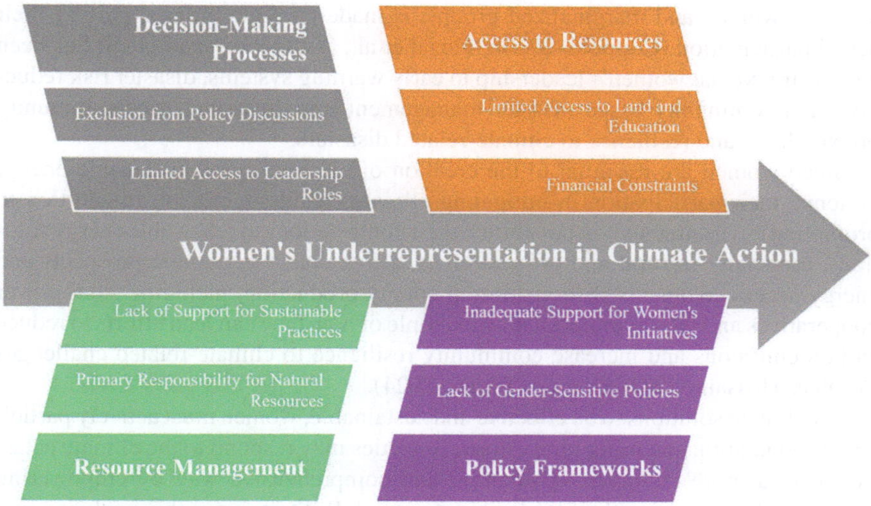

Fig. 1.9 Empowering women for climate resilience

1.8 Women as Active and Effective Agents and Promoters of Climate Change Adaptation and Mitigation

Women have continuously shown their resilience and ability to be active change agents in climate adaptation and mitigation initiatives, even though they are disproportionately vulnerable to the effects of the climate catastrophe (Alam & Ahamed, 2023; MoEFCC, 2022a). Women are essential to the day-to-day management of agriculture, family economies, and natural resources in their societies, especially in South and East Asia (Khatun & Saadat, 2021; Momtaz & Shameem, 2016a, 2016b). This emphasizes their importance as key players in climate solutions. Women's involvement in these fields gives them a distinct viewpoint on how climate change is affecting communities locally and a deep comprehension of sustainable practices that can improve resilience (Bardsley, 2018; Islam et al., 2021).

For a long time, women have been active in grassroots campaigns supporting the use of renewable energy, water management, and sustainable farming methods. Within official and informal networks, women have become leaders, advocating for culturally sensitive and regionally relevant climate adaption solutions. Their participation in these tactics, which include crop diversification, agroforestry, and water conservation, shows that they can address climate issues in a way that benefits entire communities in addition to exhibiting their understanding of environmental stewardship (Ersel, 2015).

In addition, women's voices are becoming increasingly acknowledged as crucial in discussions about climate policy and climate finance methods. The development of gender-sensitive climate policies, which may be able to address the unique needs of women and marginalized groups, is made possible in large part by their active participation (Coppola, 2020b; Faisal et al., 2021). For example, it has been demonstrated that women's leadership in early warning systems, disaster risk reduction, and community-based disaster management initiatives increases community preparedness and resilience to climate-related disasters.

Since women are essential to the creation of cleaner, more sustainable energy systems, their participation in mitigating climate change is crucial. In addition to promoting environmental sustainability, their contributions to renewable energy solutions, especially in rural and off-grid areas, also enhance economic prospects and energy access. Women's participation in energy production, including solar power cooperatives and biogas projects, is an example of how they can lead efforts to reduce carbon emissions and increase community resilience to climate-related challenges (Miah & Hossain, 2021; Singh & Tiwari, 2024).

For climate solutions to be effective and sustainable, women must actively participate in adaptation and mitigation efforts. Societies may respond to the climate catastrophe in a more inclusive, egalitarian, and comprehensive way by empowering women and recognizing their leadership potential. But to achieve this goal, ongoing assistance in the form of materials, learning opportunities, and legislative frameworks that encourage women's full and equal participation in these activities is required (Fig. 1.10). It has been shown that the impact of adaptation and mitigation techniques is increased when gender-sensitive approaches are incorporated into climate policies (Coppola, 2020a; GFDRR, 2011; Islam et al., 2013; Pelling & High, 2005; Rahman et al., 2017). Additionally, it has been demonstrated that this integration stimulates social change in the direction of greater climate justice and gender equality.

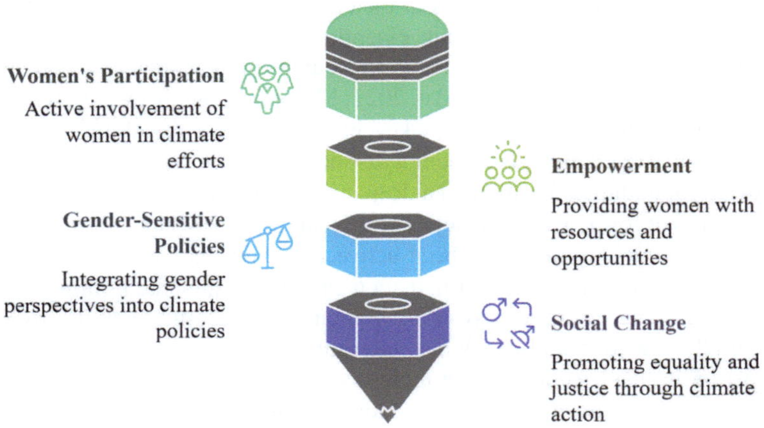

Fig. 1.10 Empowering women for climate solutions

1.9 Country Specific Gender Responsive Climate Actions

Countries all over the world are realizing the need for more gender-responsive and inclusive climate policies as the crisis continues to worsen. These initiatives aim to address the distinct effects of climate change, acknowledging that structural social, political, and economic disparities cause women to be disproportionately impacted by environmental disasters, especially in vulnerable areas (Ali, 1999; Shamsuzzoha & Al-Maruf, 2012). Even if progress is being made globally, gender views must be incorporated into national and local climate change adaptation and mitigation plans for climate action to be effective.

The execution of community-based adaptation initiatives in countries like Bangladesh is a clear example of a gender-responsive strategy to climate change (Shamsuzzoha et al., 2025b). With an emphasis on floods, cyclones, and rising sea levels, these programs highlight the leadership responsibilities that women play in disaster preparedness and recovery. By actively including women in decision-making processes and giving them the tools and training they need to become more resilient, these programs aim to overcome gender inequities in the face of climate-related concerns (Nurhidayah et al., 2025; Saunik & Shaw, 2025). Women in rural Bangladesh, for example, have received training to operate community-based early warning systems, which are essential for lowering the risk of disasters and enhancing local response systems (Fig. 1.11).

To incorporate gender into climate action, the Indian government has launched mitigation measures. The National Action Plan on Climate Change (NAPCC) is one such policy (MoEFCC, 2021; MoST, 2010). The NAPCC recognizes the need for gender-sensitive adaptation methods, especially in areas where women are disproportionately affected, like agriculture, water, and health. Initiatives like the National Mission on Sustainable Agriculture (NMSA) have taken steps to boost women's involvement in climate-resilient crop production and water management, among

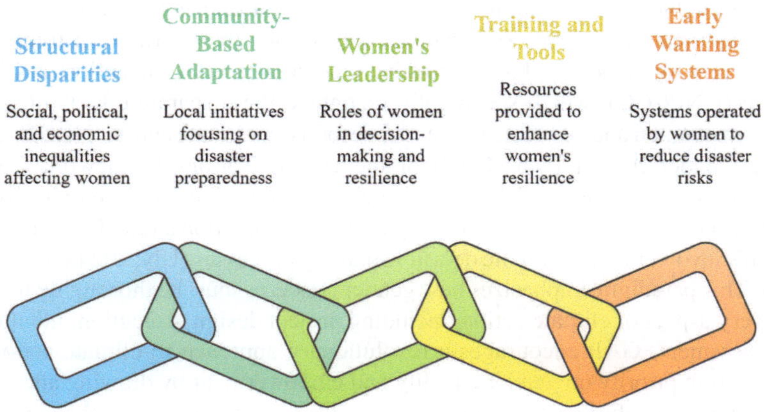

Fig. 1.11 Gender-responsive climate action

other sustainable farming methods (MoEF, 2004; MoEFCC, 2021; MoST, 2010). These steps guarantee that women have the information and resources they need to adjust to changing weather patterns.

In the Philippines, gender-responsive climate action is integrated into disaster risk reduction and management plans, demonstrating a similar strategy (Ramalho, 2019). With women represented in disaster management committees and local government entities, the nation has made great progress in encouraging women's participation in climate governance. To help women invest in sustainable livelihoods and get ready for the financial effects of climate change, the Philippines have also concentrated on expanding its access to financial resources.

These examples from nations highlight the importance of gender-responsive climate initiatives that tackle the pressing issues caused by climate change, as well as empowering women to take an active role in bringing about change in their communities. These countries are strengthening the resilience of women and marginalized groups and promoting more sustainable and effective climate solutions by giving gender equality top priority in their climate policy. Since women continue to be disproportionately impacted by climate change, other countries must follow suit and incorporate gender concerns into their climate frameworks to guarantee that no one is left behind in the global effort to combat climate change.

1.10 Global Frameworks to Promote Equitable and Sustainable Climate Solutions in the Gender Context

Gender-sensitive methods must be incorporated into global climate governance and policy frameworks to address the climate problem from an equitable and sustainable perspective. The importance of gender equality in the fight against climate change and in strengthening communities against its harmful effects has been emphasized by international agreements and frameworks. Among the well-known frameworks are the Paris Agreement and the United Nations Framework Convention on Climate Change (UNFCCC), which specifically recognize the disparate effects of climate change on women and advocate for gender-responsive climate action (NRDC, 2015).

The Gender Action Plan (GAP) of the United Nations Framework Convention on Climate Change (UNFCCC), which was created in 2017, emphasizes the need to encourage gender-sensitive mitigation and adaptation strategies and increase women's involvement in climate decision-making processes (UN, 2024a; UNFCCC, 2024). This paradigm emphasizes how gender concerns must be thoroughly included into every aspect of climate action, including project design, execution, monitoring, and assessment. GAP encourages a revolutionary approach to climate action that places a high priority on gender equality and empowerment by drawing attention to the unique vulnerabilities that women face in the context of climate change.

Similarly, the Paris Agreement, which seeks to keep global warming well below 2 °C, clearly recognizes that human rights must be respected in the fight against climate change, with a focus on gender equality. To guarantee that women and other vulnerable groups are not left behind in the shift to a low-carbon, resilient global economy, the agreement invites nations to create Nationally Determined Contributions (NDCs) that take gender issues into account.

Furthermore, regional programs like UN Women's Asia–Pacific Gender and Climate Change Initiative focus on enabling women in the area to take part in climate policies and support locally driven solutions (UNW, 2024). These frameworks acknowledge the critical role that women play in disaster preparedness, natural resource management, and sustainable farming practices, and they empower women as leaders and beneficiaries of climate efforts.

According to the results of international frameworks, there is growing recognition that advancing gender equality is not only an issue of social justice but also a crucial tactic for attaining successful climate resilience. The inclusion, impact, and sustainability of the policies and actions that come from including gender views into climate solutions are guaranteed. Whole communities gain from this, and women are allowed to take the lead in creating futures that are responsive to climate change.

1.11 Structure of the Book

This opening chapter serves as the sole contribution to the first section of the book and provides a foundational overview of the complex interrelationship between the climate crisis and gender dynamics. It critically examines how climate-induced environmental changes disproportionately impact women and gender-diverse communities, particularly in vulnerable socioeconomic and geographic contexts. Through a synthesis of global frameworks, policy analyses, and selected case studies, the chapter highlights the structural inequalities that exacerbate gender-specific vulnerabilities to climate risks. It also underscores the necessity of integrating gender equity into climate policy and practice to foster inclusive, sustainable, and resilient environmental solutions.

The book is structured into four major sections.

The first section, consisting only of this introductory chapter, sets the stage by conceptualizing the climate-gender nexus and establishing the theoretical underpinnings for subsequent discussions. The second section comprises eight chapters focused on South Asia, exploring gender-responsive approaches to drought resilience, the gendered dimensions of climate-induced disasters, multi-sectoral gender perspectives, inclusive adaptation strategies, and women-led watershed management. These chapters provide empirical insights and community-based evidence on how gender considerations can strengthen resilience and resource management in the region. The third section, also comprising eight chapters, shifts the geographical lens to East Asia and delves into gender-adaptive risk management frameworks, disaster preparedness through gender-sensitive approaches, and transformative gender strategies in climate adaptation. This section emphasizes the importance of context-specific, culturally embedded interventions that recognize the agency of women and marginalized groups. The fourth and final section includes a single concluding chapter that synthesizes the key findings across all sections. It draws critical lessons, articulates evidence-based recommendations, and proposes actionable policy directions for advancing gender-equitable climate resilience globally.

Together, these sections create a comprehensive, interdisciplinary volume that contributes to the growing field of climate and gender studies, offering both theoretical insights and practical strategies for inclusive climate action.

1.12 Concluding Remarks

One important but often overlooked aspect of the global environmental disaster is the intersection of gender issues and climate change. Existing social injustices are made worse by the climate crisis, which disproportionately affects women and other marginalized gender groups that are already at risk because they have less access to resources, decision-making authority, and legal protection. Women are disproportionately affected by climate-related disasters, such as floods, droughts, and extreme weather events, which are growing more common in places like South and East Asia. These consequences hinder women's ability to moderate or adapt to the changing environment, endangering their livelihoods while also exacerbating gendered power disparities.

However, the chapter emphasizes how crucial it is to incorporate gender-sensitive strategies into climate reactions and policy. The unique experiences, skills, and perspectives of women can play a crucial part in creating long-term solutions to climate-related problems. Developing more inclusive, egalitarian, and successful climate resilience solutions requires acknowledging and addressing the gendered aspects of climate change. Furthermore, there is potential for a significant impact on environmental sustainability and sustainable development if women are empowered to take on leadership roles in climate action, especially in vulnerable communities.

In conclusion, when developing successful mitigation efforts, it is critical to consider the gendered effects of the climate problem. Efforts to adapt and mitigate the effects of climate change must incorporate gender equality. Regardless of their social or economic standing, this integration guarantees that people alike have access to the tools, chances, and assistance required to fight and adjust to the changing climate. Implementing inclusive and intersectional methods is essential to achieving global climate goals that are fair, sustainable, and just for everyone.

References

Adger, W. N. (2003). Social capital, collective action, and adaptation to climate change. *Economic Geography, 79*(4), 387–404. http://www.jstor.org/stable/30032945

Adger, W. N. (2010). Social capital, collective action, and adaptation to climate change. In M. Voss (Ed.), *Der Klimawandel:Sozialwissenschaftliche Perspektiven* (pp. 327–345). VS Verlag für Sozialwissenschaften. https://doi.org/10.1007/978-3-531-92258-4_19

Ahamed, T., Noguchi, R., Muhsin, N., Ayu Purnamasari, R., Islam, M. A., Tasnim, F., Islam, M. Z., Islam, M. F., & Akmam, W. (2021). Sustainable agricultural development: A micro-level GIS-based study on women's perceptions of environmental protection and entrepreneurship in Japan and Bangladesh. *GeoJournal, 86*(5), 2071–2103. https://doi.org/10.1007/s10708-020-10169-5

Akter, M. M., Hosen, M. I., & Nasher, N. M. R. (2024). Assessing household resilience to climate extremes using indicator-based index in hazard-prone areas; evidence from Bangladesh. *Environmental Challenges, 15*, 100957. https://doi.org/10.1016/j.envc.2024.100957

Alam, K. F., & Ahamed, T. (2023). Climate-adaptive potential crops selection in vulnerable agricultural lands adjacent to the Jamuna River basin of Bangladesh using remote sensing and a Fuzzy expert system. *Remote Sensing, 15*(8), 2201. https://doi.org/10.3390/rs15082201

Aldrich, D. (2012). *Building resilience: social capital in post-disaster recovery*. https://doi.org/10.7208/chicago/9780226012896.001.0001

Ali, A. (1999). Climate change impacts and adaptation assessment in Bangladesh. *Climate Research, 12*, 109–116. https://api.semanticscholar.org/CorpusID:46283242

Anwarul Islam, M., Shamsuzzoha, M., Rasheduzzaman, M., Chandra Ghosh, R., & Faisal, M. (2019). Assessment on climate change adaptation: A study on coastal area of Khulna District in Bangladesh. *Australian Journal of Engineering and Innovative Technology, 6*(1), 14–20. https://doi.org/10.34104/ajeit.019.014020

Ariyaningsih, Sukmara, R. B., Fauzi, A., Wijaya, N., Latief, K., Wu, R.-S., & Shaw, R. (2025). Disaster risk reduction in Indonesia: 20 years after Aceh Tsunami. In R. Shaw, T. Izumi, R. Djalante, & F. Imamura (Eds.), *Two decades from the Indian Ocean Tsunami: Key challenges and advancements* (pp. 241–270). Singapore: Springer Nature Singapore. https://doi.org/10.1007/978-981-96-2669-4_15

Sphere Association. (2004). *Humanitarian charter and minimum standards in disaster response* (2nd ed.). Sphere Project. https://spherestandards.org/wp-content/uploads/Sphere-Handbook-2004-English.pdf

Sphere Association. (2018). *The sphere handbook: Humanitarian charter and minimum standards in humanitarian response* (4th ed.). Sphere Project. https://spherestandards.org/wp-content/uploads/Sphere-Handbook-2018-EN.pdf

Aziz, M., & Anjum, G. (2024). Transformative strategies for enhancing women's resilience to climate change: A policy perspective for low- and middle-income countries. *Women's Health, 20*, 17455057241302032. https://doi.org/10.1177/17455057241302032

Bardsley, D. K. (2018). Indigenous knowledge and practice for climate change adaptation. In *Encyclopedia of the Anthropocene* (pp. 359–367). Elsevier. https://doi.org/10.1016/B978-0-12-809665-9.09797-4

Barenstein, J. D., Phelps, P. M., Pittet, D., & Sena, S. (n.d.). *Safer Homes, stronger communities: A handbook for reconstructing after natural disasters*. Retrieved 16 Aug 2024, from. https://documents.worldbank.org/en/publication/documents-reports/documentdetail/290301468159328458/safer-homes-stronger-communities-a-handbook-for-reconstructing-after-natural-disasters

Beer, T. (2018). The impact of extreme weather events on food security. In S. Mal, R. B. Singh, & C. Huggel (Eds.), *Climate change, extreme events and disaster risk reduction: Towards sustainable development goals* (pp. 121–133). Springer International Publishing. https://doi.org/10.1007/978-3-319-56469-2_8

Binte Mostafiz, R., Noguchi, R., & Ahamed, T. (2021). Calorie-based seasonal multicrop land suitability analysis for regional food nutrition security in Bangladesh. *Asia-Pacific Journal of Regional Science, 5*(3), 757–795. https://doi.org/10.1007/s41685-021-00197-5

Bradshaw, S., & Fordham, M. (2015). Double disaster: Disaster through a gender lens (Chap. 14). In J. F. Shroder, A. E. Collins, S. Jones, B. Manyena, & J. Jayawickrama (Eds.), *Hazards, Risks, and Disasters in Society* (pp. 233–251). Academic Press. https://doi.org/10.1016/B978-0-12-396451-9.00014-7

Brady, B., Rosenberg, S., Newman, C. E., Kaladelfos, A., Kenning, G., Duck-Chong, E., & Bennett, J. (2022). Gender is dynamic for all people. *Discover Psychology, 2*(1), 41. https://doi.org/10.1007/s44202-022-00054-2

Brouwer, R., Akter, S., Brander, L., & Haque, E. (2007). Socioeconomic vulnerability and adaptation to environmental risk: a case study of climate change and flooding in Bangladesh. *Risk Analysis, 27*(2), 313–326. https://doi.org/10.1111/j.1539-6924.2007.00884.x

Bryan, E., Alvi, M., Huyer, S., & Ringler, C. (2024a). Addressing gender inequalities and strengthening women's agency to create more climate-resilient and sustainable food systems. *Global Food Security, 40*, 100731. https://doi.org/10.1016/j.gfs.2023.100731

Bryan, E., Alvi, M., Huyer, S., & Ringler, C. (2024b). Addressing gender inequalities and strengthening women's agency to create more climate-resilient and sustainable food systems. *Global Food Security, 40*, 100731. https://doi.org/10.1016/j.gfs.2023.100731

Castelo, S., Antunes, L., & Ashrafuzzaman, M. (2024). The impact of the climate crisis on gender inequality. Looking to the frontlines in search of priorities for policy. In *Frontiers in Sustainable Cities* (Vol. 6). Frontiers Media SA. https://doi.org/10.3389/frsc.2024.1304535

Coppola, D. P. (2020a). Mitigation. In *Introduction to International Disaster Management* (pp. 267–329). Elsevier. https://doi.org/10.1016/B978-0-12-817368-8.00004-X

Coppola, D. P. (2020b). Participants:—Governmental disaster management agencies. In *Introduction to International Disaster Management* (pp. 549–614.e7). Elsevier. https://doi.org/10.1016/b978-0-12-817368-8.00008-7

CPHA. (2019). *Minimum standards for child protection in humanitarian action* (2nd ed.). Alliance for Child Protection in Humanitarian Action (CPHA), Sphere Project. https://spherestandards.org/wp-content/uploads/CPMS-2019-EN.pdf

Cutter, S. L., Burton, C. G., & Emrich, C. T. (2010). Disaster resilience indicators for benchmarking baseline conditions. *Journal of Homeland Security and Emergency Management, 7*(1). https://doi.org/10.2202/1547-7355.1732

Dankelman, I., & Naidu, K. (2020). Introduction: Gender, development, and the climate crisis. *Gender & Development, 28*(3), 447–457. https://doi.org/10.1080/13552074.2020.1843830

Das, S., Hazra, S., Haque, A., Rahman, M., Nicholls, R. J., Ghosh, A., Ghosh, T., Salehin, M., & Safra de Campos, R. (2021). Social vulnerability to environmental hazards in the Ganges-Brahmaputra-Meghna delta, India and Bangladesh. *International Journal of Disaster Risk Reduction, 53*, 101983. https://doi.org/10.1016/j.ijdrr.2020.101983

Deininger, F., Woodhouse, A., Kuriakose, A. T., Gren, A., & Liaqat, S. (2023). *Placing gender equality at the center of climate action.* https://documents1.worldbank.org/curated/en/099718102062367591/pdf/IDU08c737dd00f8580412b0aed90fce874ab09b0.pdf

Erman, A., De Vries Robbe, S. A., Thies, S. F., Kabir, K., & Maruo, M. (2021). *Gender dimensions of disaster risk and resilience: Existing evidence.* https://openknowledge.worldbank.org/entities/publication/88d46d58-c4ca-53bf-82ea-4f3cc423b67e

Ersel, M. (2015). Water and sanitation standards in humanitarian action. *Turkish Journal of Emergency Medicine, 15*, 27–33. https://doi.org/10.5505/1304.7361.2015.48753

Faas, A. J., & Jones, E. C. (2017). Social Network Analysis Focused on Individuals Facing Hazards and Disasters. In E. C. Jones & A. J. Faas (Eds.), *Social network analysis of disaster response, recovery, and adaptation* (pp. 11–23). Butterworth-Heinemann. https://doi.org/10.1016/B978-0-12-805196-2.00002-9

Faisal, M., Saha, M. K., Sattar, M. A., Biswas, A. K. M. A. A., & Hossain, M. A. (2021). Evaluation of climate induced hazards risk for coastal Bangladesh: A participatory approach-based assessment. *Geomatics, Natural Hazards and Risk, 12*(1), 2477–2499. https://doi.org/10.1080/19475705.2021.1967203

FAO. (2020). *The role of gender in climate-smart agriculture.* https://www.fao.org/climate-smart-agriculture-sourcebook/enabling-frameworks/module-c6-gender/c6-overview/en/?type=111

FAO. (2022). *SCALA gender and social inclusion strategy: Scaling up climate ambition on land use and agriculture (SCALA).* https://www.adaptation-undp.org/sites/default/files/resources/scala_gender_strategy_adjusted15.pdf

FAO. (2024a). *Better life: Gender equality and women's empowerment at the Food and Agriculture Organization of the United Nations.* https://doi.org/10.4060/cd0671en

FAO. (2024b). *Gender and climate change, agroecology and biodiversity.* https://www.fao.org/gender/learning-center/thematic-areas/gender-and-climate-change-agroecology-and-biodiversity/3/en?tabInx=1

Fernandez, L. (2023). Unveiling gender dynamics: An in-depth analysis of gender realities. *Influence: International Journal of Science Review, 5*(3), 61–70. https://doi.org/10.54783/influencejournal.v5i3.182

Fernando, N., Amaratunga, D., Haigh, R., & Fernando, Y. (2025). 20 Years after: A snapshot of affected communities in the aftermath of the 2004 Indian Ocean Tsunami in Sri Lanka. In R. Shaw, T. Izumi, R. Djalante, & F. Imamura (Eds.), *Two decades from the Indian Ocean Tsunami: Key challenges and advancements* (pp. 411–429). Springer Nature Singapore. https://doi.org/10.1007/978-981-96-2669-4_22

GFDRR. (2011). *Climate risk and adaptation country profile of Bangladesh: Vulnerability, risk reduction, and adaptation to climate change.* https://www.gfdrr.org/en/publication/climate-risk-and-adaptation-country-profile-bangladesh

Ginige, K., Amaratunga, D., & Haigh, R. (2014). Tackling women's vulnerabilities through integrating a gender perspective into disaster risk reduction in the built environment. *Procedia Economics and Finance, 18*, 327–335. https://doi.org/10.1016/S2212-5671(14)00947-2

GoB. (2020). *National plan for disaster management (2021–2025): Action for disaster risk management towards resilient nation.* https://modmr.portal.gov.bd/sites/default/files/files/modmr.portal.gov.bd/page/a7c2b9e1_6c9d_4ecf_bb53_ec74653e6d05/NPDM2021-25%20DraftVer5_23032020.pdf

Graveline, M. H., & Germain, D. (2022). Disaster risk resilience: Conceptual evolution, key issues, and opportunities. *International Journal of Disaster Risk Science, 13*(3), 330–341. https://doi.org/10.1007/s13753-022-00419-0

Haque, A. K. E., Mukhopadhyay, P., Nepal, M., & Shammin, R. (2021). Climate change and community resilience: Insights from South Asia. In *Climate change and community resilience: Insights from South Asia.* Springer Nature. https://doi.org/10.1007/978-981-16-0680-9

Hirt, H., Al-Babili, S., Almeida-Trapp, M., Martin, A., Aranda, M., Bartels, D., Bennett, M., Blilou, I., Boer, D., Boulouis, A., Bowler, C., Brunel-Muguet, S., Chardon, F., Colcombet, J., Colot, V., Daszkowska-Golec, A., Dinneny, J. R., Field, B., Froehlich, K., Gardener, C. H., Gojon, A., Gomès, E., Gomez-Alvarez, E. M., Gutierrez, C., Havaux, M., Hayes, S., Heard, E., Hodges, M., Alghamdi, A. K., Laplaze, L., Lauersen, K. J., Leonhardt, N., Johnson, X., Jones, J., Kollist, H., Kopriva, S., Krapp, A., Lopez-Portillo Masson, M., McCabe, M. F., Merendino, L., Molina, A., Moreno Ramirez, J. L., Mueller-Roeber, B., Nicolas, M., Nir, I., Orduna, I. O., Pardo, J. M., Reichheld, J.-P., Rodriguez, P. L., Rouached H., Saad, M. M., Schlögelhofer, P., Singh, K. A., De Smet, I., Stanschewski, C., Stra, A., Tester, M., Walsh, C., Weber, A. P. M., Weigel, D., Wigge, P., Wrzaczek, M., Wulff, B. B. H., Young, I. M. (2023). PlantACT!—How to tackle the climate crisis. *Trends in Plant Science, 28*(5), 537–543. https://doi.org/10.1016/j.tplants.2023.01.005

Huq, N., Hugé, J., Boon, E., & Gain, A. K. (2015). Climate change impacts in agricultural communities in rural areas of coastal Bangladesh: A tale of many stories. *Sustainability (Switzerland), 7*(7), 8438–8460. https://doi.org/10.3390/su7078437

Ikeda, K. (1995). Gender differences in human loss and vulnerability in natural disasters: A case study from Bangladesh. *Indian Journal of Gender Studies, 2*(2), 171–193. https://doi.org/10.1177/097152159500200202

IPCC. (2023). Climate change 2022: Impacts, adaptation and vulnerability. In *Intergovernmental Panel on Climate Change (IPCC), Contribution of Working Group II to the Sixth Assessment Report of the Intergovernmental Panel on Climate Change*. Cambridge University Press. https://doi.org/10.1017/9781009325844

Islam, A., Shaw, R., & Mallick, F. (2013). National adaptation programme of action. In R. Shaw, F. Mallick, & A. Islam (Eds.), *Climate change adaptation actions in Bangladesh* (pp. 93–106). Springer Japan. https://doi.org/10.1007/978-4-431-54249-0_6

Islam, Md. N., Tamanna, S., van Amstel, A., Noman, Md., Ali, Md. S. S., Aparajita, D. M., Roy, P., Tanha, S. R., Sarkar, N., Ashiquzzaman, Md., Ghosh, S. K., Hasnat, S. R., Bhuiyan, Md. F., Sayef, Md., Saha, P., Hasan, G. R., & Ghosh, A. (2021). Climate change impact and comprehensive disaster management approach in Bangladesh: A review. In Md. N. Islam & A. van Amstel (Eds.), *Bangladesh II: Climate change impacts, mitigation and adaptation in developing countries* (pp. 1–39). Springer International Publishing. https://doi.org/10.1007/978-3-030-71950-0_1

Kadetz, P., & Mock, N. B. (2018). Problematizing vulnerability: Unpacking gender, intersectionality, and the normative disaster paradigm (Chap. 9). In M. J. Zakour, N. B. Mock, & P. Kadetz (Eds.), *Creating Katrina, rebuilding resilience* (pp. 215–230). Butterworth-Heinemann. https://doi.org/10.1016/B978-0-12-809557-7.00009-0

Kempa, K., & Tyagi, A. (2025). Climate and environmental impacts of green recovery: Evidence from the financial crisis. *World Development Sustainability, 6*, Article 100194. https://doi.org/10.1016/j.wds.2024.100194

Khan, S. A. (2024). *The role of innovation and adaptation technologies in climate change-induced disaster risk reduction: Enhancing resilience in satkhira communities of Bangladesh* (University of Manitoba). https://umanitoba.ca/environment-earth-resources/sites/environment-earth-resources/files/2024-04/MNRM%20Thesis%20Khan%2C%20Sabbir.pdf

Khatun, F., & Saadat, S. Y. (2021). *Climate change in Bangladesh: A sustainable development perspective*. https://www.un.org/ldc5/sites/www.un.org.ldc5/files/revised_climate_change_in_bangladesh_a_sustainable_development_perspective-22_september_2021.pdf

Kim, H., Lazurko, A., Linney, G., Maskell, L., Díaz-General, E., Březovská, R. J., Keune, H., Laspidou, C., Malinen, H., Oinonen, S., Raymond, J., Rounsevell, M., Vaňo, S., Venâncio, M. D., Viesca-Ramirez, A., Wijesekera, A., Wilson, K., Ziliaskopoulos, K., & Harrison, P. A. (2024). Understanding the role of biodiversity in the climate, food, water, energy, transport and health Nexus in Europe. *Science of The Total Environment, 925*, 171692. https://doi.org/10.1016/j.scitotenv.2024.171692

Miah, M. D., & Hossain, A. (2021). Carbon Concentration in the Coastal Afforestation Sites of Cox's Bazar Bangladesh. *Forestist, 71*(2), 84–92. https://doi.org/10.5152/forestist.2020.20012

Mirza, M. M. Q. (2003). Climate change and extreme weather events: Can developing countries adapt? *Climate Policy, 3*(3), 233–248. https://doi.org/10.3763/cpol.2003.0330

MoEF. (2004). *India's initial national communication to the United Nations framework convention on climate change*. Ministry of Environment and Forests (MoEF), Government of India. https://unfccc.int/resource/docs/natc/indnc1.pdf

MoEFCC. (2021). *National action plan on climate change (NAPCC)*. https://dst.gov.in/climate-change-programme

MoEFCC. (2022a). *Climate change initiatives of Bangladesh: Achieving climate resilience*. https://doe.portal.gov.bd/sites/default/files/files/doe.portal.gov.bd/npfblock/2022-11-02-08-08-ade27c3a48eeeedbf1394e5fa527edd2.pdf

MoEFCC. (2022b). *National adaptation plan of Bangladesh (2023–2050)*. https://www.undp.org/bangladesh/publications/national-adaptation-plan-bangladesh-2023-2050

Momtaz, S., & Shameem, M. I. M. (2016a). Adaptation in climate change discourse. In *Experiencing Climate Change in Bangladesh* (pp. 7–28). Elsevier. https://doi.org/10.1016/b978-0-12-803404-0.00002-8

Momtaz, S., & Shameem, M. I. M. (2016b). Livelihood adaptation to climate change. In *Experiencing Climate Change in Bangladesh* (pp. 123–139). Elsevier. https://doi.org/10.1016/b978-0-12-803404-0.00008-9

Mortoja, M. G., & Yigitcanlar, T. (2020). How does peri-urbanization trigger climate change vulnerabilities? An investigation of the dhaka megacity in Bangladesh. *Remote Sensing, 12*(23), 1–41. https://doi.org/10.3390/rs12233938

MoST. (2010). *National mission on strategic knowledge for climate change under national action plan on climate change*. https://dst.gov.in/sites/default/files/NMSKCC_mission%20document%201.pdf

Naim, C., Pokorny, J., Uyen, A., Shortall, C., Farra, A., & Moussally, K. (2024). *The role of humanitarian actors in global governance for AMR*. The Lancet Global Health. https://doi.org/10.1016/S2214-109X(24)00319-X

Nasreen, M. (2021, March 26). *Fifty years of disaster risk governance in Bangladesh*. The Daily Star. https://www.thedailystar.net/supplements/celebrating-50-years-bangladesh/news/fifty-years-disaster-risk-governance-bangladesh-2067109

Naz, F., Doneys, P., & Saqib, S. E. (2018). Adaptation strategies to floods: A gender-based analysis of the farming-dependent char community in the Padma floodplain, Bangladesh. *International Journal of Disaster Risk Reduction, 28*, 519–530. https://doi.org/10.1016/j.ijdrr.2017.12.016

Naz, F., & Doneys, P. (2022). Gender-based differences in access to and use of loans from rural credit programs for flood adaptation in the farming-dependent char communities of Bangladesh. In *Women's Studies International Forum*, (Vol. 95). https://doi.org/10.1016/j.wsif.2022.102651

NRDC. (2015). *Issue brief: The Paris agreement on climate change*. https://www.nrdc.org/sites/default/files/paris-climate-agreement-IB.pdf

Nurhidayah, L., Djalante, R., Suntoro, A., & Warsilah, H. (2025). Key progress and challenges in tsunami risk reduction in the ASEAN region. In R. Shaw, T. Izumi, R. Djalante, & F. Imamura (Eds.), *Two decades from the Indian ocean tsunami: Key challenges and advancements* (pp. 201–218). Springer Nature Singapore. https://doi.org/10.1007/978-981-96-2669-4_13

Pal, I., & Bhatia, S. (2018). Disaster risk governance and city resilience in Asia-Pacific region. In R. Shaw, K. Shiwaku, & T. Izumi (Eds.), *Science and technology in disaster risk reduction in Asia* (pp. 137–159). Academic Press. https://doi.org/10.1016/B978-0-12-812711-7.00009-2

Paul, C. J., Weinthal, E. S., Bellemare, M. F., & Jeuland, M. A. (2016). Social capital, trust, and adaptation to climate change: Evidence from rural Ethiopia. *Global Environmental Change, 36*, 124–138. https://doi.org/10.1016/j.gloenvcha.2015.12.003

Pelling, M., & High, C. (2005). Understanding adaptation: What can social capital offer assessments of adaptive capacity? *Global Environmental Change, 15*(4), 308–319. https://doi.org/10.1016/j.gloenvcha.2005.02.001

Pothiawala, S. (2015). Food and shelter standards in humanitarian action. *Turkish Journal of Emergency Medicine 15*, 34–39). https://doi.org/10.5505/1304.7361.2015.98360

Rahman, M. H., Rahman, Md. S., & Rahman, M. M. (2017). Disasters in Bangladesh: Mitigation and management. *Barisal University Journal Part 1, 1*(1), 139–163. https://bu.ac.bd/uploads/BUJ1V4I1/13.%20Hashinoor.pdf

Rahman, M. A., Afridi, S., Hossain, M. B., Rana, M., Al Masum, A., Rahman, M. M., & Al-Maruf, A. (2024). Nexus between heat wave, food security and human health (HFH): Developing a framework for livelihood resilience in Bangladesh. *Environmental Challenges, 14*(100802), 1–15. https://doi.org/10.1016/j.envc.2023.100802

Ramalho, J. (2019). Empowerment in the era of resilience-building: Gendered participation in community-based (disaster) risk management in the Philippines. *International Development Planning Review, 41*(2), 129–148. https://doi.org/10.3828/idpr.2018.25

Rao, N., Lawson, E. T., Raditloaneng, W. N., Solomon, D., & Angula, M. N. (2019). Gendered vulnerabilities to climate change: Insights from the semi-arid regions of Africa and Asia. *Climate and Development, 11*(1), 14–26. https://doi.org/10.1080/17565529.2017.1372266

Ravera, F., Iniesta-Arandia, I., Martín-López, B., Pascual, U., & Bose, P. (2016). Gender perspectives in resilience, vulnerability and adaptation to global environmental change. *Ambio, 45*(3), 235–247. https://doi.org/10.1007/s13280-016-0842-1

Sadia, H., Iqbal, M. J., Ahmad, J., Ali, A., & Ahmad, A. (2016). Gender-sensitive public health risks and vulnerabilities' assessment with reference to floods in Pakistan. *International Journal of Disaster Risk Reduction, 19*, 47–56. https://doi.org/10.1016/j.ijdrr.2016.08.024

Saha, M. K., Biswas, A. A. A., & Faisal, M. (2024). Livelihood vulnerability of coastal communities in context of the climate change: A index-based assessment. *World Development Sustainability, 4*, 100152. https://doi.org/10.1016/j.wds.2024.100152

Saunik, S., & Shaw, R. (2025). India's progress on disaster risk reduction over the past 20 years. In R. Shaw, T. Izumi, R. Djalante, & F. Imamura (Eds.), *Two decades from the Indian Ocean tsunami: Key challenges and advancements* (pp. 305–316). Springer Nature Singapore. https://doi.org/10.1007/978-981-96-2669-4_18

Shamsuzzoha, M., Noguchi, R., & Ahamed, T. (2021). Damaged area assessment of cultivated agricultural lands affected by cyclone bulbul in coastal region of Bangladesh using landsat 8 OLI and TIRS datasets. *Remote Sensing Applications: Society and Environment, 23*, 100523. https://doi.org/10.1016/j.rsase.2021.100523

Shamsuzzoha, M., Noguchi, R., & Ahamed, T. (2022). Rice yield loss area assessment from satellite-derived NDVI after extreme climatic events using a fuzzy approach. *Agricultural Information Research, 31*(1), 32–46. https://doi.org/10.3173/air.31.32

Shamsuzzoha, Md., & Al-Maruf, A. (2012). Post SIDR life strategy: Adaptation scenario of settlements of the South. *Institute of Bangladesh Studies (IBS) Journal, 19*, 207–222. https://www.ru.ac.bd/ibs/wp-content/uploads/sites/56/2022/07/JIBS-Bangla-19.pdf

Shamsuzzoha, M., Shaw, R., & Ahamed, T. (2024, January 17). Country Perspective on Disaster Risk Governance in Bangladesh. In *Rethinking Disaster Risk Management in South Asia, Panel-1: A Synergistic Approach to Disaster Risk Governance*. https://www.researchgate.net/publication/379946968

Shamsuzzoha, Md., Biswas, A. K. M. A. A., Hasan, Md. M., Faisal, Md., Hazra, P., Ghosh, R. C., Rasheduzzaman, Md., Debnath, P., Rahman, Md. A., & Shaw, R. (2025). Determination of constraints in existing DRR coordination mechanisms among different humanitarian actors of coastal Bangladesh. In R. Shaw, T. Izumi, R. Djalante, & F. Imamura (Eds.), *Two decades from the Indian ocean tsunami: Key challenges and advancements* (pp. 379–409). Springer Nature Singapore. https://doi.org/10.1007/978-981-96-2669-4_21

Shamsuzzoha, Md., Hasan, Md. F., Rasheduzzaman, Md., Talukder, Md. F., Ahamed, T., & Shaw, R. (2025). Bangladesh's progress on disaster risk reduction over the past 20 years. In R. Shaw, T. Izumi, R. Djalante, & F. Imamura (Eds.), *Two decades from the Indian Ocean tsunami: Key challenges and advancements* (pp. 317–358). Springer Nature Singapore. https://doi.org/10.1007/978-981-96-2669-4_19

Shanta, M. A., Suman, S. U., & Islam, M. M. (2025). Role of formal aquaculture industry for empowering women: A study in the southern coastal region of Bangladesh. *Women's Studies International Forum, 109*, 103046. https://doi.org/10.1016/j.wsif.2024.103046

Singh, H., & Tiwari, S. C. (2024). Carbon stock and carbon sequestration potential of forest growing stock around national thermal power plant, Bilaspur, Chhattisgarh, Central India. *Forestist, 74*(3), 289–297. https://doi.org/10.5152/forestist.2024.23052

Smith, K., & Smith, K. (2013). *Environmental hazards: Assessing risk and reducing disaster* (6th ed.). Routledge. https://doi.org/10.4324/9780203805305

Suresh, D., Saunik, S., Izumi, T., & Shaw, R. (2025). All hazards approach and its importance in the risk reduction. In R. Shaw, T. Izumi, R. Djalante, & F. Imamura (Eds.), *Two decades from the Indian Ocean tsunami: Key challenges and advancements* (pp. 29–46). Springer Nature Singapore. https://doi.org/10.1007/978-981-96-2669-4_3

Tewari, N., Bush, A., Butt, M. N., Stevens, E., & Zafar, S. (2023). *Gendered dimensions of loss and damage in Asia*. https://doi.org/10.21201/2023.000005

Tozier de la Poterie, A., Castro, E., Rahaman, H., Heinrich, D., Clatworthy, Y., & Mundorega, L. (2023). Anticipatory action to manage climate risks: Lessons from the Red Cross Red Crescent in Southern Africa, Bangladesh, and Beyond. *Climate Risk Management*, *39*, 100476. https://doi.org/10.1016/j.crm.2023.100476

UN. (2024a). *The gender action plan*. https://unfccc.int/topics/gender/workstreams/the-gender-action-plan

UN. (2024b). *The sustainable development goals report 2024*. https://unstats.un.org/sdgs/report/2024/The-Sustainable-Development-Goals-Report-2024.pdf

UNFCCC. (2024). *Decision-CP. 29: Gender and climate change*. https://unfccc.int/documents/644433

UNW. (2024). *Gender and climate change*. https://asiapacific.unwomen.org/en/focus-areas/humanitarian-action-and-disaster-risk-reduction/gender-and-climate-change

WB. (2021, March 3). *Gender dynamics of disaster risk and resilience*. World Bank (WB) Group. https://www.worldbank.org/en/topic/disasterriskmanagement/publication/gender-dynamics-of-disaster-risk-and-resilience

UN Women. (2021). *Training manual on gender and climate change resilience*. https://wrd.unwomen.org/practice/resources/training-manual-gender-and-climate-resilience-module-1-gender-and-climate-change

Wright, H., Fenton, A., Huq, S., Stott, C., Taub, J., & Chow, J. (2019). Introduction: Bangladesh Responds to Climate Change. In S. Huq, J. Chow, A. Fenton, C. Stott, J. Taub, & H. Wright (Eds.), *Confronting climate change in Bangladesh: Policy strategies for adaptation and resilience* (pp. 1–6). Springer. https://doi.org/10.1007/978-3-030-05237-9_1

Wyles, K. J., Pahl, S., & Thompson, R. C. (2014). Perceived risks and benefits of recreational visits to the marine environment: Integrating impacts on the environment and impacts on the visitor. *Ocean & Coastal Management*, *88*, 53–63. https://doi.org/10.1016/j.ocecoaman.2013.10.005

Yumarni, T., Amaratunga, D., & Haigh, R. (2014). Assessing gender vulnerability within post-earthquake reconstruction: Case study from Indonesia. *Procedia Economics and Finance*, *18*, 763–771. https://doi.org/10.1016/S2212-5671(14)01000-4

Part II
Climate Crisis and Gender in South-Asia

Chapter 2
Gender Responsive Drought Resilience Assessment in Afghanistan Using Geospatial Techniques

Sara Tokhi Arab, Md Shamsuzzoha, Ahmad Shekib Arab, Md. Mazharul Anwar, Md. Anwarul Abedin, and Rajib Shaw

Abstract Climate change, deforestation, and unsustainable agricultural practices contribute significantly to drought conditions. Afghanistan's steep terrain and arid plains make the country particularly vulnerable to water scarcity and extreme drought events. A large portion of the population relies on rainfall and groundwater for irrigated agriculture. Droughts can lead to crop failures, food shortages, and economic instability. Prolonged droughts have caused widespread food insecurity and malnutrition, particularly affecting children, the elderly, and women. Climate change has further intensified the frequency and severity of droughts, making adaptation increasingly difficult for impoverished and gender-vulnerable communities. Considering Afghanistan's chronic water shortages and the critical role of women in agriculture, household economics, water management, and sustainable practices—each essential to community resilience—evaluating gender-responsive drought resilience is vital. This study utilized the normalized difference vegetation index (NDVI), land surface temperature (LST), and Climate Hazards Group InfraRed Precipitation with Station data (CHIRPS) to derive drought indices, including the vegetation condition

S. T. Arab (✉)
Division of Intelligent Agriculture Machinery Research, Institute of Agriculture Machinery, National Agriculture and Food Research Organization, Kannondai, Tsukuba, Ibaraki, Japan
e-mail: tokhiarab.sara359@naro.go.jp

M. Shamsuzzoha (✉) · Md. Mazharul Anwar
Department of Emergency Management, Faculty of Environmental Science and Disaster Management, Patuakhali Science and Technology University, Dumki, Patuakhali, Bangladesh
e-mail: zohageo@pstu.ac.bd

A. S. Arab
National Agriculture and Food Research Organization, Tsukuba, Ibaraki, Japan

Md. Anwarul Abedin
Department of Soil Science, Faculty of Agriculture, Bangladesh Agricultural University, Mymensingh, Bangladesh

R. Shaw
Graduate School of Media and Governance, Keio University, Fujisawa, Japan

Md. A. Abedin et al. (eds.), *Climate Crisis in South and East Asia*, Sustainability Solutions, https://doi.org/10.1007/978-3-032-01012-4_2

index (VCI), temperature condition index (TCI), and the standardized precipitation index (SPI). A drought severity map was developed using the random forest algorithm. Finally, gender-specific data, particularly related to gender-based population and population involved in the agriculture sector, were integrated with the drought severity map to delineate gender-based drought impact zones in the study area. Results indicated that in 2022, approximately 1.1%, 41.2%, and 46.9% of the study area experienced extreme, severe, and moderate drought, respectively. In 2023, these figures shifted to 2.5%, 70.5%, and 24.2%, demonstrating a substantial increase in drought severity over time. The drought-affected communities map revealed that 37.2% of the population in 2022 and 39.4% in 2023 were impacted by extreme to severe drought conditions within the study area. Therefore, assessing gender-responsive drought resilience in Afghanistan is crucial, considering the country's vulnerability to water scarcity and the impact of climate change on women's contributions to agriculture, home economics, water management, and sustainable practices, all of which contribute to community resilience. This research was conducted to gather reliable evidence for enhancing women's access to water resources in Afghanistan through gender-sensitive policies. The results of the study effectively documented sustainable collaboration between Afghan government agencies, NGOs, local communities, and global collaborators in disaster risk reduction (DRR) and drought emergency management.

Keywords Drought assessment · Gender · Geospatial techniques · VCI TCI SPI

2.1 Introduction

In Afghanistan, drought is an ongoing and escalating environmental issue that has significant ramifications for food security, agriculture, water resources, and overall socioeconomic stability (FAO, 2019; Komino & Safi, 2024). As climate change accelerates, there is an observed increase in the frequency and intensity of droughts, which has the potential to increase the vulnerability of certain regions in Afghanistan (IPCC, 2022; Saha et al., 2018). The agrarian economy of Afghanistan is dependent on natural water supplies. Prolonged droughts have a detrimental effect on access to drinking water, food production, and rural livelihoods (Mohanty et al., 2019). Nonetheless, the consequences of drought are contingent on underlying social, economic, and gender-based inequalities, which vary among communities and are not uniform (FAO, 2020a). In Afghanistan, the issue of gender is a salient yet often overlooked aspect of drought resilience. Women experience a disproportionate impact from climate-related stress due to their socioeconomic limitations, including restricted access to land, financial resources, education, and decision-making platforms, particularly in rural and agrarian regions (Deininger et al., 2023; FAO, 2019). Conventional gender norms have been shown to impede women's capacity for mobility and engagement in the administration of the country's resources. This, in turn, hampers their capacity to employ adaptable strategies during periods of environmental emergency (UN Women, 2021). The demographic is particularly vulnerable to the repercussions of drought, which are exacerbated by their involvement in agricultural pursuits, domestic water collection, and food security. It is unfortunate that these individuals are frequently excluded from the formulation of policies and disaster response initiatives (Aziz & Anjum, 2024). Therefore, in order to cultivate inclusive and effective gender-responsive drought resilience and disaster risk reduction (DRR), it is imperative to address these gender disparities (Erman et al., 2021; FAO, 2024a; WB, 2021).

This study employs geospatial methodologies, including satellite remote sensing and geographic information systems (GIS), to examine drought patterns and their socioeconomic ramifications. A gendered analysis was employed to assess the resilience to drought in Afghanistan (Alam et al., 2024; Arab et al., 2022a). The efficacy of geospatial instruments in monitoring environmental shifts, identifying high-risk areas, and informing governmental interventions has been thoroughly documented (Hassan et al., 2025). The primary objective of this research was to assess the spatial variation in drought vulnerability and resilience across Kandahar Province, Afghanistan (RRERS, 2007). To achieve this objective, the study integrates remote sensing-based drought indices with gender-disaggregated socioeconomic indicators. The primary objective of the study was to generate a drought severity map using remote sensing datasets and overlay it with demographic data, particularly focusing on gender and agricultural involvement. The objective of this research is to provide policy recommendations that are grounded in empirical evidence and incorporate gender-related considerations into strategies for adapting to climate change. A gender-responsive approach is imperative to ensure equitable participation and benefit-sharing among men and women in drought resilience initiatives. The objective of this study is to furnish development agencies, policymakers, and humanitarian organizations with the findings necessary to formulate inclusive and sustainable drought mitigation strategies. The integration of gender-sensitive policy frameworks with geospatial analysis presents a novel methodology for understanding the role of gender dynamics in drought resilience in Afghanistan.

2.2 Material and Method

2.2.1 Geographical Scope of the Study

Kandahar Province (KP) is in the southernmost region of Afghanistan (Fig. 2.1). It is the second-largest province in Afghanistan, following Helmand Province, with a total area of 54,844.5 km^2 (8.4% of the total area) and a total population of approximately 1,399,594. The province's male population is approximately 713,352, while the female population is approximately 686,242 (ACSO, 2020, 2021). The region is comprised of the following districts: Khakrez, Ghorak, Arghandab, Marof, Spenboldak, Kandahar, Dand, Daman, Arghistan, Shorabak, Panjwayi, Maywand, Shah Wali Kot, Zhar, Miyanshen, and Reg (ACSO, 2020). Kandahar Province is classified as semi-arid, with an average annual rainfall 176 mm (IFAD, 2019). The southern portion of the province is predominantly desertic (RRERS, 2007). A flat, sandy plain, with a prominent ridge of dunes in the Western portion of the valley, characterizes most of the inhabited area. It has been documented that temperatures in this region can reach as high as 55 °C, with an average altitude of approximately 1075 m above sea level (UN, 1990).

Kandahar is predominantly an agricultural province; the agriculture sector is divided into two primary categories: cereal grains and fruits. The primary income crops for farmers in Kandahar, excluding opium, are wheat, maize, corn, and saffron. In rain-fed regions, melons and watermelons are also cultivated. The region is particularly renowned for its fruit production, such as grapes (*kishmish*), pomegranates, dates, and apricots. In 2020, the province's total grape yield was approximately 234,377 tons. The region's diverse grape cultivars are utilized in traditional and local methods for further processing, such as drying, resulting in the production of raisins (*kishmish khana*). These raisins are subsequently marketed, contributing to the region's economic activities. The total production of pomegranates in the province was approximately 71,227 tons (ACSO, 2020). From 2021 to 2024, a persistent drought had a substantial impact on agricultural production in this province (FAO, 2023; OCHA, 2024a).

The ongoing drought, coupled with natural disasters, population displacement, escalating living costs, food insecurity, and pervasive unemployment, have converged to create a nexus of challenges that have severely impacted nutrition across the country. Among all the provinces, Kandahar and Paktika are severely impacted by drought. The 2024 Integrated Phase Classification (IPC) report for Afghanistan indicates that KP, with a projected population of 2,067,723, has a Global Acute Malnutrition (GAM) rate of 15.5, classified as "critical." This encompasses 103,386 children under the age of five who are affected by severe acute malnutrition and 41,354 expectant and lactating mothers experiencing moderate acute malnutrition (OCHA, 2024b).

The ongoing drought has had a significant impact on the agricultural sector of the province. The scarcity of surface water has led to the destruction of vineyards and orchards, as well as a decline in wheat cultivation. The scarcity of water has led to a situation where individuals must pay for each liter of water, as the only remaining option is to extract water from deep bore wells using diesel-powered water pumps, a costly process. Consequently, farmers have resorted to poppy cultivation, which is regarded as the sole viable cash crop capable of covering production expenses under the prevailing internal irrigation methods in the region encompassing wells, deep wells, karezes, and rivers. The province of Kandahar is endowed with numerous rivers, including the Dorai, Tarnak, Arghistan, and Arghandab (IOM, 2022; RRERS, 2007).

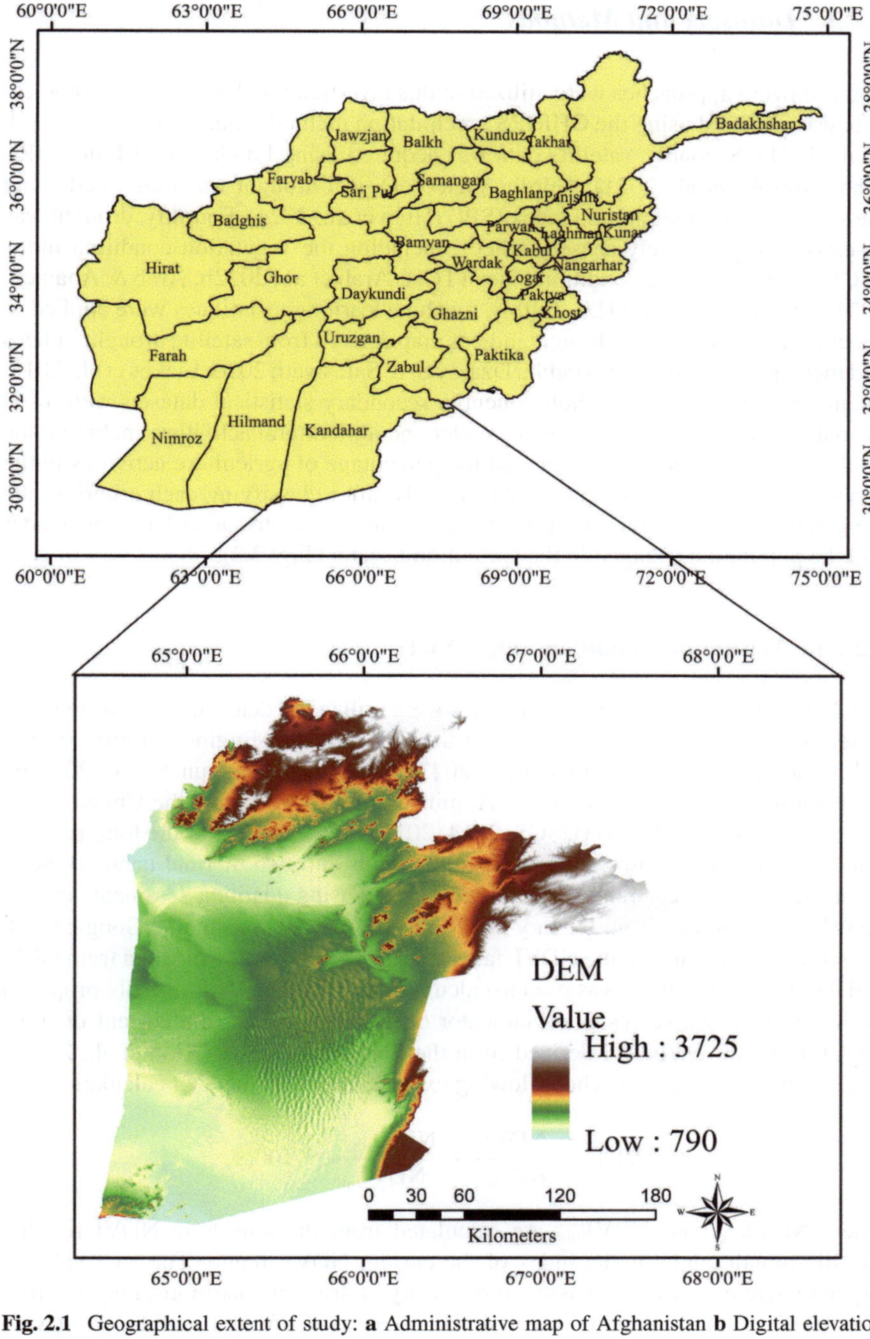

Fig. 2.1 Geographical extent of study: **a** Administrative map of Afghanistan **b** Digital elevation map (DEM) of Khandahar Province (KP)

2.2.2 Datasets and Methods

The following approaches were utilized in this investigation. Firstly, meteorological data was gathered using the CHRIPS precipitation method (Alam et al., 2024; Arab et al., 2021). Secondly, satellite data was acquired using Landsat 8 OLI and TIRS (Shamsuzzoha et al., 2021). Thirdly, meteorological drought was monitored using the standardized precipitation index (SPI) (Htun et al., 2024). Fourthly, drought was assessed using remotely sensed indices, including the vegetation condition index (VCI) and temperature condition index (TCI) (Arab et al., 2022b; Arab & Ahamed, 2023; Shahfahad et al., 2023). Fifthly, machine learning techniques were applied to generate a comprehensive drought indices map derived from satellite drought indices and meteorological datasets (Faliha Dzakiyah & Saraswati, 2020; Liakos et al., 2018; Shamsuzzoha et al., 2024). Subsequently, secondary statistical datasets were used to analyze the effect of droughts on gender and agricultural activities, including the total male and female population and the percentage of agriculture activities in the economic sector (Tewari et al., 2023). Finally, after classifying each criterion, the weighted overlay method was applied to generate the affected areas based on gender and the population engaged in the agricultural sector (Fig. 2.2).

2.2.2.1 Vegetation Condition Index (VCI)

The Landsat 8 OLI multispectral datasets were utilized to determine the normalized difference vegetation index (NDVI) on the Google Earth Engine platform (Gorelick et al., 2017; Shamsuzzoha et al., 2022). Landsat 8 was launched in 2013 by the National Aeronautics and Space Administration (NASA) and the United States Geological Survey (USGS) (USGS, 2014, 2019, 2025). As part of the long-running Landsat program, it possesses nine spectral bands and two thermal bands (USGS, 2014). Landsat 8 scene pixels were cloud-masked using quality assessment, and the NDVI was computed from January 2022 to December 2023 using the Google Earth Engine platform. Finally, the NDVI data for 2022–2023 was downloaded from GEE, and the ArcGIS platform was used to calculate the VCI. The VCI, initially proposed by Kogan in 1995, serves as an indicator of the proximity of the current month's NDVI to the lowest NDVI derived from the long-term dataset (Arab et al., 2022a, 2022b; Htun et al., 2024). The following expression shows the VCI calculation.

$$\mathrm{VCI} = \frac{\mathrm{NDVI}_J - \mathrm{NDVI}_{\min}}{\mathrm{NDVI}_{\max} - \mathrm{NDVI}_{\min}} \times 100,$$

where $\mathrm{NDVI}_{\max}$ and $\mathrm{NDVI}_{\min}$ are calculated from the long-term NDVI for that specific month, and J is the index of the current NDVI month. The VCI value is a quantitative metric used to assess the severity of drought conditions, ranging from 0 to 100% of the maximum potential. This scale is delineated in Table 2.1 (Arab et al., 2022a, 2022b)

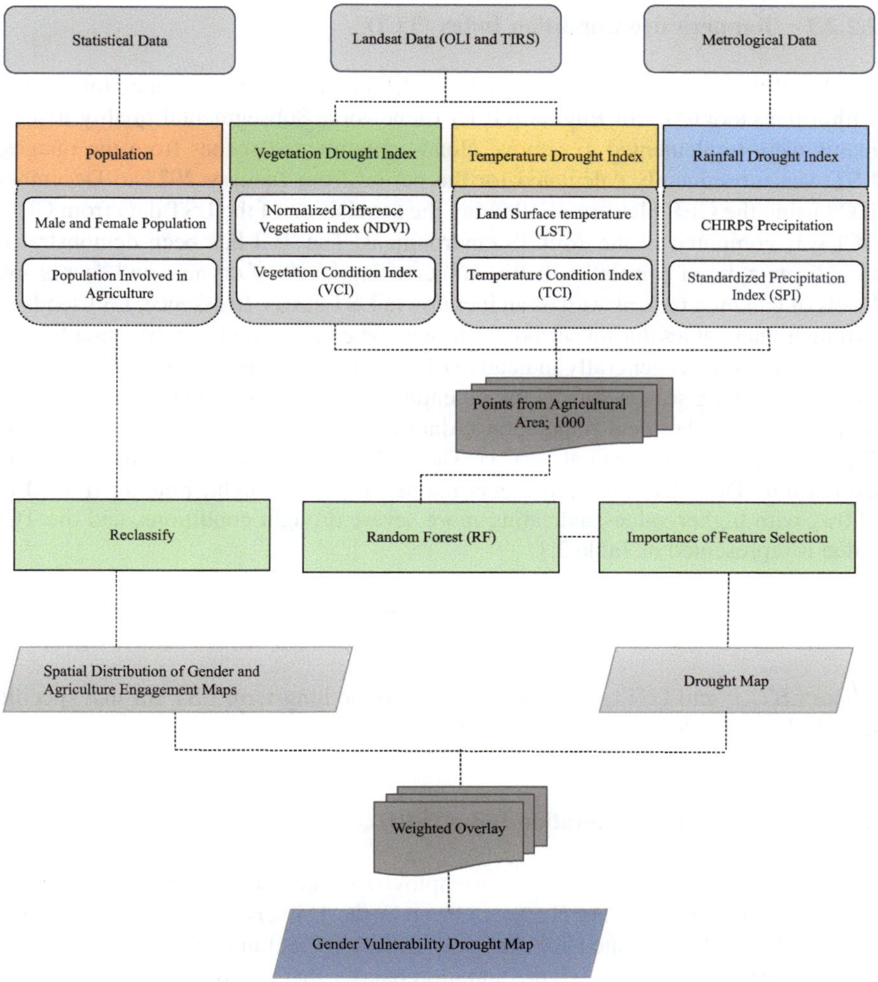

Fig. 2.2 Illustration of the framework applied in the study

Table 2.1 VCI and TCI-based drought categories in percentage (Kogan, 1995)

Drought categories	VCI (value in %)
Extremely drought	0–10
Severe drought	10–20
Moderately drought	20–30
Near normal	30–40
Moderately wet	40–50
Very wet	50–60
Extremely wet	70–100

2.2.2.2 Temperature Condition Index (TCI)

Landsat 8 thermal bands were utilized to calculate land surface temperature (LST) within the Google Earth Engine (GEE) framework. Subsegmental quality assessments were implemented to remove cloud-contaminated scenes from the images. LST was subsequently calculated for the period from January 2022 to December 2023 using the GEE platform. Following the acquisition of the LST data from GEE, TCI was computed in the ArcGIS environment. The TCI has been demonstrated to indicate a direct correlation between increasing levels of dryness and decreasing levels of soil moisture, as well as an increase in LST stress. In essence, LST tends to exhibit higher values during periods of dryness when compared to typical conditions. While a low LST is generally indicative of favorable conditions, a high LST during the crop-growing season signals the potential for adverse or drought conditions for both plants and the local population (Alam & Rahman, 2014; Alam et al., 2024; Bento et al., 2018; Hang et al., 2024). The TCI was calculated using the following expression: The TCI value represents the severity of drought, ranging from 0 to 100%, with higher values indicating more severe drought conditions, and the TCI value is represented in Table 2.1.

$$\mathrm{TCI} = \frac{\mathrm{LST}_{\mathrm{max}} - \mathrm{LST}_J}{\mathrm{LST}_{\mathrm{max}} - \mathrm{LST}_{\mathrm{min}}} \times 100,$$

where $\mathrm{LST}_{\mathrm{max}}$ and $\mathrm{LST}_{\mathrm{min}}$ are calculated from the long-term LST for that specific month, J is the index of the current LST month.

2.2.2.3 Standard Precipitation Index (SPI)

The CHIRPS precipitation dataset was employed to calculate the SPI-1. This product originates from the Climate Hazards Group at the University of California, Santa Barbara. CHIRPS (Climate Hazards Group InfraRed Precipitation with Station data) is a high-resolution, extensive precipitation dataset that combines satellite-derived infrared data with ground station data to estimate precipitation worldwide. Rainfall data from January 2022 to December 2023 was collected to generate the standardized precipitation index (SPI) for drought evaluation. The SPI is a widely utilized metric for monitoring drought conditions. The categorization of drought severity is achieved through the measurement of precipitation anomalies, defined as deviations from the mean rainfall over a designated period. These anomalies are then compared to a comprehensive historical record to derive a quantitative assessment of drought intensity (Arab & Ahamed, 2023; McFeeters, 1996; Mckee et al., 1993). The following expression was used for SPI calculation.

$$\mathrm{SPI} = \frac{P - \mu}{\sigma},$$

Table 2.2 SPI-based drought categories based on McKee classification (Mckee et al., 1993)

Drought category	SPI values	Probability %
Extreme drought	≤ − 2.00	2.3
Severe drought	− 1.50 to − 1.99	4.4
Moderate drought	− 1.00 to − 1.49	9.2
Mild drought	0 to − 0.99	34.1
Near normal	1 to 0.99	34.1
Moderately wet	1.0 to 1.49	9.2
Very wet	1.5 to 1.99	4.4
Extremely wet	2 and above	2.3

where P represents the monthly mean precipitation, μ is the average precipitation during 2022–2023, and σ standard deviation of precipitation over the entire historical period. Table 2.2 indicates the drought classes, SPI values, and the percentage of probability of drought (Arab & Ahamed, 2023; Mckee et al., 1993).

2.2.2.4 Statistical Dataset

This study employed a statistical dataset to elucidate the percentage of agriculture's contribution to the economy, thereby assessing the impact of drought. The analysis focuses specifically on the effects of drought on gender dynamics within the province. The objective is to comprehend how drought conditions influence both male and female populations, with a particular emphasis on the agricultural sector's role in the region's economy, a pivotal component of the province's economic structure. (ACSO, 2020; IOM, 2022).

2.2.2.5 Radom Forest (RF)

Random forest (RF) is a machine learning method that has gained considerable popularity for its use in classification and regression applications (Breiman, 2001; Park et al., 2019). This study employed the RF model to assess the significance of various variables, including the SPI, TCI, and VCI, in relation to agricultural systems experiencing drought. The model's hyperparameters were determined by essential elements, including the quantity of trees in the forest, the minimum samples necessary to bifurcate an internal node, the minimum samples required at a leaf node, and the maximum depth of each tree. The model's performance was assessed by partitioning the dataset into a training set of 70% and a testing set of 30%. This partitioning of the dataset is paramount for ensuring that the model is adequately trained on enough data while being evaluated on novel data, thereby assessing its generalization capability.

The optimal hyperparameter combinations for the random forest models were determined through a process of testing and adjustment (Etikan, 2017). The following critical hyperparameters were selected for this study: The quantity of trees in the forest (n_estimators) was established at 1000, the maximum depth of the trees (max_depth) was fixed at 40, the minimum number of samples necessary to bifurcate an internal node (min_samples_split) was determined to be 2, and the minimum number of samples required at a leaf node (min_samples_leaf) was set to 5. The selection of these hyperparameters was driven by the need to strike a balance between model complexity and performance. The primary objective was to enhance the assessment of feature significance while reducing the likelihood of overfitting (Park et al., 2019).

2.3 Results

2.3.1 VCI

As demonstrated in Fig. 2.3, the VCI outcome illustrates the spatial and temporal distribution of drought in KP from January 2022 to December 2023, as assessed by the NDVI. The findings indicate that the severity of the drought was significantly high for several months, particularly in April, May, and June 2022, as well as in May, July, August, September, and October 2023. These periods of heightened drought severity underscore the pervasive and recurrent dry conditions that afflicted the region over the study period (Fig. 2.3). The graph in Fig. 2.4 offers supplementary insight by presenting the mean average values for specific months. For the year 2022, the values were 10.8%, 10%, and 9% for April, May, and June, respectively, reflecting moderate to severe drought conditions during the spring and early summer months. In 2023, the values were 29%, 19%, 16%, 13%, and 18% for April, June, July, August, and October, respectively, indicating that drought conditions were not only persistent but, in some cases, more severe compared to the previous year (Fig. 2.4). These findings highlight the persistent vulnerability of KP to drought, particularly during the warmer months, and underscore the necessity for proactive measures to mitigate the impact of protracted dry conditions on agriculture, water resources, and overall livelihoods. The data further indicate that monitoring and intervention strategies should prioritize these critical periods to optimize drought risk management and bolster resilience in the region.

2.3.2 TCI

As demonstrated in Fig. 2.5, the TCI exhibited considerable variations in drought conditions across multiple months in 2022 and 2023 within the Kandahar region. In 2022, the TCI values were found to be significantly low in February (8.6%), March (12%), April (5.8%), June (33.7%) July (33%), and September (4.5%). These findings suggest that these months experienced varying degrees of drought intensity,

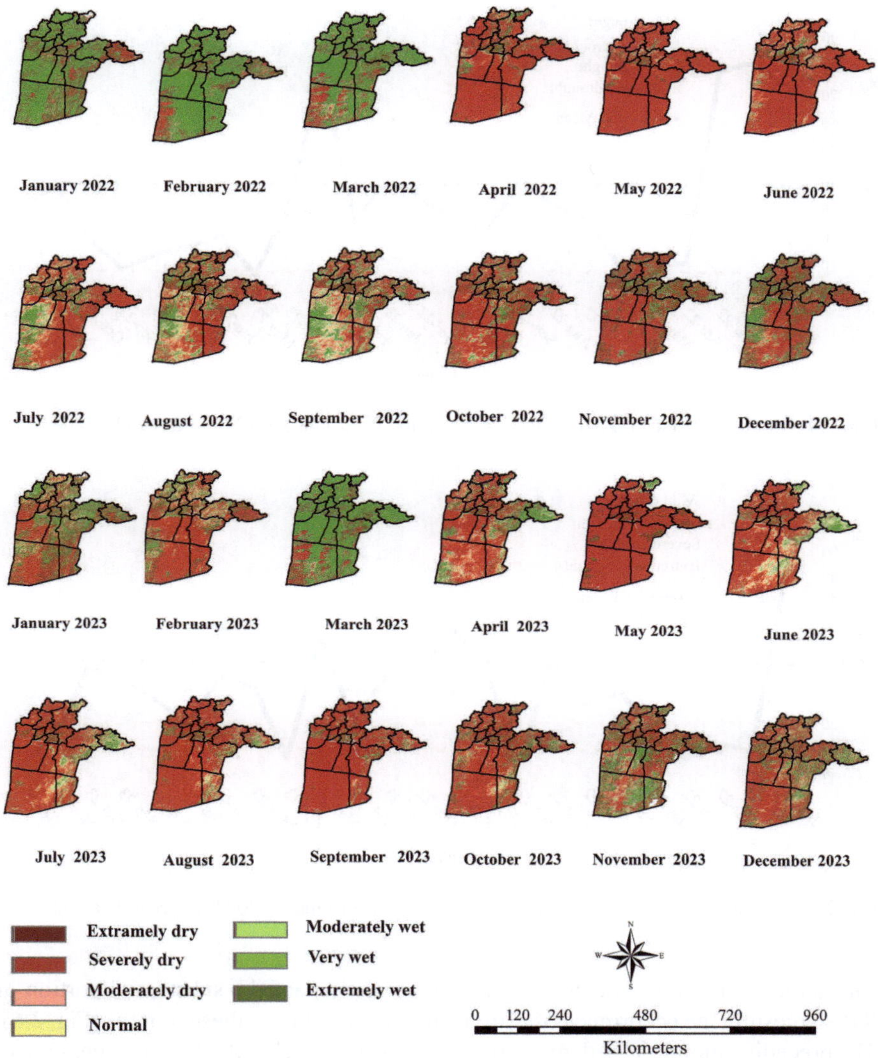

Fig. 2.3 Spatial distribution of VCI in KP of Afghanistan during the 2022–2023

ranging from moderate to intense. In 2023, the TCI values demonstrate the presence of substantial drought effects in February (20.9%), April (5%), July (20%), August (30%), September (20%), November (14%) and December (3.09%) (Fig. 2.5). As illustrated in Fig. 2.6, the TCI map reveals that temperatures in KP exhibited significant increases during specific intervals in both 2022 and 2023. Specifically, the months of March, April, May, partially June, November, and December 2022, along with February, April, partially June, July, November and December 2023, exhibited

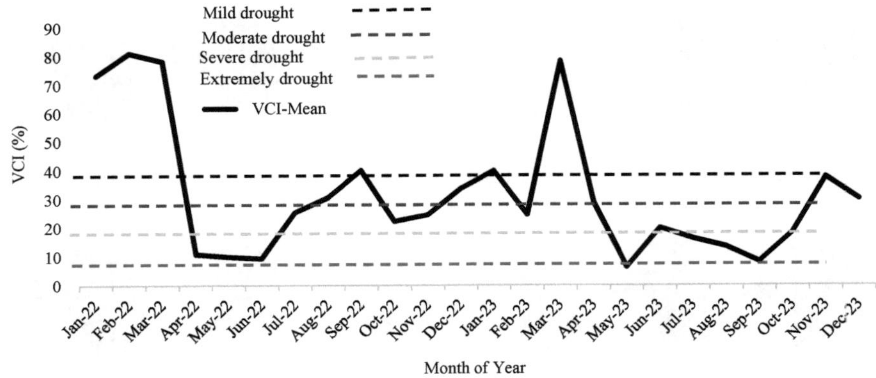

Fig. 2.4 Monthly VCI average for KP over the period from January 2022 to December 2023

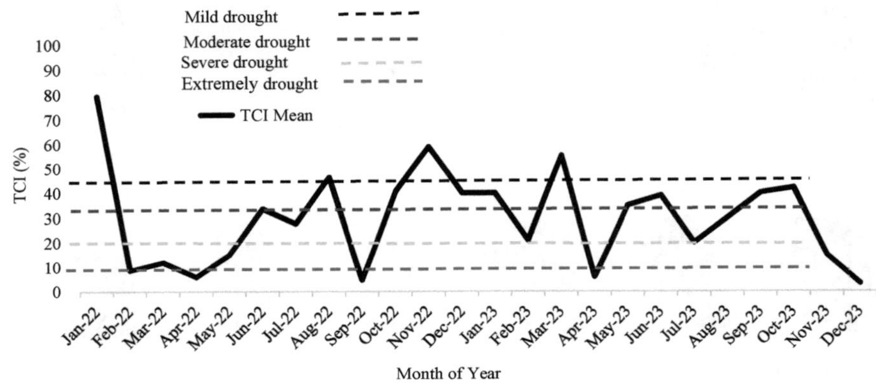

Fig. 2.5 Monthly TCI average for KP over the period from January 2022 to December 2023

substantial disruption. The elevated temperatures impacted a substantial portion of KP, suggesting the occurrence of extreme heat events during these months (Fig. 2.6). The preceding months also demonstrated circumstances classified as extreme, severe, or moderate drought. The findings underscore the persistent and cyclical nature of drought in KP, attributable to low TCI values indicative of inadequate vegetation and diminished moisture availability.

2.3.3 SPI

As demonstrated in Fig. 2.6, instances of low and extremely low drought conditions were observed in various months throughout 2022 and 2023. Specifically, these

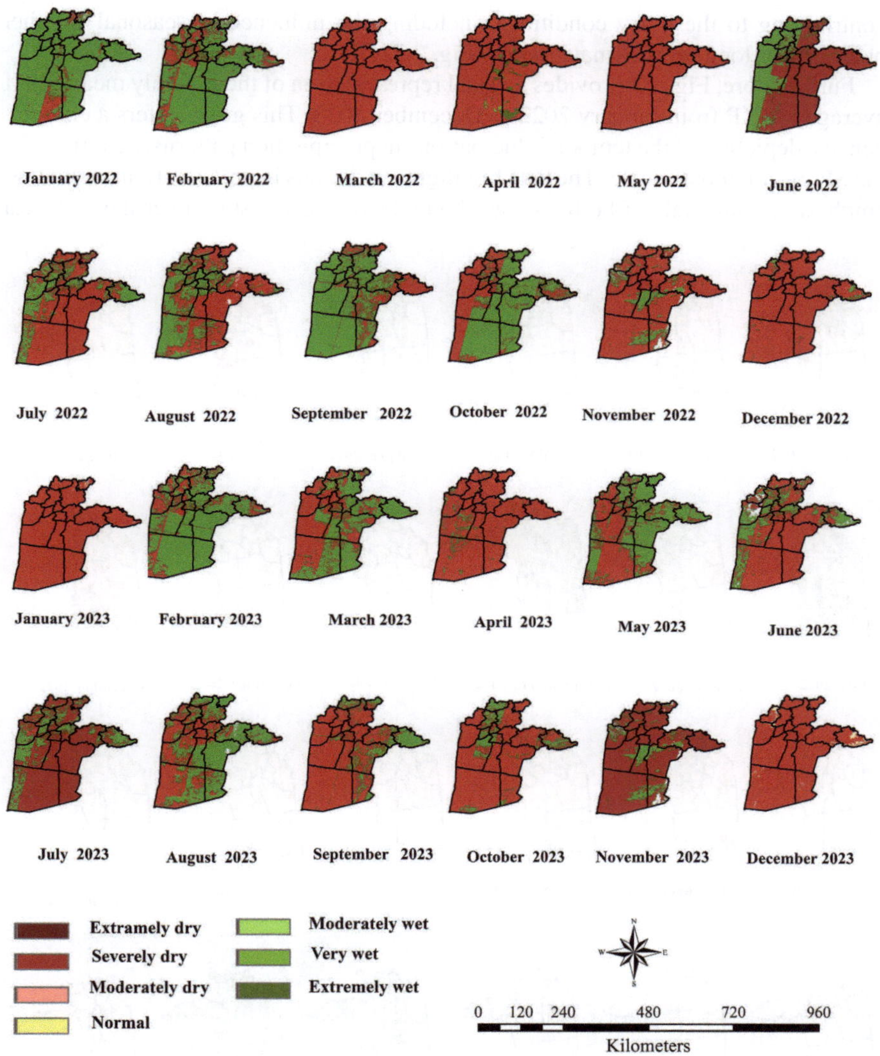

Fig. 2.6 Spatial distribution of TCI in the KP of Afghanistan during the 2022–2023

conditions were prevalent in February, March, April, May, November, and December 2022, as well as in January, February, April, May, June and July 2023. The recurrent presence of low and extremely low drought severity during these months suggests a consistent pattern of insufficient precipitation and elevated environmental stress. These prolonged periods of reduced precipitation may have significant implications for agricultural productivity, water resources, and the overall health of ecosystems in the region. Consequently, further analysis is needed to identify the underlying factors contributing to these dry conditions, including the influence of seasonal weather patterns and long-term climate trends (Fig. 2.7).

Furthermore, Fig. 2.7 provides a visual representation of the monthly mean SPI-1 average for KP from January 2022 to December 2023. This graph offers a comprehensive depiction of the temporal fluctuations in precipitation patterns, quantified by the SPI on a 1-month scale. The data highlights variations in precipitation anomalies, emphasizing intervals of both increased and decreased moisture over the two-year

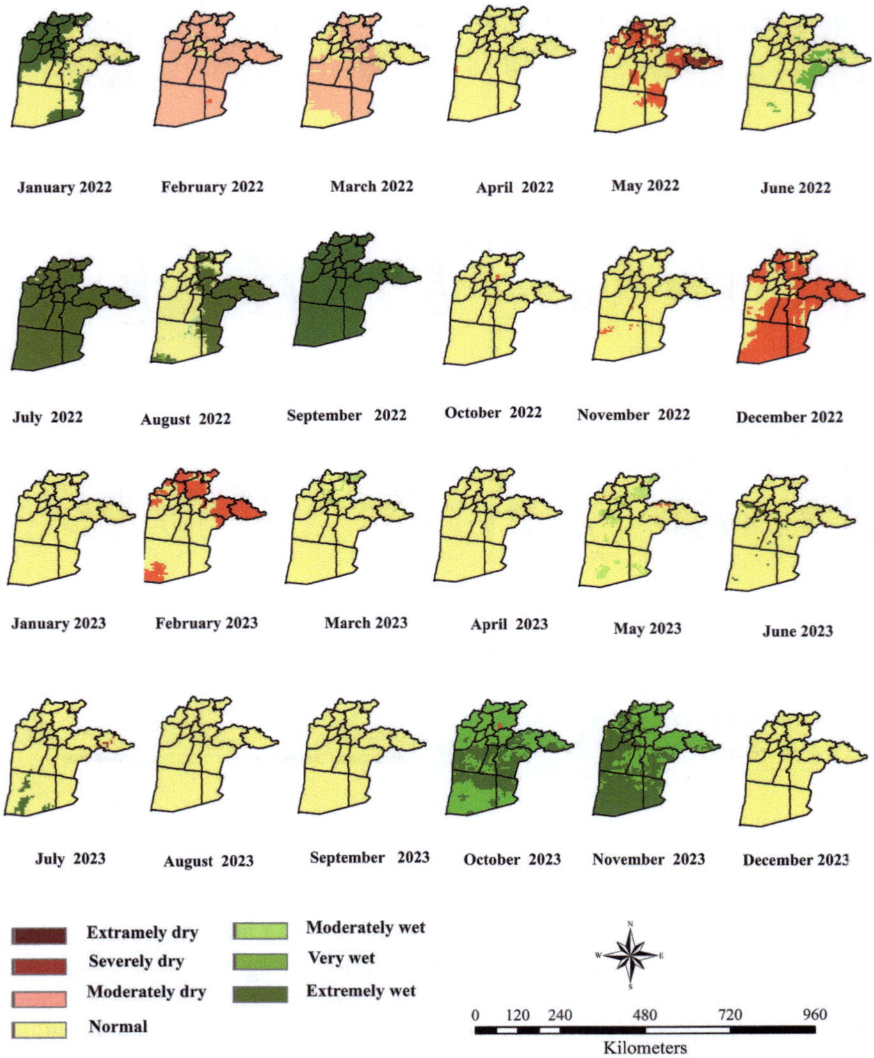

Fig. 2.7 Spatial distribution of SPI in the KP of Afghanistan during the 2022–2023

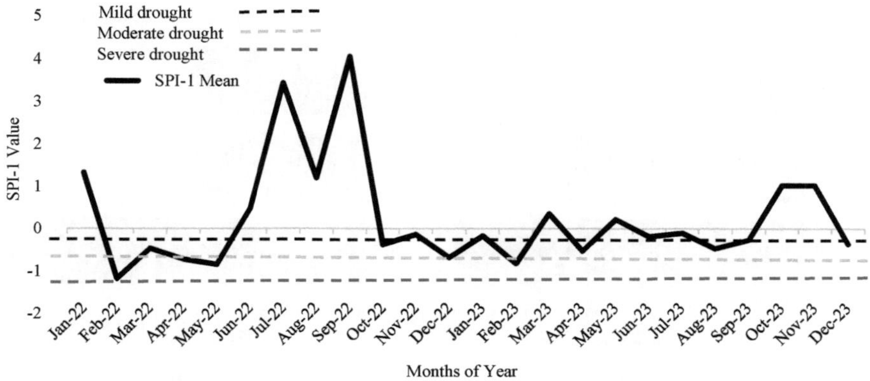

Fig. 2.8 Monthly SPI-1 average for KP over the period from January 2022 to December 2023

period. For instance, months such as February (− 1.33), March (− 0.46), April (− 0.72), May (− 0.85), October (− 0.14), November (− 0.13), and December (− 0.67) in 2022, and January (− 0.16), February (− 0.18), April (− 0.52), June (− 0.19), July (− 0.11), August (− 0.47), September (− 0.27), and December (− 0.36) in 2023. As demonstrated in Fig. 2.8, these months exhibit significant deviations from the long-term mean precipitation. A comprehensive understanding of these patterns is imperative for evaluating the ramifications of drought and precipitation anomalies on regional ecosystems, agriculture, and water resources. The SPI-1 values presented in this figure are instrumental in identifying instances of meteorological dryness and informing effective drought management strategies in KP.

2.3.4 *Random Forest*

The findings of the RF model indicated that SPI consistently exhibited a significantly higher level of importance in predicting the target variable compared to the other features, namely VCI and TCI, in both 2022 and 2023 (Figs. 2.9 and 2.10). In 2022, SPI exhibited the highest level of significance, with a score of 0.49, followed by VCI at 0.34 and TCI at 0.17. In 2023, SPI's importance score remained the highest at 0.47, while VCI's score declined to 0.30. Despite this decline, TCI's score increased, reaching 0.23 in 2023 (Figs. 9a and b). This finding is consistent with established knowledge that SPI is significantly associated with precipitation patterns and drought intensity, which are critical factors in determining agricultural drought, particularly in regions prone to drought. Consequently, SPI's greater significance relative to VCI and TCI underscores its pivotal role in evaluating drought effects and the importance of precipitation for water availability in Afghanistan (Figs. 9a and b).

The drought maps generated using the RF model's important features indicate that KP experienced significant drought in both 2022 and 2023. The analysis of

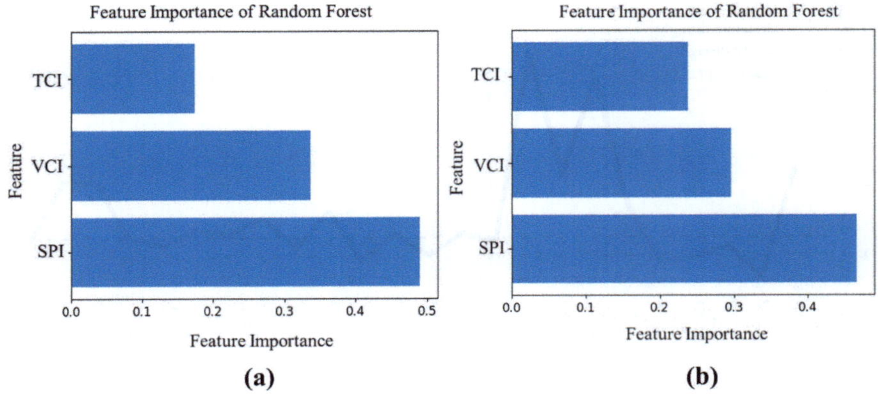

Fig. 2.9 Importance of features based on the RF model for SPI, VCI, and TCI **a** 2022 and **b** 2023

affected areas reveals that in 2022, 41% of the province was severely affected by drought, while 47% experienced moderate drought conditions (Table 2.3, Fig. 2.10). In 2023, the severity of drought escalated, with 70% of the province experiencing severe conditions and 24% exhibiting moderate impacts. These findings highlight a worrisome trend of escalating drought conditions over time. This trend shows the critical need for proactive measures, such as improved water resource management and climate adaptation strategies, to mitigate the impacts of drought in this drought-prone region (Table 2.3, Fig. 2.10).

2.3.5 Gender Distribution and Agricultural Contribution by District

The demographic data of the male and female population indicates that the most highly populated area is the center of Kandahar Province, specifically Kandahar District, with a total population of 614,254. Following this, the districts of Spin Boldak, Panjwayi, and Zheray also have significant populations, totaling 112,768, 96,754, and 95,316, respectively. However, in terms of involvement in agricultural activities, the highest participation is observed in the districts of Maruf (68%), Maywand (66%), Zheray (61%), and Arghandab (61%) (Fig. 2.11).

The gender-based population maps and the level of involvement in the agriculture sector were classified based on agricultural intensity. All three variables were then integrated with the RF drought map to develop the final drought-affected zones based on socioeconomic data.

The integrated drought impact map for 2022 indicated that some areas, such as Registan, Shorabak, Ghorak, and Miyanishin, experienced minimal effects, with some parts not affected by drought at all. However, Daman and Nesh were

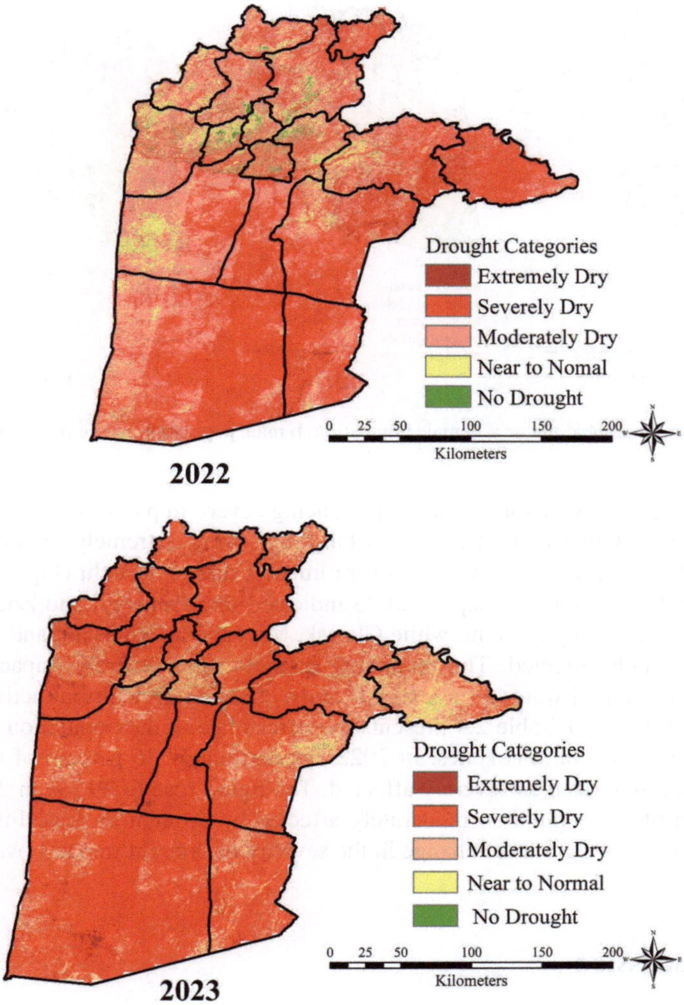

Fig. 2.10 Drought severity maps for 2022 and 2023 of the KP, Afghanistan

Table 2.3 Drought-affected area by percentage in KP during 2022 and 2023

Drought classes	2022	2023
	Area affected (%)	Area affected (%)
Extremely dry	1.1	2.5
Severely dry	41.2	70.5
Moderately dry	46.9	24.2
Near to normal	10.1	2.8
No drought	0.7	0.1
Total	100.0	100.0

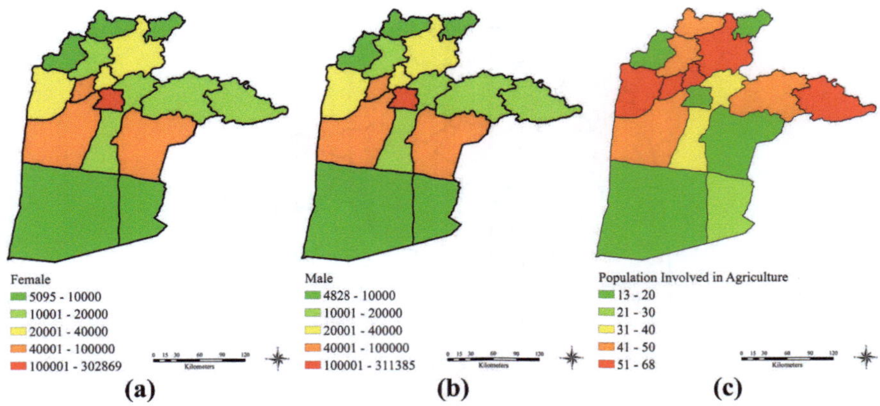

Fig. 2.11 Demographic maps: **a** Female population, **b** male population, and **c** population

partially affected, with some areas experiencing severe impacts and others moderately affected. Other districts, such as Panjwayi, were extremely affected, while the rest of the districts experienced severe impacts due to drought (Fig. 2.12). The integrated drought impact map for 2023 indicated that Panjwayi and Zheray were extremely affected by drought, while Ghorak, Miyanishin, Registan, and Shorabak were moderately affected. The remaining districts were severely impacted, with drought and limited water access significantly affecting agricultural activities and daily life (Fig. 2.12). Table 2.4 presents the percentage of the population impacted by drought across various zones. In 2022, approximately 36 percent of the population was categorized as severely affected. The figure rose to 39.3% in 2023. The proportion of the population moderately affected by drought increased to 48% in 2023, reflecting a significant change in the severity of drought impacts over time.

2.4 Discussion

2.4.1 The Role of Geospatial Techniques in Drought Assessment

This study reveals the substantial impact of drought on agricultural and water resources in Afghanistan, especially in Kandahar Province, where conditions have declined from 2022 to 2023. The implementation of the random forest (RF) model for the generation of drought impact maps has led to the discernment of a marked escalation in drought severity. The data indicates a substantial increase in the percentage of the province experiencing severe impacts, from 41 to 70%, underscoring the escalating challenges faced by populations reliant on surface water and precipitation. The intensified drought conditions pose a multifaceted challenge, particularly with

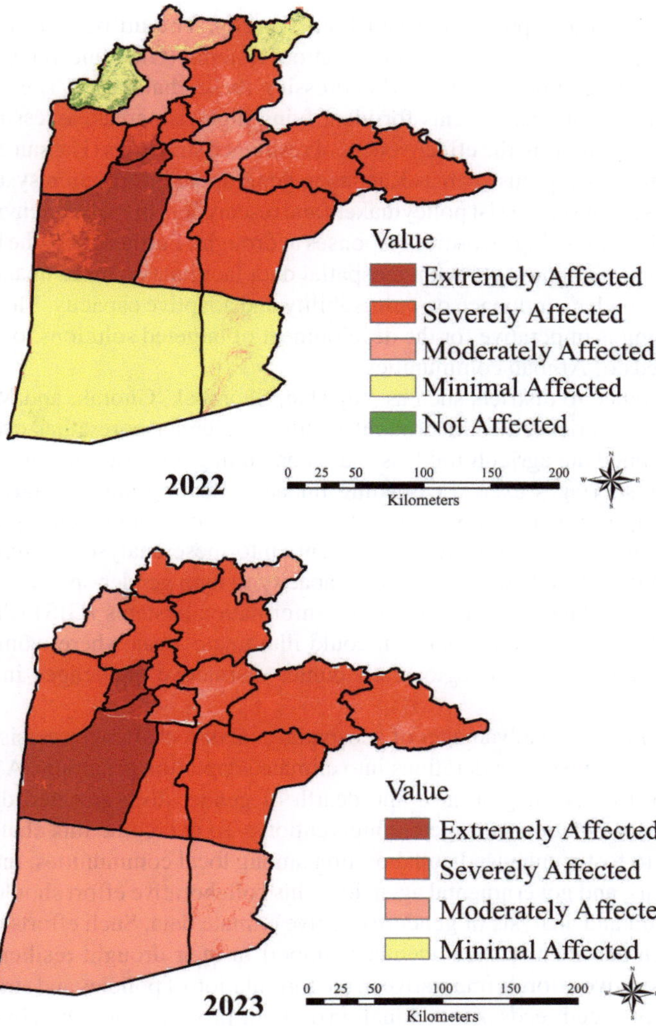

Fig. 2.12 Integrated drought impact map by district (2022–2023) for KP, Afghanistan

Table 2.4 Drought-affected area by percentage in KP during 2022 and 2023

Categories	2022	2023
	Affected zones (%)	Affected zones (%)
Extremely affected zones	0.8	0.1
Severely affected zones	36.4	39.3
Moderately affected zones	12.0	48.4
Minimal affected zones	40.9	12.1
Not affected zones	10.0	0.0

regard to agricultural production and food security. The utilization of geospatial techniques, including geographic information systems (GIS) and remote sensing, is imperative in comprehending and addressing these challenges. These technologies offer invaluable instruments for identifying high-risk areas, assessing drought severity, and monitoring the effectiveness of resilience strategies (Hassan et al., 2025; Li et al., 2024). In regions such as Kandahar, geographic information systems (GIS) and remote sensing can assist policymakers and researchers in tracking environmental changes and formulating informed responses to drought. Furthermore, the integration of socioeconomic indicators with geospatial data facilitates a more nuanced understanding of gender's influence on vulnerability and adaptive capacity. This enhanced understanding is imperative for the development of targeted solutions to address the specific needs of Afghan communities.

The reference to districts such as Registan, Shorabak, Ghorak, and Miyanishin, which have experienced mild drought conditions, underscores that even regions with less immediate agricultural losses are contending with significant water shortages. These shortages exert a cascading impact, affecting not just agriculture but also everyday life, potentially leading to wider socioeconomic consequences. The integration of gender-responsive assessments into these analyses is imperative, as evidenced by studies demonstrating the capacity to map gender-specific vulnerabilities to climate change using geographic information systems (GIS) (Sadia et al., 2016). In Afghanistan, this approach could illuminate areas where women, particularly in drought-affected regions, encounter distinctive challenges in achieving resilience.

Despite the evident advantages of geospatial analysis, challenges persist in the full integration of gender considerations into climate adaptation programs. A significant impediment to this integration is the dearth of gender-disaggregated data, which hinders the formulation of targeted interventions. To overcome this challenge, it is imperative to foster enhanced collaboration among local communities, international organizations, and governmental agencies. This collaborative effort should prioritize the collection and analysis of gender-sensitive climate data. Such efforts ensure that both women and men receive adequate support in their drought resilience efforts. This collaborative effort is imperative in the formulation of policies and strategies that address the nuanced needs of individuals in drought-prone regions, thereby enhancing overall resilience and sustainability. Given Afghanistan's arid, landlocked region, it is imperative to address water scarcity and enhance resilience to droughts to protect food security and livelihoods in the area (FAO, 2023). The findings of this study, in conjunction with the incorporation of geospatial and gender-sensitive data, are imperative for the formulation of policy decisions and the development of strategies for drought mitigation, resource management, and sustainable agricultural practices, particularly in regions susceptible to climatic extremes (FAO, 2020b, 2022).

2.4.2 Gendered Dimensions of Drought Resilience in Afghanistan

Drought has long been a recurring environmental threat in Afghanistan, disproportionately affecting disadvantaged groups, especially women. A multitude of socioeconomic and cultural factors, including restricted access to land, water, financial resources, and decision-making processes, contribute to the heightened vulnerability of women to the impacts of drought (UN Women, 2021). Despite their limited or non-existent participation in decision-making processes related to agricultural resources and adaptation plans, women in many rural areas are often entrusted with crucial responsibilities, such as collecting water and overseeing household food supplies (FAO, 2020b, 2024b). This incongruity hinders their capacity to utilize sustainable coping mechanisms and reduces their resilience to stress induced by drought. The observed incongruity hinders the implementation of sustainable coping mechanisms and reduces their resilience to stress induced by drought (Fig. 2.13).

A substantial body of research has highlighted the exacerbating impact of drought on women, particularly in contexts characterized by gender disparities in access to opportunities and resources. For instance, female-headed households in South Asia are more likely to experience food insecurity during droughts because they have less access to financial aid and irrigation systems. (Aziz & Anjum, 2024; UN Women, 2021). A similar trend is also observed in Afghanistan, where women are frequently excluded from formal agricultural decision-making processes due to the influence of sociocultural norms (Bryan et al., 2024; FAO, 2020b). Consequently, drought interventions that neglect to consider gender-specific vulnerabilities risk excluding a substantial segment of the population from initiatives designed to enhance resilience. It is imperative that drought interventions incorporate a gender-specific perspective to ensure the efficacy of initiatives aimed at enhancing resilience.

2.4.3 Policy Implications and Recommendations

The results of this study highlight the urgent need for gender-responsive drought resilience strategies in Afghanistan. To this end, policymakers must adopt a multi-dimensional approach that incorporates geospatial analysis and socioeconomic assessments to develop inclusive adaptation policies. The following recommendations are deemed to be of the utmost importance:

i. The necessity of enhancing gender-inclusive data collection is paramount (Fig. 2.14). The collection of gender-disaggregated data on subjects such as drought vulnerability, resource access, and adaptation capability should be prioritized by government organizations and academic institutions. This approach will empower policymakers to develop evidence-based programs that address the unique needs of women, as asserted by UN Women in 2021 (UN Women, 2021).

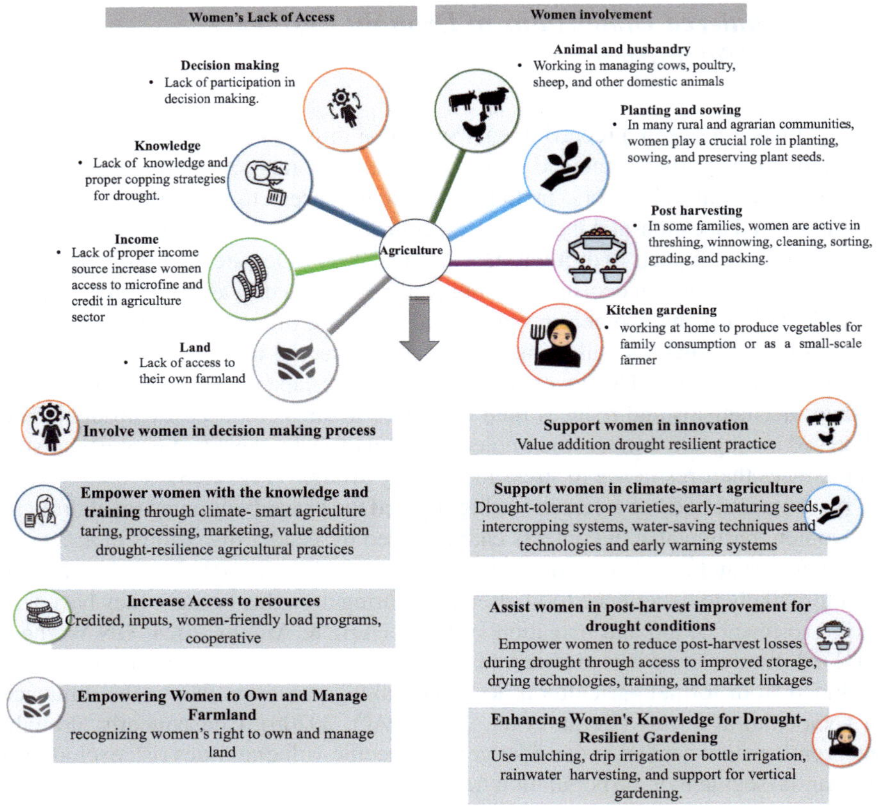

Fig. 2.13 Building drought resilience through Women's Involvement in Agriculture and Livestock Production (Author, 2025)

ii. The creation of initiatives is imperative to ensure that women have equitable access to resources such as agricultural inputs, financial loans, land ownership, and training in CCA techniques. Empirical evidence suggests that women may possess the aptitude to utilize sustainable agricultural methods more proficiently and to a greater extent, thereby enhancing their resilience to the repercussions of drought through appropriate policy and initiatives (Fig. 2.13) (FAO, 2019, 2020a, 2020b).

iii. This study examines the potential benefits of increasing women's participation in decision-making processes. Women should play an active role in national and local policy debates concerning climate resilience. Addressing existing disparities necessitates the establishment of climate committees led by women and the incorporation of gender-sensitive frameworks into drought response plans (WB, 2021).

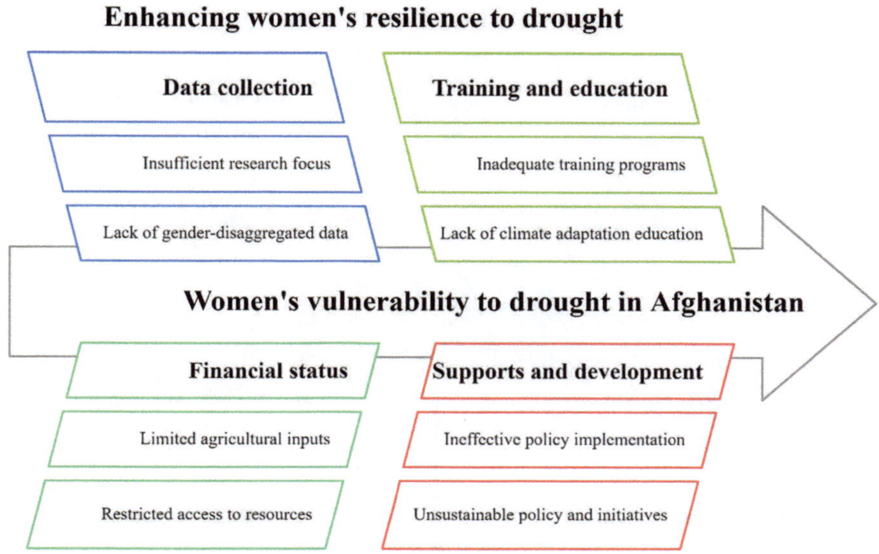

Fig. 2.14 Limitations of enhancing vulnerable women's resilience to drought in Afghanistan (Author, 2025)

iv. It is imperative that gender-responsive planning using geospatial technologies, such as geographic information systems (GIS) and remote sensing, be used to assess drought vulnerability from a gender perspective. The integration of gender-sensitive geospatial analysis into climate adaptation planning has been demonstrated to enhance resource allocation, ensuring that interventions reach the most vulnerable groups (Hassan et al., 2025; Kim et al., 2025).

By adhering to these recommendations, Afghanistan stands to make significant strides in formulating a more sustainable and inclusive strategy for drought resilience. A proposed gender-responsive drought resilience framework for Afghanistan is presented, utilizing a range of methodologies. This framework employs a combination of CCA perspectives, gender-sensitive approaches, and geospatial technologies to address the challenges posed by drought resilience in Afghanistan (Fig. 2.15). A comprehensive understanding of the interplay between gender and environmental concerns is imperative to ensure that adaptation initiatives are inclusive of all segments of society, particularly the most vulnerable populations. (Fig. 2.15).

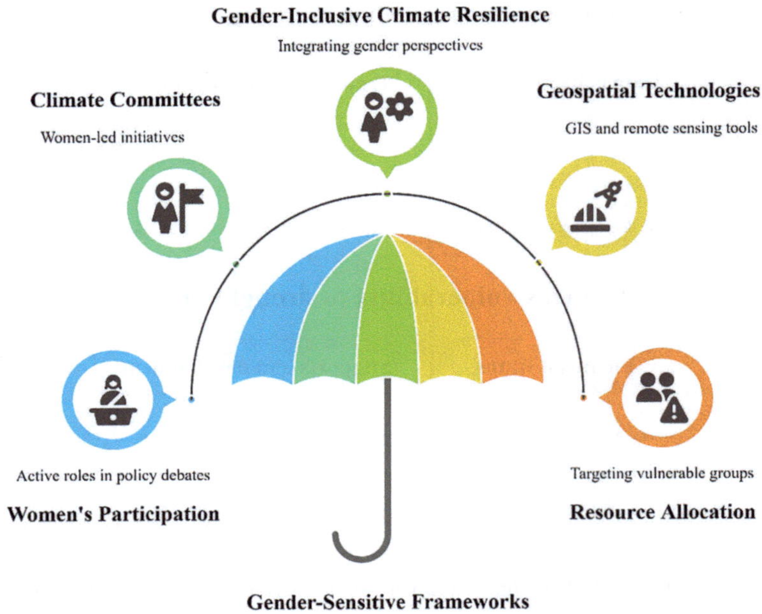

Fig. 2.15 Proposed gender-responsive drought resilience framework for Afghanistan (Author, 2025)

2.5 Conclusions

The study revealed that Kandahar Province (KP) experienced a range of drought severity levels, from minimal to extreme, during the years 2022 and 2023. The deteriorating drought conditions exerted substantial influence on agricultural activities, water resources, and livelihoods, particularly in rural regions. Gender disparities in vulnerability and resilience were evident, as structural injustices, limited access to resources, and exclusion from decision-making processes disproportionately affected women. In Afghanistan, drought continues to be a significant environmental and socioeconomic problem that exerts considerable influence on livelihoods, agriculture, and water resources. However, given that gender is a salient factor in determining resilience and vulnerability, its impacts are not equally dispersed throughout society. The pervasive influence of structural injustices, the limitation of access to resources, and the exclusion from decision-making processes engender a state of vulnerability that is especially pronounced among women, particularly in rural areas. Developing effective and equitable solutions to support drought resilience requires a comprehensive understanding of these disparities in KP. This study underscores the significance of employing gender-responsive geospatial methodologies in the assessment of drought resilience in KP. The integration of remote sensing, GIS-based analysis, and socioeconomic data provides novel insights into the regional and demographic variations in drought risk. The findings emphasize the necessity for targeted interventions that address gender-specific concerns to ensure women have equitable access to climate change adaptation (CCA) as development organizations and policymakers must prioritize gender-responsive planning strategies for climate resilience in the future. The enhancement of women's access to resources, the expansion of their participation in decision-making processes, and the integration of gender-sensitive data into drought management plans have the potential to render climate adaptation more equitable and sustainable. This study provides a foundation for a more robust and equitable future for all Afghan communities grappling with the repercussions of the drought by integrating gender-inclusive policymaking with geospatial analysis.

Acknowledgements The authors express their profound gratitude to the respondents and relevant officials of Kandahar Province (KP) in Afghanistan for their invaluable assistance in facilitating the research team's collection of critical data for the study. Additionally, we extend our appreciation to the Environmental Systems Research Institute (ESRI) for providing the ArcGIS® 10.8.2 software utilized for map preparation in this study. The authors are grateful for the LANDSAT/LC08/C01/T1 8DAY NDVI product from the United States Geological Survey and the Google Earth Engine and Google Colab developers. Additionally, we acknowledge the CHIRPS precipitation dataset from the Climate Hazards Group and the Afghanistan Geodesy and Cartography Head Office (AGCHO) for providing the map of Afghanistan's administrative boundaries. Finally, our heartfelt thanks goes to the ministries and organizations that provided the necessary open-source remote sensing data, reports, and literature, which were instrumental in conducting this research.

Author Contributions Sara Tokhi Arab & Md Shamsuzzoha: Conceived and designed the research, conducted the literature review, developed the methodology, prepared geospatial maps, performed the analysis, wrote and finalized the manuscript, and revised it based on reviewers' comments. Md. Mazharul Anwar, Md. Anwarul Abedin, and Arab Ahmed Shekib contributed to writing and editing. Rajib Shaw: Provided feedback for manuscript development.

Funding This research did not receive any external funding.

Data Availability Statement Upon submission of a formal request, the corresponding author is obligated to make the dataset created and analyzed for this study available to the public. However, there are restrictions on the data's repeatability for reuse and commercially reliable details.

Conflicts of Interest Regarding the research, the authors claim to have no conflicts of interest.

References

ACSO. (2020). *Statistical yearbook of 2019–2020.*

ACSO. (2021). *Statistical yearbook of 2018–2019.* Kabul.

Alam, K., Rahman, Md. H. (2014). Women in natural disasters: A case study from southern coastal region of Bangladesh. *International Journal of Disaster Risk Reduction 8*, 68–82. https://doi.org/10.1016/j.ijdrr.2014.01.003

Alam, K. F., Shamsuzzoha, M., Arab, S. T., Pangaribuan, I. J. P., Ahamed, T. (2024). Application of remote sensing in the analysis of climate extremes due to global climate change. In: Ahamed, T. (Ed.), *Remote sensing application II* (pp. 1–21). New Frontiers in Regional Science: Asian Perspectives. Springer, Singapore. https://doi.org/10.1007/978-981-97-1188-8_1

Arab, S. T., & Ahamed, T. (2023). Near-real-time drought monitoring and assessment for vineyard production on a regional scale with standard precipitation and vegetation indices using Landsat and chirps datasets. *Asia-Pacific Journal of Regional Science, 7*, 591–614. https://doi.org/10.1007/s41685-023-00286-7

Arab, S. T., Noguchi, R., Matsushita, S., & Ahamed, T. (2021). Prediction of grape yields from time-series vegetation indices using satellite remote sensing and a machine-learning approach. *Remote Sensing Applications: Society and Environment, 22*, 100485. https://doi.org/10.1016/j.rsase.2021.100485

Arab, S. T., Islam, Md. M., Shamsuzzoha, Md., Alam, K. F., Muhsin, N., Noguchi, R., Ahamed, T. (2022a). A review of remote sensing applications in agriculture and forestry to establish big data analytics. In: Ahamed, T. (Ed.), *Remote sensing application* (pp. 1–24). New Frontiers in Regional Science: Asian Perspectives. Springer, Singapore. https://doi.org/10.1007/978-981-19-0213-0_1

Arab, S. T., Noguchi, R., Ahamed, T. (2022b). Yield loss assessment of grapes using composite drought index derived from Landsat OLI and TIRS datasets. *Remote Sensing Applications: Society and Environment 26*. https://doi.org/10.1016/j.rsase.2022.100727

Aziz, M., & Anjum, G. (2024). Transformative strategies for enhancing women's resilience to climate change: A policy perspective for low- and middle-income countries. *Women's Health, 20*, 17455057241302032. https://doi.org/10.1177/17455057241302032

Bento, V. A., Gouveia, C. M., DaCamara, C. C., & Trigo, I. F. (2018). A climatological assessment of drought impact on vegetation health index. *Agricultural and Forest Meteorology, 259*, 286–295. https://doi.org/10.1016/j.agrformet.2018.05.014

Breiman, L. (2001). Random forests. *Machine Learning, 45*, 5–32. https://doi.org/10.1023/A:1010933404324

Bryan, E., Alvi, M., Huyer, S., Ringler, C. (2024). Addressing gender inequalities and strengthening women's agency to create more climate-resilient and sustainable food systems. *Global Food Security 40*, 100731. https://doi.org/10.1016/j.gfs.2023.100731

Deininger, F., Woodhouse, A., Kuriakose, A. T., Gren, A., Liaqat, S. (2023). *Placing gender equality at the center of climate action.*

Erman, A., De Vries Robbe, S. A., Thies, S. F., Kabir, K., Maruo, M. (2021). *Gender dimensions of disaster risk and resilience: Existing evidence.* Washington.

Etikan, I. (2017). Combination of probability random sampling method with non probability random sampling method (sampling versus sampling methods). *Biometrics & Biostatistics International Journal 5*, 210–213. https://doi.org/10.15406/bbij.2017.05.00148

Faliha Dzakiyah, I., Saraswati, R. (2020). Drought area of agricultural land using Tasseled Cap Transformation (TCT) method in Ciampel Subdistrict Karawang Regency. In *E3S Web of Conferences* (Vol. 211, p. 02005). https://doi.org/10.1051/e3sconf/202021102005

FAO. (2019). *Afghanistan: Drought risk management strategy.*

FAO. (2020a). *The role of gender in climate-smart agriculture.* Rome.

FAO. (2020b). *The state of food security and nutrition in the world 2020, the state of food security and nutrition in the world 2020.* FAO, IFAD, UNICEF, WFP and WHO, Rome. https://doi.org/10.4060/ca9692en

FAO. (2022). *SCALA gender and social inclusion strategy: Scaling up climate ambition on land use and agriculture (SCALA).* Rome.

FAO. (2023). *Assessing of agricultural drought severity in Afghanistan.* Rome. https://doi.org/10.5067/MODIS/MOD13Q1.061

FAO. (2024a). *Better life: Gender equality and women's empowerment at the Food and Agriculture Organization of the United Nations.* Rome. https://doi.org/10.4060/cd0671en

FAO. (2024b). *Gender and climate change, agroecology and biodiversity.* Rome.

Gorelick, N., Hancher, M., Dixon, M., Ilyushchenko, S., Thau, D., & Moore, R. (2017). Google earth engine: Planetary-scale geospatial analysis for everyone. *Remote Sensing of Environment, 202*, 18–27. https://doi.org/10.1016/j.rse.2017.06.031

Hang, Q., Guo, H., Meng, X., Wang, W., Cao, Y., Liu, R., De Maeyer, P., & Wang, Y. (2024). Optimizing the vegetation health index for agricultural drought monitoring: Evaluation and application in the yellow river Basin. *Remote Sensing (Basel), 16*, 4507. https://doi.org/10.3390/rs16234507

Hassan, S. S., Goheer, M. A., Farah, H., Hafeez, F., Sheraz, K., Fahad, J., & Tariq, A. (2025). Geospatial assessment of climate variability and drought patterns: A case study from Pakistan. *Theoretical and Applied Climatology, 156*, 185. https://doi.org/10.1007/s00704-025-05414-1

Htun, A. M., Arab, S. T., Ahamed, T. (2024). Drought monitoring and rice yield loss assessment using various drought indices and the SPI by an integrated approach of the AHP in the central dry zone of Myanmar. In: Ahamed, T. (Ed.), *Remote sensing application II* (pp. 295–320). New Frontiers in Regional Science: Asian Perspectives. Springer, Singapore. https://doi.org/10.1007/978-981-97-1188-8_11

International Fund for Agricultural Development (IFAD). (2019). Arghandab integrated water resources development programme design report. *The Asian Development Bank*. https://webapps.ifad.org/members/eb/128/docs/EB-2019-128-R-20-Design-report.pdf

IOM. (2022). *Provincial overview Kandahar Province.*

IPCC. (2022). *Climate change 2022: Impacts, adaptation and vulnerability.* New York. https://doi.org/10.1017/9781009325844

Kim, H.-J., Rahman, M., Hammad, Z., Kim, H.-S., Lee, S.-J., Kim, T.-E., & Jung, S.-H. (2025). Comprehensive analysis of coastal flood susceptibility, drought severity, and crop water stress using data fusion. *Geomatics, Natural Hazards and Risk, 16*, 2483809. https://doi.org/10.1080/19475705.2025.2483809

Kogan, F. N. (1995). Application of vegetation index and brightness temperature for drought detection. *Advances in Space Research, 15*, 91–100. https://doi.org/10.1016/0273-1177(95)00079-T

Komino, T., Safi, N. K. (2024). Complex disaster risk management in Afghanistan. In: Izumi, T., Abe, M., Fujita, K., Shaw, R. (Eds.), *All-hazards approach: Towards resilience building* (pp. 141–151). Springer Nature Singapore, Singapore. https://doi.org/10.1007/978-981-97-1860-3_10

Li, P., Jia, L., Lu, J., Jiang, M., Zheng, C., & Menenti, M. (2024). Investigating the response of vegetation to flash droughts by using cross-spectral analysis and an evapotranspiration-based drought index. *Remote Sensing (Basel), 16*, 1564. https://doi.org/10.3390/rs16091564

Liakos, K. G., Busato, P., Moshou, D., Pearson, S., & Bochtis, D. (2018). Machine learning in agriculture: A review. *Sensors (Switzerland).* https://doi.org/10.3390/s18082674

McFeeters, S. K. (1996). The Use of the Normalized Difference Water Index (NDWI) in the Delineation of Open Water Features. *International Journal of Remote Sensing, 17*, 1425–1432. https://doi.org/10.1080/01431169608948714

Mckee, T. B., Doesken, N. J., Kleist, J. (1993). The relationship of drought frequency and duration to time scales, In: *Eighth Conference on Applied Climatology* (pp. 179–183). California. https://climate.colostate.edu/pdfs/relationshipofdroughtfrequency.pdf

Mohanty, A., Hussain, M., Mishra, M., Kattel, D. B., Pal, I. (2019). Exploring community resilience and early warning solution for flash floods, debris flow, and landslides in conflict prone villages of Badakhshan, Afghanistan. *International Journal of Disaster Risk Reduction 33*, 5–15. https://doi.org/10.1016/j.ijdrr.2018.07.012

OCHA. (2024a). *Afghanistan: Drought—2021–2025.*

OCHA. (2024b). *Afghanistan Malnutrition Crisis 2024—DREF Operation (MDRAF017).*

Park, H., Kim, K., & Lee, D. K. (2019). Prediction of severe drought area based on random forest: Using satellite image and topography data. *Water (Switzerland), 11*, 705. https://doi.org/10.3390/w11040705

RRERS. (2007). *Provincial profile for-Kandahar Province-executive summary.*

Sadia, H., Iqbal, M. J., Ahmad, J., Ali, A., Ahmad, A. (2016). Gender-sensitive public health risks and vulnerabilities' assessment with reference to floods in Pakistan. *International Journal of Disaster Risk Reduction 19*, 47–56. https://doi.org/10.1016/j.ijdrr.2016.08.024

Saha, A., Patil, M., Goyal, V. C., Rathore, D. S. (2018). Assessment and impact of soil moisture index in agricultural drought estimation using remote sensing and GIS techniques. In: *ECWS-3. MDPI* (p. 2), Basel Switzerland. https://doi.org/10.3390/ECWS-3-05802

Shahfahad, Bindajam, A. A., Naikoo, M. W., Horo, J. P., Mallick, J., Rihan, M., Malcoti, M. D., Talukdar, S., Rahman, R., Rahman, A. (2023). Response of soil moisture and vegetation conditions in seasonal variation of land surface temperature and surface urban heat island intensity in sub-tropical semi-arid cities. *Theoretical and Applied Climatology, 153*(1), 367–395. https://doi.org/10.1007/s00704-023-04477-2

Shamsuzzoha, M., Noguchi, R., & Ahamed, T. (2021). Damaged area assessment of cultivated agricultural lands affected by cyclone bulbul in coastal region of Bangladesh using Landsat 8 OLI and TIRS datasets. *Remote Sensing Applications, 23*, 100523. https://doi.org/10.1016/j.rsase.2021.100523

Shamsuzzoha, M., Noguchi, R., & Ahamed, T. (2022). Rice yield loss area assessment from satellite-derived NDVI after extreme climatic events using a fuzzy approach. *Agricultural Information Research, 31*, 32–46. https://doi.org/10.3173/air.31.32

Shamsuzzoha, M., Shaw, R., & Ahamed, T. (2024). Machine learning system to assess rice crop change detection from satellite-derived RGVI due to tropical cyclones using remote sensing dataset. *Remote Sensing Applications, 35*, Article 101201. https://doi.org/10.1016/j.rsase.2024.101201

Tewari, N., Bush, A., Butt, M. N., Stevens, E., Zafar, S. (2023). *Gendered dimensions of loss and damage in Asia.* https://doi.org/10.21201/2023.000005

UN. (1990). *Maruf Mission Report.*

UN Women. (2021). *Training manual on gender and climate change resilience.*

USGS. (2014). *Landsat 8 bands (WWW Document).* United States Geological Survey. URL https://landsat.gsfc.nasa.gov/satellites/landsat-8/landsat-8-bands/. Accessed 8 April 23.

USGS. (2019). *Landsat 8 (L8) data users handbook.* United States Geological Survey (USGS).

USGS. (2025). Earth Explorer (WWW Document). URL https://earthexplorer.usgs.gov/. Accessed 28 Feburary 25.

WB. (2021). *Gender dynamics of disaster risk and resilience (WWW Document).* World Bank (WB) Group. URL https://www.worldbank.org/en/topic/disasterriskmanagement/publication/gender-dynamics-of-disaster-risk-and-resilience. Accessed 8 September 24.

Chapter 3
Gender Dynamics in Climate Change-Induced Disasters in Bangladesh: Early Warning to Build Back Better

Md. Shamsuzzoha, Sanjida Hossain Setu, Kamrunnahar Ishana, A. K. M. Abdul Ahad Biswas, Mohammad Kabirul Islam, Md. Saiful Islam, Md. Rasheduzzaman, Md. Jashim Uddin, Md. Anwarul Abedin, and Rajib Shaw

Abstract Climate change-induced disasters are gender-neutral; however, the effects are not. Males, females, children, disabled persons, or any other marginalized groups face separate levels of vulnerabilities, risks, and emergencies to extreme climatic disasters, driven by different humanitarian standards, gender-based interactions, and lack of protections, socioeconomic factors, and discriminations in the society. This study explored gender dynamics in preparedness to response phases on alerts and early warning stages in Bangladesh. In the emergency stage, due to various types

Md. Shamsuzzoha (✉) · S. H. Setu · Md. Rasheduzzaman
Department of Emergency Management, Faculty of Environmental Science and Disaster Management, Patuakhali Science and Technology University, Patuakhali, Bangladesh
e-mail: zohageo@pstu.ac.bd

K. Ishana
Department of Soil Science, Government Brojomohun College, National University, Barishal, Bangladesh

A. K. M. Abdul Ahad Biswas
Department of Disaster Risk Management, Faculty of Environmental Science and Disaster Management, Patuakhali Science and Technology University, DumkiPatuakhali, Bangladesh

M. K. Islam · Md. Saiful Islam
Department of Soil Science, Faculty of Agriculture, Patuakhali Science and Technology University, DumkiPatuakhali, Bangladesh

Md. Jashim Uddin
Department of Social Work, Government Brojomohun College, National University, Barishal, Bangladesh

Md. Anwarul Abedin
Department of Soil Science, Faculty of Agriculture, Bangladesh Agricultural University, Mymensingh, Bangladesh

R. Shaw
Graduate School of Media and Governance, Keio University, Fujisawa, Japan

Md. A. Abedin et al. (eds.), *Climate Crisis in South and East Asia*, Sustainability Solutions, https://doi.org/10.1007/978-3-032-01012-4_3

of risks, all catastrophic incidents are always gender-sensitive in place, on time, and for people. Steps for completing evacuation, search, rescue, relief, and short-term rehabilitation are also necessarily synchronized by gender dynamics from the phases like response and recovery to build back better. Gender dynamics are also important to maintain humanitarian principles such as humanity, neutrality, impartiality, and independence, which are vital to humanitarian action. To recognize the gender dynamics due to climate-induced disaster risks, the minimum humanitarian standards for disaster response are WASH (WASH: water supply, sanitation, and hygiene promotion), food security and nutrition, shelter and settlement, and health action. Therefore, in the reduction phase, to build back better to mitigate disaster risks, the governmental and non-governmental disaster risk reduction (DRR) plans, policies, and institutional frameworks should be gender and child sensitive. Gender dynamics are successfully regulated and affected by women's empowerment, capacity development, need-based education, child protection, gender friendly employment, stopping violence, child marriage, child labour, climate displacement, and so on. As one of the most disaster-prone countries, Bangladesh also faces gender-focused challenges by climate change-induced disasters in several natural outcomes, including life expectancy, corruption, unemployment, inequalities, labour force re-entry, and relative asset losses. Therefore, this research review aims to summarize the gender dynamics and it's prospective in climate change-induced disasters in Bangladesh.

Keywords Bangladesh · Build back better · Climate change · Early warning · Gender dynamics

3.1 Introduction

In general, disasters seem to be gender-neutral catastrophes affecting people and their livelihoods, regardless of caste, class, race, or gender (Khan, 2022). Bangladesh is acknowledged globally as one of the most vulnerable countries to climate change due to its geographical position, low-lying elevation, and dense population (Hossen et al., 2019). Several kinds of extreme events, like floods, tropical cyclones, storm surges, land erosion, heat waves, excessive rainfall, droughts, tornadoes, salinity intrusion, etc., are occurring more frequently or with greater intensity (Alam et al., 2024; Shamsuzzoha & Al-Maruf, 2012; Shamsuzzoha et al., 2022). These events result in livestock losses, pastureland damage, increased fodder scarcity, destroyed shelters, decreased production, increased management costs, increased disease incidence, etc., in Bangladesh (Hossain et al., 2011). Climate change affects everyone, yet not everyone is equally affected. The home location of an individual, along with other socioeconomic factors like race, gender, age, culture, and ethnicity, might influence that individual's susceptibility to the effects of climate change. Vulnerability to an event is not based on sex or biological differences between men and women, but rather on how society constructs what it means to be a man or woman, what roles they should play, and how they should behave in the context of disasters (Bradshaw &

Fordham, 2015). The gender perspective on the effects of climate change is among the most important since gender relations set the norms for men's and women's roles and status in a given social, religious, and cultural context. Men and women face different risks of exposure and vulnerability to natural hazards, which leads to differentiated impacts of disasters (Sophie et al., 2021).

In the coastal regions of Bangladesh, women are vulnerable to climate change-induced disasters because of the gender inequality and socio-cultural norms that exist in Bangladesh (Shamsuddoha et al., 2024). Women become more susceptible both in the short-term, i.e., to major natural disasters, cyclones, floods, and long-term, i.e., sea level rise, salinity intrusion in water and soil, land erosion, droughts, and climatic events, as these disasters enhance gender inequalities (Md et al., 2022). Disasters like floods, cyclones, and riverbank erosion brought on by climate change force many coastal populations to relocate (Alam & Ahamed, 2023; Arab et al., 2022; Shamsuzzoha et al., 2021). Women had greater workloads, hygienic issues, and the possibility of sexual harassment or violence while living in temporary housing. Women are especially vulnerable to the effects of climate change-induced natural hazards due to their limited access to economic opportunities, healthcare, and education due to these hazards (Shamsuddoha et al., 2024). Due to disasters, women face partial and full loss in their lives and livelihoods. Partial losses include their family members' education, life security, and loss of homestead-based occupation (Alam & Rahman, 2014a, 2014b). Immediately following a calamity, there is a tendency for violence to rise. Females and children are at greater risk of physical and sexual violence in emergency settings (Gennari et al., 2015). In the developing world, adolescent girls' specific needs are mostly invisible and remain unfulfilled as they are being conceptualized neither as women nor children (Bradshaw & Fordham, 2015). In a society characterized by male domination and the subjugation of women, the vulnerabilities of specific intersectional groups become evident. These groups include the economically disadvantaged, the elderly, the disabled, minors, widows, and those left behind due to the migration of dependent family members.

Women are more vulnerable than men to disasters; thus, it is important to acknowledge their unique responsibilities, talents, vulnerabilities, and demands and consider them in disaster risk reduction in the built environment. Gender mainstreaming as a way of bringing a gender perspective into disaster risk reduction can be applied to recognize women's varying needs and capacities and integrate them into disaster risk reduction in the built environment (Ginige et al., 2014a, 2014b). Socially, women have very little decision-making power; therefore, families and communities seldom pay attention to what women have to say. Because they lack the necessary information or experience, they cannot apply risk reduction-related skills or knowledge in practical settings. In the coastal areas of Bangladesh, women have almost no opportunity for income and cannot invest to reduce disaster-induced risks (Alam & Rahman, 2014a, 2014b). Women can play a major role in community mobilization during crises due to gendered societal roles, yet their contributions are frequently disregarded and marginalized (Lafrenière et al., 2019). Thus, community resilience may improve more, if the local women can be engaged at the local operational level of disaster management (Islam et al., 2014).

3.2 Materials and Methods

3.2.1 Study Area

A total of five coastal districts have been selected from the southern coastal region of Bangladesh for this study. These study districts are Bagerhat, Khulna, Satkhira, Barguna, and Patuakhali, respectively. The basemap shapefiles were collected from the National Spatial Data Infrastructure (NSDI) website of the Government of Bangladesh (GoB) through the link https://new.nsdi.gov.bd/ as open-source data (Fig. 3.1). The five districts are particularly vulnerable to a range of adverse effects related to climate change, including floods, tropical cyclones, storm surges, land erosion, heat waves, excessive rainfall, droughts, tornadoes, and salinity intrusion (Shamsuzzoha, 2023). The extreme events have been shown to result in a variety of detrimental outcomes, including decreased crop yield and livestock losses, land use damage, increased fodder scarcity, destroyed shelters, decreased production, increased management costs, and increased disease incidence. The climatic conditions in these districts are marked by elevated temperatures, substantial precipitation, and high humidity levels. The coastal regions are particularly vulnerable to the impacts of climate change, including rising sea levels, increased frequency and intensity of cyclones, and salinity intrusion in Bangladesh (Hossen & Sultana, 2023; Shamsuzzoha & Ahamed, 2023, 2024). Women residing in these regions face heightened vulnerability to climate change-induced disasters due to the exacerbating effects of gender inequality and the constraints imposed by socio-cultural norms (Ahamed et al., 2021; Naz & Doneys, 2022a, 2022b). They shoulder greater responsibilities in their work and face challenges related to hygiene (Alam & Rahman, 2014a, 2014b; Shanta et al., 2025). Besides, they are vulnerable to sexual harassment and violence while residing in temporary shelters (Alam & Rahman, 2019; Ginige et al., 2014a, 2014b). Furthermore, women encounter constrained access to economic prospects, healthcare services, and educational opportunities, thereby amplifying their preexisting vulnerabilities (Bryan et al., 2024; Naz & Doneys, 2022b; WB, 2021).

3.2.2 Literature Review and Data Collection

A comprehensive review of the existing literature, including books, reports, and published documents, was conducted to identify relevant sources of information. This research drew data from numerous government websites in Bangladesh, including those of the Bangladesh Bureau of Statistics (BBS), the Ministry of Environment, Forest and Climate Change (MoEFCC), and the Ministry of Women and Child Affairs (MoWCA), the Ministry of Disaster Management and Relief (MoDMR), among others. Furthermore, another systematic review of pertinent research articles published in academic journals was conducted to support the study. However, the

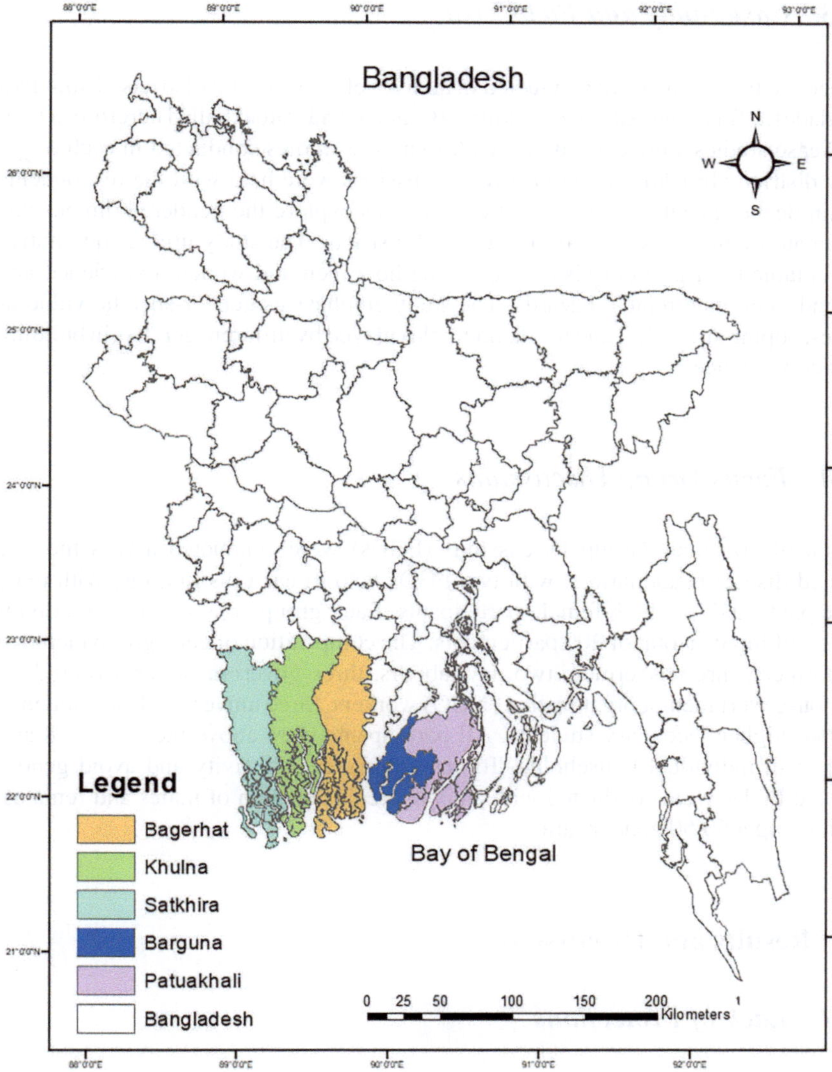

Fig. 3.1 Climate change-induced disaster-prone coastal study districts in Bangladesh (Author, 2025)

research faced certain limitations. Due to limitations in time and funding, we were unable to conduct a questionnaire-based survey.

3.2.3 Case Study and Field Visit

Respective field visits were conducted in all five selected coastal districts of southern Bangladesh: Bagerhat, Khulna, Satkhira, Barguna, and Patuakhali. Therefore, a total of 20 case studies were carried out, with four case studies conducted in each of the study districts. In addition, descriptive discussions were held with the respondents to gain deeper insights. The present case studies explore the gendered dimensions of climate-induced disasters in the selected districts. The study utilizes qualitative data obtained during field visits to examine how men and women experience and respond to climate-related hazards. The study emphasizes gender-specific vulnerabilities, coping strategies, and the distinct roles played by different genders in building disaster resilience.

3.2.4 Focus Group Discussions

A total of 10 Focus Group Discussions (FGDs) were conducted across the five selected district headquarters, with two FGDs held in each district, one with male participants and one with female participants. Each group consisted of 20 individuals, resulting in a total of 200 participants. The composition of each group included five farmers, three fishermen, two day-labours, three government service holders, two non-governmental organization (NGO) workers, three university-level students, and two higher secondary students. All participants were above the age of 18 and came from individual households. To ensure gender inclusivity and avoid gender bias, the FGDs were conducted with an equal representation of males and females, maintaining a 50:50 gender ratio.

3.3 Results and Discussion

3.3.1 Lack of Protections

In everyday life, women who experience social and economic disadvantages are likely to be more vulnerable to or affected more thoroughly by a disaster. Rather, a higher-risk woman is more vulnerable to living in hazardous areas (Yumarni et al., 2014). Due to current power imbalances, external shocks that generate financial and environmental crises may disproportionately affect women and girls. Women and girls have particular reproductive health concerns, even while their lesser access to necessities like food and water may make them more vulnerable to health risks. There are specific health risks for expectant mothers and women who give birth shortly before, during, or right after an event. Nursing mothers, for example, are particularly susceptible to malnutrition and dehydration and this might be assumed to cause higher mortalities and morbidity (Bradshaw & Fordham, 2015). However, many parents in Bangladesh consider cyclone shelters to be unsafe for girls and prefer to leave them at home (Swarup et al., 2011).

3.3.2 Identification of Gender-Sensitive Socioeconomic Factors

Gender equality and the ability to bounce back from shocks in future are determined by the current socioeconomic conditions, which are influenced by gender differences in disaster outcomes (Sophie et al., 2021). Gendered norms directly impact women's ability to respond to natural hazards (Bradshaw & Fordham, 2015). Among social factors, gender is a significant determinant and a major dimension of social difference. The social dimensions of gender vulnerability are not equally distributed; some are more prevalent than others. These include women who are heads of households, women who live alone, women who are widows with numerous dependents, women who are homeless, women who lack skills, women who lack access to education and training, women who are illiterate, and women who experience violence against women (Yumarni et al., 2014). Because women's employment typically pays less than men's, they may be less wealthy and/or dependent on men for cash, which limits their influence at home and in the community (Bradshaw & Fordham, 2015) (Fig. 3.2).

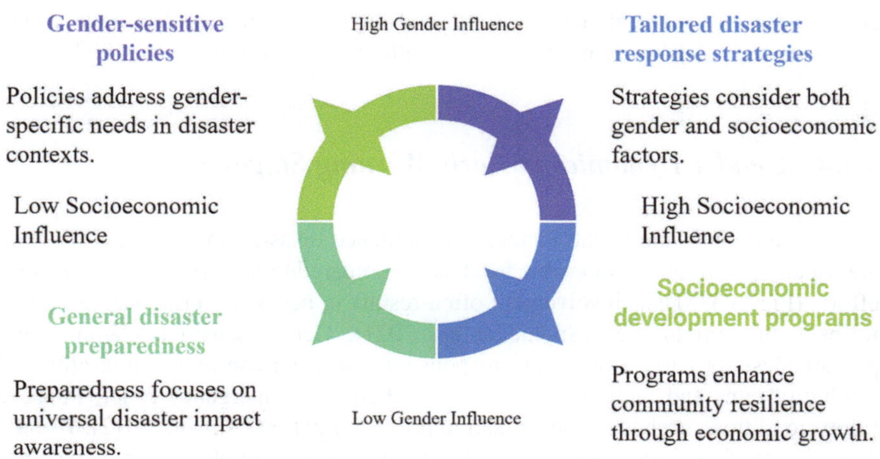

Fig. 3.2 Gender and socioeconomic influence on disaster vulnerability (Author, 2025)

3.3.3 Discrimination in Society

Illiteracy, food shortages poor health, patriarchy, religious taboos, and traditional standards are further examples of gender inequality. Furthermore, while family authority relations and gender-based socioeconomic, cultural, and institutional barriers persist, women's mobility, income, and economic prospects are shifting (Md et al., 2022). Sometimes disasters strengthen existing male domination of resources and discrimination against women within a patriarchal culture and it leaves inequality in access and control to power, structure, and resources within the disaster management process (Yumarni et al., 2014). Women and girls experience disasters differently from men, suffering longer-term and more intangible impacts such as a rise in violence or greater insecurity in employment (Bradshaw & Fordham, 2015). During disasters, women in Bangladesh must deal with difficulties that are distinct from those faced by men, leading to a type of functional dysfunction. Women must deal with sexual harassment, a lack of resources for aid, and lost livelihood prospects. Instead, they have limited opportunities to participate in any management or response operations. Moreover, they also suffer from inverse care law after a disaster (Alam & Rahman, 2014a, 2014b). In Bangladesh, women are even more susceptible to the adverse impacts of climate variability and changes as they are often not allowed to participate in the public sphere and are therefore, less likely to receive critical information for emergency preparedness (CCC, 2009). Rather, the traditional subordinate position of women in society and limited access to economic resources results in higher risk which becomes crucial during any disaster in Bangladesh (Alam & Collins, 2010). In developing countries like Bangladesh, women experience inadequate access to medical support, food, and physical protection and have scanty resources and rights resulting in a higher death rate (Chowdhury et al., 1993).

3.3.4 Gender Dynamics in Early Warning Stages

Gender can contribute to the factors that influence disaster preparedness. Women are usually potentially responsible for disaster preparedness, response, and recovery efforts (Fig. 3.3). Their involvement often results in better performance and transforms their communities (Sophie et al., 2021). Gender-sensitive planning and preparedness for early warnings or response to a disaster can reduce mortality and morbidity rates and facilitate impartial distribution of emergency relief, improve safety conditions in shelter houses, and improve mitigation (Ahmad, 2012). Women differ widely from men in the preparedness case where they play a complementary rather than an independent role (Alam & Rahman, 2014a, 2014b). Participating in local-level DRR helps women realize that they can play a leading role in all phases, including saving men. Women's engagement in early warning systems and other preparedness activities can minimize the necessity for them and others to be rescued (Buvinic et al., 1999). Women play crucial roles pre, during, and after disaster events.

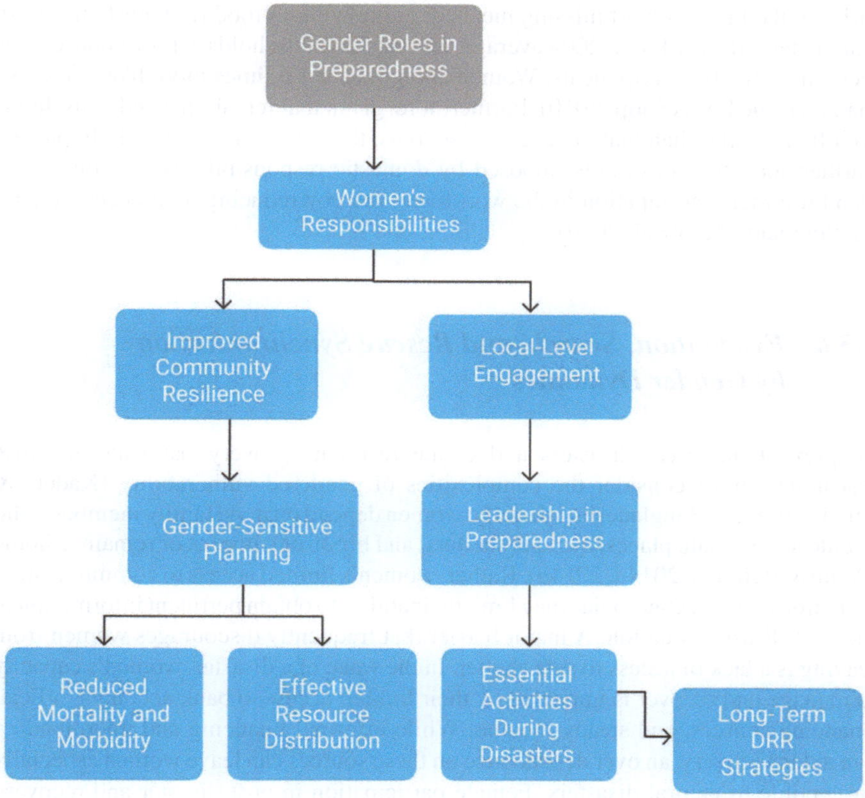

Fig. 3.3 Gender dynamics in disaster preparedness in Bangladesh (Author, 2025)

They prepare portable stoves, gather firewood, store dry food (Chira, Muri, and Khoi, biscuits and, molasses), and essential medicines. Whenever possible, they set aside a certain amount from their family's expenses. To keep utensils hanging above the floor during floods, they make reticulate sacks called Chhika, which are made of jute and hang from the wall or ceiling. They also plant trees specifically fruit, banana, and Ipomoea carnea around their houses as part of the disaster risk reduction (DRR) approach (Alam & Rahman, 2014a, 2014b).

3.3.5 Gender-Sensitive Emergency Stage Evaluation

Gendered differences in disaster impacts can also influence resilience to future disasters, creating a negative feedback loop (Sophie et al., 2021). An indirect gender-sensitive change of catastrophe can be household consumption patterns. When there is a shortage of food, as frequently occurs in the wake of a natural disaster, women and girls are the first to report missing meals or going without food for their families. In Bangladesh, during Flood 2004 average, among 326 households 5.1% of households were more likely to skip meals. Women and girls are 1.6 times more likely to do so than men and boys (Shoji, 2010). Furthermore, given that female-headed households are often smaller than male-headed households, they are more negatively impacted. Furthermore, the constraints imposed by domestic responsibilities on women can also limit their participation in the workforce, thereby reducing their access to paid employment. (Koolwal, 2019).

3.3.6 Evacuation, Search, and Rescue Synchronization by Gender Dynamics

To prevent and treat disasters and create resilient recovery pathways, disaster researchers must consider the complexities of gendered vulnerability (Kadetz & Mock, 2018). In Bangladesh, generally, women depend on male family members who decide to go to safe places, cyclone shelters, and big strong houses or remain at home (Alam & Rahman, 2014a, 2014b). Rather, women's limited access to communication and information technologies may limit their ability to obtain pertinent information in the post-disaster situation. A major barrier that frequently discourages women from leaving is a lack of access to safe shelter. In the wake of a disaster, women's capacity to manage and recover is impacted by their limited access to bank accounts, official financing sources, and steady income. While informal financing and microlending can aid in recovery, an over dependence on these sources can leave women especially susceptible to natural disasters. Female participation in post-disaster and recovery phases may transform gender dynamics. (Sophie et al., 2021). Bangladesh during the 1991 storm suggests that inadequate conditions for women in cyclone shelters may be a significant factor in the decision of female household members to remain at home. They also find that fear of looting is an important factor keeping families from evacuating; but in this case, men tend to stay behind to protect the property (Haque, 1995).

3.3.7 Gender Dynamics to Maintain Humanitarian Principles

The objectives of humanitarian action are twofold: to safeguard human dignity and to preserve life and health. It is imperative for humanitarian agents to refrain from engaging in political activities, considering factors such as race, religion, or ideology, and to avoid taking sides in conflicts. Humanitarian action must be prioritized based only on need, with no discrimination committed based on nationality, race, gender, religion, class, or political beliefs. When it comes to places where humanitarian action is being carried out, humanitarian activity must be independent of any actor's political, economic, military, or other goals. Gender dynamics for minimum humanitarian standards for disaster response in Bangladesh:

i. WASH (WASH: water supply, sanitation, and hygiene promotion): Women in rural regions face societal, domestic, and intrapersonal harassment and abuse as a result of inadequate or nonexistent water and sanitation infrastructure because of intersectional social dynamics.
ii. Food security and nutrition: Food security remains a critical global issue in terms of climate change effects. Gender mainstreaming is a necessary component of adaptation techniques aimed at addressing food insecurity because it prevents women and girls from bearing an excessive amount of the burden of ensuring intra-household food security (Alston & Akhter, 2016).
iii. Shelter and settlement: Social structures like gender systems have a significant influence on how people think and make decisions that contribute to their immobility during a disaster. In Bangladesh, evacuation is occasionally impossible, and previous disasters have claimed the lives of more women than males (Ayeb-Karlsson, 2020). Rather, women, children, the elderly, and the disabled get trapped in vulnerable areas due to a lack of migration options (Begum, 2017).
iv. Health action: Various socio-cultural elements, particularly cultural perspectives on gender roles and relationships, have a big impact on women's access to healthcare during and after disasters. These effects are amplified in times of economic hardship (Rezwana, 2017).

3.3.8 Gender Dynamics in the Reduction Phase

The risk reduction phase of any disaster is associated with the coping strategies of the affected community. When a community experiences a disaster, its capacity to recover will depend on whether it has coping strategies that can help it secure or diversify its sources of revenue, adjust to new technologies, or endure income shocks. The ability to change sources of income, government assistance, financial accessibility, and migratory adaption are examples of coping mechanisms. Women tend to rely on informal finance more than men, which can make it more difficult for them to access funds in case of a disaster (Sophie et al., 2021). In Bangladesh, in the event of flooding, both men and women accumulated jewellery as a precautionary savings strategy to mitigate potential financial losses (Quisumbing et al., 2017). Following a disaster, individuals move in search of better economic possibilities, to rebuild their livelihoods, and to reduce the risk of another one. In Bangladesh, around 5–20% of flood exposure at the sub-district level leads to increased short-distance mobility within the district, primarily among women and the poor (Gray & Mueller, 2012). Gender dynamics in disaster risk reduction (DRR):

i. **Plans**: Ensuring that both men and women are involved in the planning process can lead to more comprehensive and effective DRR strategies. Conducting gender analysis helps identify the different needs, capacities, and vulnerabilities of men and women.

ii. **Policies**: Policies must explicitly address gender inequalities and aim to empower women. This includes ensuring equal access to resources, information, and decision-making processes.

iii. **Institutional frameworks**: Training and capacity-building programmes should be designed to enhance the skills of both men and women in DRR. Appointing gender focal points within DRR institutions can help ensure that gender considerations are integrated into all aspects of disaster management. Implementing gender-sensitive monitoring and evaluation systems can help track the effectiveness of DRR initiatives and ensure that they are meeting the needs of both men and women.

3.3.9 Build Back Better to Mitigate Disaster Risks

Reducing disaster risk for women demands an unequivocal focus on reducing gender inequalities (Bradshaw & Fordham, 2015). Addressing gender inequalities by enhancing the participation of women in decision-making is essential for building communities' resilience to natural disasters (Sophie et al., 2021). However, women better understand their necessity, so their involvement and leadership in disaster decision-making are crucial (Tanner et al., 2018). In Bangladesh, it has worked well to involve women in community mobilization initiatives aimed at addressing cultural barriers to shelter access. This could be because they can recognize women's needs with ease or because, in certain situations, women are more likely to trust other women.

Women's contributions to resilience and disaster risk reduction are supported and legitimized when they are represented more frequently in civil protection and disaster risk management (Fig. 3.4). In addition to strengthening their standing in the community, providing women's groups with emergency response and preparedness training and resources would guarantee that more women receive information on disaster readiness and response. Reconstruction after a disaster presents an opportunity to break down the barriers that women encounter. Moreover, after a disaster, reconstruction can strengthen preparedness by, for example, turning schools and other public buildings into multi-purpose shelters that can be used in future disaster events. (Sophie et al., 2021). If the infrastructure, such as utilities, roads, community assets, and cultural sites, are maintained, then the community will have better resources to recover (Eburn, 2015).

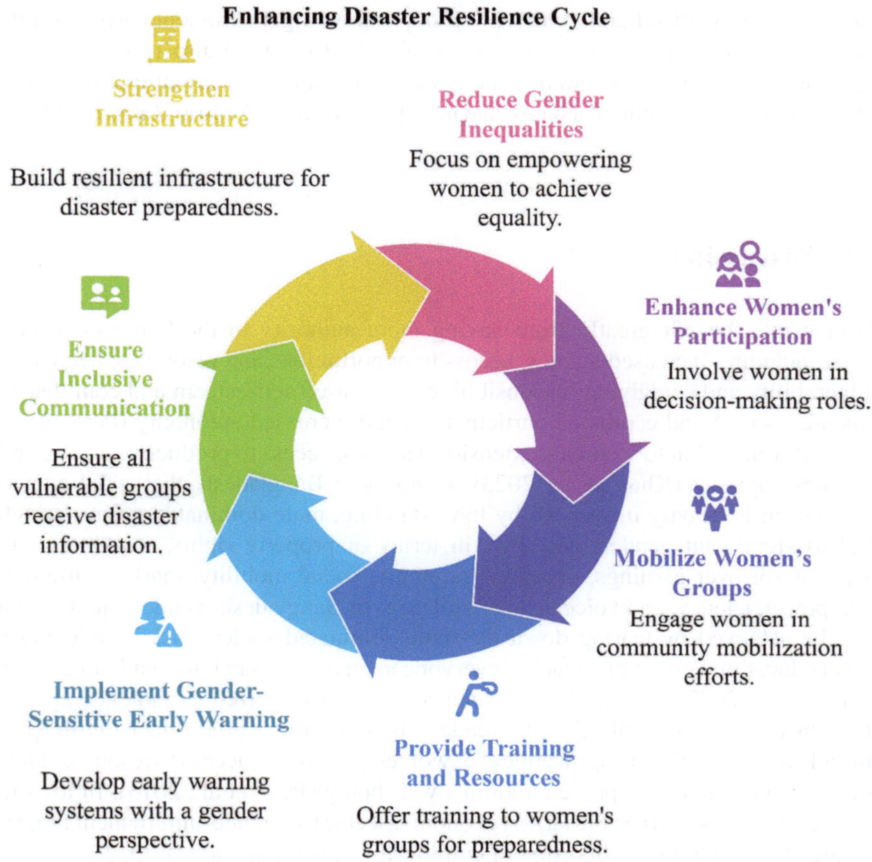

Fig. 3.4 Gender-inclusive disaster resilience cycle (Author, 2025)

3.3.10 Early Warning to Build Back Better

An early warning system is one of the most crucial aspects of emergency preparedness (Shah et al., 2022). A good early warning system necessitates communicating with the most susceptible group, particularly those who have been excluded from the development phase, such as underprivileged children and women (Handapangoda & Kumara, 2013). Planning for disaster risk reduction (DRR) from a gender perspective emphasizes the necessity for gender sensitivity in early warning and response. Early warning systems can function more effectively when they take a gender perspective into account because it makes it possible to gather data that is more accurate and distinct and to perform analyses that are more precise and thorough, which improves readiness and preventative actions combined with more accurate and quantitative answers, leading to a reduced possible loss. Affected individuals, notably women and other marginalized and vulnerable groups, should get sufficient early warning messages. In Bangladesh, primary sources of early warning of disasters in the local community are word-of-mouth from neighbours or relatives, information from radio, television, local government, NGOs, and newspapers rarely (Alam & Rahman, 2014a, 2014b).

3.4 Discussion

Women may benefit greatly from having more authority in the home in several ways, including increased agency, access to opportunities and resources, decreased vulnerability and caregiving responsibilities, increased self-esteem and confidence, expanded social and economic participation, and increased autonomy (Patel et al., 2013). Pay attention to women's extension services, access to productive assets, and skill development (Khan et al., 2023). In northern Bangladesh, the social education system is heavily influenced by long-standing, male-dominated norms, which lead to significant gender inequality in terms of property rights, workload, pay rate, control over earnings, decision autonomy, social mobility, market participation, power, agency, and voice against violence. In Bangladesh, gender equality can only be achieved by tearing down the male-dominated society and introducing a social education system that teaches everyone to respect others fairly and impartially (Khan et al., 2023). Gender-based violence is also recognized as violence against children (Callaghan et al., 2018). Gender impacts child welfare decision-making (Kobulsky et al., 2020). Empowerment of women and greater access to resources have been demonstrated to improve children's well-being (Patel et al., 2013). In low- to middle-income countries, the largest economic sector for female employment is agriculture, where a higher proportion of employed women than men (ILO, 2016). One industry that is particularly susceptible to natural calamities is agriculture. Agriculture accounted for 23% of all natural disaster-related losses and damages from 2006 to 2016 (FAO, 2018). According to Sophie et al., (2021), women are more likely to experience unemployment, return to the workforce, or transition to self-employment following a disaster.

Systemic inequity between men and women is reflected in gender-based violence. During an emergency, it gets worse (Abiona & Koppensteiner, 2018). Women and girls are at greater danger in refugee and displacement camps, where privacy and protection are frequently insufficient (Sophie et al., 2021). For females, the tendency of child marriage is influenced by the loss of a parent in a disaster, whereas for boys, it is reduced. The impact of shocks on the rate of child marriage among girls is determined by the economic function of marriage and prevailing cultural norms. The age of marriage declines due to disasters (Sophie et al., 2021). There are differences in the effects of natural disasters on school enrolment between males and girls. Natural catastrophes may compel parents to pull their children out of school if they are unable to pay the tuition or if they require more financial assistance from home or a second source of income. Boys are typically more impacted than girls when families affected by disasters require more money and labour (Takasaki, 2017). Conversely, girls are usually more impacted when parents are unable to pay for tuition or require extra help at home (Björkman-Nyqvist, 2013). Due to extreme weather patterns and gradual occurrences like desertification and sea level rise, climate change will cause at least 150 million people to be displaced by the year 2050. Extreme weather events and disasters linked to climate change are more likely to occur when power dynamics and relationships are in place. The impact of women and girls becoming climate refugees exacerbates this structural vulnerability. Of all climate displacement, 80% is attributed to women and girls (UNDP & GGCA, 2016). Throughout the migration process, there is a state of multifaceted injustice and insecurity, which is especially problematic for migrant women and girls that regulating gender dynamics (Table 3.1) (UNHCR, 2022).

Table 3.1 Regulating factors of gender dynamics

Regulating factors	Description	References
Women's empowerment	Increased power for women in households enhances agency, access to resources, confidence, and autonomy	Patel et al. (2013); Khan et al. (2023)
Education	Deep-rooted male-dominated systems create disparities in property rights, wages, decision-making, and mobility. Gender equality requires systemic educational change	Khan et al. (2023)
Child protection	Gender-based violence impacts children, affecting welfare decisions. Women's empowerment improves child well-being	Callaghan et al. (2018); Kobulsky et al. (2020); Patel et al. (2013)
Employment	Women in low-income countries rely on agriculture, which is vulnerable to disasters. Women face unemployment or self-employment shifts post-disaster	ILO (2016); FAO (2018); Sophie et al. (2021)
Stopping violence	Gender-based violence worsens during emergencies, especially in refugee camps where women face greater risks	Abiona & Koppensteiner (2018); Sophie et al. (2021)
Child marriage	For girls, child marriage is more common when a parent dies in a disaster. Economic hardship and cultural norms drive early marriages	Sophie et al. (2021)
Child labour	Natural disasters force children out of school, with boys more likely to work and girls more likely to stay home due to financial hardship	Takasaki (2017); Björkman-Nyqvist (2013)
Climate refugees	Women and girls make up 80% of climate refugees, facing increased insecurity and injustice during migration	UNDP & GGCA (2016); UNHCR (2022)

3.4.1 Disaster-Prone Countries, Bangladesh also Faces Gender-Focused Challenges

Though women in Bangladesh have made considerable progress in attaining social and economic advances in almost five decades of independence, they still face many challenges against many handicaps and social stigmas in a male-dominated society (UNDP, 2022). One of the most prominent gender-focused challenges is women's safety (Swarup et al., 2011). After a disaster, the impacted households are frequently compelled to relocate to shelters or refuges, where women are left to carry out their gender-specific domestic duties under trying conditions. When males relocate in search of employment, women must take on both male and female responsibilities to provide for their families with the resources at hand (Nasreen, 2022). Depending on their position in society, women experience different outcomes from subordination. Women from poor economic classes have insufficient access to their inherited property ownership due to the socio-cultural settings and lack of consciousness in northern Bangladesh (Khan et al., 2023). Thus, poverty, starvation, malnutrition, economic crises, environmental degradation, insecurity, and health-related issues, notably those on adolescent and reproductive health, affect women and girls more than males do (Nasreen, 2022).

3.4.2 Climate Change-Induced Disasters Outcomes

Climate change-induced natural disasters have direct impacts on mortality and life expectancy, such as death from injury or drowning (Table 3.2). Moreover, climate change-induced natural disasters can increase morbidity and may have economic impacts on life expectancy. Natural disasters disproportionately negatively affect women's life expectancy (Sophie et al., 2021). Corruption can be brought about by disasters and disaster risks, which can worsen the number of fatalities and damages while decreasing a nation's ability to react. Several resource windfalls brought about by disasters may be abused by elites and governments (Zafar et al., 2023). Climate change-induced disasters usually destroy small rural communities, disrupt their sources of income, and destroy local infrastructure, which raises poverty rates and keeps these communities from making catastrophe investments.-lives that are resilient (Ashrafuzzaman, 2022). Countries that have experienced extreme weather events more frequently and have higher levels of inequality could be caught in a vicious cycle. Adopting adaptation strategies would likely be hampered by increasingly frequent disasters, further undermining resilience, unless governments can re-establish a fair distribution of income and resources together with wider access to health care and preventative measures (Cappelli et al., 2021). Since climate change typically threatens the provision of ecosystem services and increases the frequency of extreme weather events, it may result in losses in jobs and productivity at work. Catastrophes cause job and production losses by destroying infrastructure and taking lives

(Sangeetha & Usha, 2022). Financial authorities are growing increasingly concerned that climate change-induced shifts in the frequency and distribution of catastrophic weather events could endanger the stability of the world's financial system (Mandel et al., 2021).

Table 3.2 Climate change-induced disasters outcomes in Bangladesh

Key factors	Findings in Bangladesh context	References
Health impacts	In Bangladesh, women face elevated mortality risks during disasters, attributable to constrained access to information, mobility restrictions, and social norms that disadvantage them. A study of the 1991 cyclone reveals a disproportionately high mortality rate for women. Concurrently, children are afflicted with ailments such as pneumonia, measles, chickenpox, colds, and fevers, further exacerbating their already vulnerable state. Anaemia, nutritional imbalances, headaches, vomiting, and other health concerns are prevalent among pregnant women. Furthermore, there has been an observed increase in the prevalence of diarrhoea and other waterborne infections. Concurrently, cutaneous ailments, including discomfort, abscesses, and heightened allergic responses, are exhibiting exacerbated symptoms	Ikeda (1995), Md et al. (2022)
Mental health	The presence of menstrual irregularities in women has been identified as a contributing factor to the development of depression	Md et al. (2022)
Gender-based violence	Post-disaster, domestic violence and sexual exploitation have been shown to increase, especially in displacement settings where women lack security and privacy	UNDP (2022)
Early development of children	The consequences of exposure to tornadoes in 2005 on female newborns are negligible. For male newborns, the administration of vitamin A supplements has been demonstrated to mitigate the adverse health consequences of prenatal exposure to tornadoes. In contrast, female newborns in both control and treatment areas appear to be predominantly unimpacted by tornadoes	Gunnsteinsson et al. (2019)
Gender disparities	According to surveys conducted in the southern coastal areas of Bangladesh, women place less value than men on information about climate change and disasters (including emergency information and training programmes), domestic and public administration procedures, financial resources including economic instruments such as microcredit, land ownership, and employment	Md et al. (2022)

3.4.3 Gender Dynamics in Climate Change-Induced Disasters in Bangladesh

In the context of disasters, gender dynamics have been shown to influence a variety of factors, including exposure and vulnerability, preparedness, coping mechanisms, and recovery (Tables 3.1 and 3.2). While gender dynamics exert an influence on the consequences of disasters, the impact varies between men and women, as well as between boys and girls. The regulation of current socioeconomic circumstances by gender-differentiated consequences has the potential to exacerbate existing gender inequality. Consequently, this dynamic can influence the elements that determine the disaster impacts of future hazards (Sophie et al., 2021). Climate-related disasters have had a profound impact on various demographic groups in Bangladesh. Women constitute the most affected demographic. The food security, water usage, health, and general standard of living of women and their families are all directly impacted by climate change, which heightens their socioeconomic vulnerabilities. In contrast to many patriarchal societies, Bangladeshi women often face challenges related to unequal social relations and hierarchies, which enforce gender-specific vulnerabilities (Tanjeela, 2022). In Bangladesh, the mortality risk for women over the age of 19 during Cyclone 1991 was three to five times higher than average (Ikeda, 1995). Similarly, the mortality rate during the 2005 tornado event was 1.24 times higher than average. (Sugimoto et al., 2010).

3.5 Conclusions

The present study posits that men and women play equal roles in society. When women are marginalized and disadvantaged, men, children, and other women who are close to those women also suffer, perhaps not equally, but at least somewhat (Khan et al., 2023). Disaster-based interventions, on their own, will prove insufficient to ameliorate the effects of calamities on women and girls. It is imperative to recognize that injustices have the potential to amplify the risks faced by these groups, thereby diminishing their prospects for success, not only during catastrophic events but also in the aftermath (Bradshaw & Fordham, 2015). The assessment of gender vulnerability within the context of post-disaster reconstruction is a critical step in identifying the dimensions and determinants of gender vulnerability. By conducting such an assessment, key stakeholders can develop a comprehensive understanding of the challenges faced by different genders during the reconstruction phase. This understanding is essential for formulating effective strategies to promote gender equality in the post-disaster reconstruction process (Yumarni et al., 2014). It is imperative to recognize the linkages between women's empowerment and socioeconomic status in the context of climate change adaptation. Empowering women and promoting their involvement in eco-friendly stoves, rural electrification, and the growth of renewable energy, microfinance, and handicrafts will reduce the vulnerability of women in

Bangladesh's coastal regions to climate change (Md et al., 2022). In formulating policies and initiatives, effective disaster risk management must consider how gender dynamics affect the impact of disasters in each location. In instances where the impact of disasters varies significantly between female and male demographics, policymakers bear the responsibility of leveraging available resources to mitigate these disparities and fortify societal resilience, thereby narrowing the gender gap in disaster-related consequences (Sophie et al., 2021).

3.6 Data Availability Statement

By receiving a formal request, the corresponding author is obligated to make the dataset created and analysed for this study available to the public. However, limitations are imposed on the reusability of the data, including restrictions on its reproducibility and the disclosure of commercially sensitive information.

3.7 Conflicts of Interest

With regard to the research, the authors assert that they have no conflicts of interest.

Acknowledgements The authors express their profound gratitude to the respondents and relevant officials of Bagerhat, Khulna, Satkhira, Barguna, and Patuakhali Districts respectively in Bangladesh for their invaluable assistance in facilitating the research team's collection of critical data for the study. The authors would also like to acknowledge the Environmental Systems Research Institute (ESRI) for providing the ArcGIS® 10.8.2 software used for study area map of the study. The authors would like to express their profound gratitude to their respective ministries and organizations, as well as the district administration, for providing the necessary open-source data, reports, and works of literature that were instrumental in the conduct of this study.

Author Contributions Md Shamsuzzoha, Sanjida Hossain Setu, Kamrunnahar Ishana & Md. Jashim Uddin: Conceived and designed the research, literature review, methodology development, geospatial maps preparation, wrote and finalized the manuscript, revised the manuscript based on reviewers' comments. AKM Abdul Ahad Biswas, Mohammad Kabirul Islam, Md. Saiful Islam & Md. Rasheduzzaman: Contributed materials, analysed and interpreted the data, and wrote the paper. Md. Anwarul Abedin & Rajib Shaw: Supervised and reviewed the manuscript. Following a thorough review of the final manuscript, all authors consented to its publication.

Funding The authors did not receive any internal or external funding for the research.

References

Abiona, O., & Koppensteiner, M. F. (2018). The impact of household shocks on domestic violence: Evidence from Tanzania. *SSRN Electronic Journal.* https://doi.org/10.2139/ssrn.3301756

Ahamed, T., Noguchi, R., Muhsin, N., Ayu Purnamasari, R., Islam, M. A., Tasnim, F., Islam, M. Z., Islam, M. F., Akmam, W. (2021). Sustainable agricultural development: A micro-level GIS-based study on women's perceptions of environmental protection and entrepreneurship in Japan and Bangladesh. *GeoJournal 86*, 2071–2103. https://doi.org/10.1007/s10708-020-10169-5

Ahmad, N. (2012). *Gender and climate change in Bangladesh: The role of institutions in reducing gender gaps in adaptation program.* Washington, DC. https://hdl.handle.net/10986/27416

Alam, K. F., & Ahamed, T. (2023). Erosion vulnerable area assessment of Jamuna River system in Bangladesh using a multi-criteria-based geospatial fuzzy expert system and remote sensing. *Asia-Pacific Journal of Regional Science, 7*, 433–454. https://doi.org/10.1007/s41685-023-00292-9

Alam, E., & Collins, A. E. (2010). Cyclone disaster vulnerability and response experiences in coastal Bangladesh. *Disasters, 34*(4), 931–954. https://doi.org/10.1111/j.1467-7717.2010.01176.x

Alam, K., & Rahman, M. H. (2019). Post-disaster recovery in the cyclone Aila affected coastline of Bangladesh: Women's role, challenges and opportunities. *Natural Hazards, 96*, 1067–1090. https://doi.org/10.1007/s11069-019-03591-7

Alam, K., & Rahman, M. H. (2014a). Women in natural disasters: A case study from southern coastal region of Bangladesh. *International Journal of Disaster Risk Reduction, 8*, 68–82. https://doi.org/10.1016/j.ijdrr.2014.01.003

Alam, K., Rahman, Md. H. (2014b). Women in natural disasters: A case study from southern coastal region of Bangladesh. *International Journal of Disaster Risk Reduction 8*, 68–82. https://doi.org/10.1016/j.ijdrr.2014.01.003

Alam, K. F., Shamsuzzoha, M., Arab, S. T., Pangaribuan, I. J. P., Ahamed, T. (2024). Application of remote sensing in the analysis of climate extremes due to global climate change. In Ahamed, T. (Ed.), *Remote sensing application II*. New Frontiers in Regional Science: Asian Perspectives. Singapore: Springer. https://doi.org/10.1007/978-981-97-1188-8_1

Alston, M., & Akhter, B. (2016). Gender and food security in Bangladesh: The impact of climate change. *Gender Place & Culture, 23*(10), 1450–1464. https://doi.org/10.1080/0966369x.2016.1204997

Arab, S. T., Islam, Md. M., Shamsuzzoha, Md., Alam, K. F., Muhsin, N., Noguchi, R., Ahamed, T. (2022). A review of remote sensing applications in agriculture and forestry to establish big data analytics. In Ahamed, T. (Ed.), *Remote sensing application* (pp. 1–24) New Frontiers in Regional Science: Asian Perspectives. Singapore: Springer. https://doi.org/10.1007/978-981-19-0213-0_1

Ashrafuzzaman, M. (2022). Climate change driven natural disasters and influence on poverty in the South Western Coastal Region of Bangladesh (SWCRB). *SN Social Sciences, 2*(7). https://doi.org/10.1007/s43545-022-00351-6

Ayeb-Karlsson, S. (2020). 'I do not like her going to the shelter': Stories on gendered disaster (im)mobility and wellbeing loss in coastal Bangladesh. *International Journal of Disaster Risk Reduction, 50*, Article 101904. https://doi.org/10.1016/j.ijdrr.2020.101904

Begum, A. (2017). *Review of migration and resettlement in Bangladesh: effects of climate change and its impact on gender roles.* Bangladesh Institute of Development Studies, Dhaka. http://www.deccma.com/

Björkman-Nyqvist, M. (2013). Income shocks and gender gaps in education: Evidence from Uganda. *Journal of Development Economics, 105*, 237–253. https://doi.org/10.1016/j.jdeveco.2013.07.013

Bradshaw, S., & Fordham, M. (2015). Double disaster: Disaster through a Gender Lens. In *Hazards, Risks and, Disasters in Society* (pp. 233–251). Elsevier Inc. https://doi.org/10.1016/B978-0-12-396451-9.00014-7

Bryan, E., Alvi, M., Huyer, S., Ringler, C. (2024). Addressing gender inequalities and strengthening women's agency to create more climate-resilient and sustainable food systems. *Global Food Security 40*, 100731. https://doi.org/10.1016/j.gfs.2023.100731

Buvinic, M., Vega, G., Bertrand, M., Urban, A., & Nakata, G. T. (1999). *Hurricane Mitch: Women's needs and contributions*. https://doi.org/10.18235/0008902

Callaghan, J. E., Alexander, J. H., Sixsmith, J., & Fellin, L. C. (2018). Beyond "witnessing": Children's experiences of coercive control in domestic violence and abuse. *Journal of Interpersonal Violence, 33*(10), 1551–1581. https://doi.org/10.1177/0886260515618946

Cappelli, F., Costantini, V., & Consoli, D. (2021). The trap of climate change-induced "natural" disasters and inequality. *Global Environmental Change, 70*, 102329. https://doi.org/10.1016/j.gloenvcha.2021.102329

CCC. (2009). *Climate Change, Gender and Vulnerable Groups in Bangladesh*. Climate Change Cell, DoE, MoEF; Component 4b, CDMP, MoFDM. Month 2009, Dhaka.

Chowdhury, A. M., Bhuyia, A. U., Choudhury, A. Y., & Sen, R. (1993). The Bangladesh cyclone of 1991: Why so many people died. *Disasters, 17*(4), 291–304. https://doi.org/10.1111/j.1467-7717.1993.tb00503.x

Eburn, M. (2015). Disaster risk reduction in the shadow of the law. In *Hazards, Risks and, Disasters in Society* (pp. 255–271). Elsevier Inc. https://doi.org/10.1016/B978-0-12-396451-9.00015-9

FAO. (2018). *The Impact of Disasters and Crises On Agriculture and Food Security 2017*. Food and Agriculture Organization of the United Nations. http://www.fao.org/3/I8656EN/i8656en.pdf

Gennari, F., Arango, D., Hidalgo, N., & McCleary-Sills, J. (2015). *Violence against women and girls resource guide : Disaster risk management brief* (pp. 1–18). Washington, DC: World Bank. http://hdl.handle.net/10986/21093

Ginige, K., Amaratunga, D., & Haigh, R. (2014a). Tackling women's vulnerabilities through integrating a gender perspective into disaster risk reduction in the built environment. *Procedia Economics and Finance, 18*, 327–335. https://doi.org/10.1016/s2212-5671(14)00947-2

Ginige, K., Amaratunga, D., Haigh, R. (2014b). Tackling women's vulnerabilities through integrating a gender perspective into disaster risk reduction in the built environment. *Procedia Economics and Finance 18*, 327–335. https://doi.org/10.1016/S2212-5671(14)00947-2

Gray, C. L., & Mueller, V. (2012). Natural disasters and population mobility in Bangladesh. *Proceedings of the National Academy of Sciences of the United States of America, 109*(16), 6000–6005. https://doi.org/10.1073/pnas.1115944109

Gunnsteinsson, S., Adhvaryu, A., Christian, P., Labrique, A., Sugimoto, J., Shamim, A. A., & West, K. (2019). *Protecting infants from natural disasters: The case of vitamin A supplementation and a Tornado in Bangladesh*. https://doi.org/10.3386/w25969

Handapangoda, W. S., & Kumara, A. S. (2013). The world at her fingertips?: Examining the empowerment potential of mobile phones among poor housewives in Sri Lanka. *Gender Technology and Development, 17*(3), 361–385. https://doi.org/10.1177/0971852413498742

Haque, C. E. (1995). Climatic hazards warning process in Bangladesh: Experience of, and lessons from, the 1991 April cyclone. *Environmental Management, 19*(5), 719–734. https://doi.org/10.1007/bf02471954

Hossain, M. A., Reza, M. I., Rahman, S., & Kayes, I. (2011). Climate change and its impacts on the livelihoods of the vulnerable people in the southwestern coastal zone in Bangladesh. In *Climate change management* (pp. 237–259). https://doi.org/10.1007/978-3-642-22266-5_15

Hossen, M. A., Chowdhury, M. A., Hans, A., Tagoe, C. A., Allan, A., Nelson, W., Patel, A., Mondal, M. S., Salehin, M., Quaye, R. M., & Das, S. (2019). Governance challenges in addressing climatic concerns in coastal Asia and Africa. *Sustainability (Switzerland), 11*(7). https://doi.org/10.3390/su11072148

Hossen, M. F., & Sultana, N. (2023). Shoreline change detection using DSAS technique: Case of Saint Martin Island, Bangladesh. *Remote Sensing Applications: Society and Environment, 30*, 100943. https://doi.org/10.1016/j.rsase.2023.100943

Ikeda, K. (1995). Gender differences in human loss and vulnerability in natural disasters: A case study from Bangladesh. *Indian Journal of Gender Studies, 2*(2), 171–193. https://doi.org/10.1177/097152159500200202

ILO. (2016). *Women at work trends 2016*. Geneva: International Labour Office. https://www.ilo.org/wcmsp5/groups/public/---dgreports/---dcomm/---publ/documents/publication/wcms_457317.pdf

Islam, M. R., Ingham, V., Hicks, J., & Manock, I. (2014). The changing role of women in resilience, recovery and economic development at the intersection of recurrent disaster: A case study from Sirajgang, Bangladesh. *Journal of Asian and African Studies, 52*(1), 50–67. https://doi.org/10.1177/0021909614560244

Kadetz, P., & Mock, N. B. (2018). Problematizing vulnerability: Unpacking gender, intersectionality, and the normative disaster paradigm. In *Creating Katrina, Rebuilding Resilience: Lessons from New Orleans on Vulnerability and Resiliency* (pp. 215–230). Elsevier Inc. https://doi.org/10.1016/B978-0-12-809557-7.00009-0

Khan, M. S., Haque, S., Sarkar, M. A. R., Hoque, M. N., Noman, S. M. M. H., & Wahid, T. (2023). Thinking out of the 'Man box': An intersectional exploration of gender dynamics in northern Bangladesh via gender tracking framework. *World Development Sustainability, 3*, 100100. https://doi.org/10.1016/j.wds.2023.100100

Khan, M. M. (2022). Introduction: gender and women in understanding disasters. In *Disaster and Gender in Coastal Bangladesh: Women's Changing Roles, Risk and Vulnerability* (pp. 1–42). Singapore: Springer Nature Singapore. https://doi.org/10.1007/978-981-19-3284-7_1

Kobulsky, J. M., Wildfeuer, R., Yoon, S., & Cage, J. (2020). Distinguishing characteristics and disparities in child protective services-investigated maltreatment by fathers. *Child Maltreatment, 26*(2), 182–194. https://doi.org/10.1177/1077559520950828

Koolwal, G. B. (2019). *Improving the measurement of rural women's employment: Global momentum and survey research priorities*. Washington, DC: World Bank. http://documents.worldbank.org/curated/en/938101556812458877/Improving-the-Measurement-of-Rural-Womens-Employment-Global-Momentum-and-Survey-Research-Priorities

Lafrenière, J., Sweetman, C., & Thylin, T. (2019). Introduction: Gender, humanitarian action and crisis response. *Gender & Development, 27*(2), 187–201. https://doi.org/10.1080/13552074.2019.1634332

Mandel, A., Tiggeloven, T., Lincke, D., Koks, E., Ward, P., & Hinkel, J. (2021). Risks on global financial stability induced by climate change: The case of flood risks. *Climatic Change, 166*(1–2). https://doi.org/10.1007/s10584-021-03092-2

Md, A., Gomes, C., Dias, J. M., & Cerdà, A. (2022). Exploring gender and climate change nexus, and empowering women in the south western coastal region of Bangladesh for adaptation and mitigation. *Climate, 10*(11). https://doi.org/10.3390/cli10110172

Md Shamsuddoha, M. A. (2024). Impacts of climate change-induced natural hazards on women and their human rights implications: A study in the southwest coast of Bangladesh. *Journal of Migration and Health, 9*.

Nasreen, M. (2022). Gender and disaster in Bangladesh. In *Oxford research Encyclopedia of natural hazard science*. https://doi.org/10.1093/acrefore/9780199389407.013.380

Naz, F., Doneys, P., (2022a). Gender-based differences in access to and use of loans from rural credit programs for flood adaptation in the farming-dependent char communities of Bangladesh. In *Womens Studies International Forum* (Vol. 95, p. 102651). https://doi.org/10.1016/j.wsif.2022.102651

Naz, F., Doneys, P. (2022b). Gender-based differences in access to and use of loans from rural credit programs for flood adaptation in the farming-dependent char communities of Bangladesh. In *Womens Studies International Forum* (Vol. 95, p. 102651). https://doi.org/10.1016/j.wsif.2022.102651

Patel, L., Hochfeld, T., & Moodley, J. (2013). Gender and child sensitive social protection in South Africa. *Development Southern Africa, 30*(1), 69–83. https://doi.org/10.1080/0376835x.2012.755872

Quisumbing, A. R., Kumar, N., & Behrman, J. A. (2017). Do shocks affect men's and women's assets differently? Evidence from Bangladesh and Uganda. *Development Policy Review, 36*, 3–34. https://doi.org/10.1111/dpr.12235

Rezwana, N. (2017). Gender and healthcare access after disasters. In *Routledge eBooks* (pp. 113–142). https://doi.org/10.4324/9780203701546-6

Sangeetha, S., & Usha, S. (2022). Impact of climate change on the labor market. *ECS Transactions, 107*(1), 11287–11294. https://doi.org/10.1149/10701.11287ecst

Shah, A. A., Ullah, A., Khan, N. A., Pal, I., Alotaibi, B. A., & Traore, A. (2022). Gender perspective of flood early warning systems: People-centered approach. *Water (Switzerland), 14*(14). https://doi.org/10.3390/w14142261

Shamsuzzoha, M., & Ahamed, T. (2023). Shoreline change assessment in the coastal region of Bangladesh delta using tasseled cap transformation from satellite remote sensing dataset. *Remote Sensing (Basel), 15*, 295. https://doi.org/10.3390/rs15020295

Shamsuzzoha, M., Noguchi, R., & Ahamed, T. (2021). Damaged area assessment of cultivated agricultural lands affected by cyclone bulbul in coastal region of Bangladesh using Landsat 8 OLI and TIRS datasets. *Remote Sensing Application: Society and Environment, 23*, 100523. https://doi.org/10.1016/j.rsase.2021.100523

Shamsuzzoha, M., Noguchi, R., & Ahamed, T. (2022). Rice yield loss area assessment from satellite-derived NDVI after extreme climatic events using a fuzzy approach. *Agricultural Information Research, 31*, 32–46. https://doi.org/10.3173/air.31.32

Shamsuzzoha, M., Ahamed, T. (2024). Assessment of shoreline and agricultural land use changes in the onshore coastal region of Bangladesh delta using satellite remote sensing and GIS. In Ahamed, T. (Ed.), *Remote sensing application II* (pp. 85–119). New Frontiers in Regional Science: Asian Perspectives. Singapore: Springer. https://doi.org/10.1007/978-981-97-1188-8_4

Shamsuzzoha, Md., Al-Maruf, A. (2012). Post SIDR life strategy: Adaptation scenario of settlements of the south. *Institute of Bangladesh Studies (IBS) Journal 19*, 207–222.

Shamsuzzoha, Md. (2023). *Assessment of agricultural land use change in onshore coastal regions due to extreme climatic events using machine learning and remote sensing* (Ph.D. Thesis). University of Tsukuba, Tsukuba.

Shanta, M. A., Suman, S. U., Islam, M. M. (2025). Role of formal aquaculture industry for empowering women: A study in the southern coastal region of Bangladesh. In *Women's Studies International Forum* (Vol. 109, p. 103046). https://doi.org/10.1016/j.wsif.2024.103046

Sophie, A. E., de Vries, A., Stephan, R., Thies, F., Kabir, K., & Maruo, M. (2021). *Gender dimensions of disaster risk and resilience existing evidence.*

Sugimoto, J. D., Labrique, A. B., Ahmad, S., Rashid, M., Shamim, A. A., Ullah, B., Klemm, R. D., Christian, P., & West, K. P. (2010). Epidemiology of tornado destruction in rural northern Bangladesh: Risk factors for death and injury. *Disasters, 35*(2), 329–345. https://doi.org/10.1111/j.1467-7717.2010.01214.x

Swarup, A., Dankelman, I., Ahluwalia, K., & Hawrylysgym, K. (2011). *Weathering the storm: Adolescent girls and climate change 2011.* http://repository.ubn.ru.nl/handle/2066/92554

Takasaki, Y. (2017). Do natural disasters decrease the gender gap in schooling? *World Development, 94*, 75–89. https://doi.org/10.1016/j.worlddev.2016.12.041

Tanjeela, M. (2022). Understanding the struggles of Bangladeshi women in coping with climate change through a gender analysis. *Gender Technology and Development, 27*(2), 250–265. https://doi.org/10.1080/09718524.2022.2144101

Tanner, L., Markek, D., & Komuhangi, C. (2018). *Women's leadership in disaster preparedness.* https://disasterpreparedness.ngo

UNDP & GGCA. (2016). *Overview of linkages between gender and climate change*, (pp. 1–8). New York: UNDP.

UNDP. (2022). *New threats to human security in the Anthropocene: Demanding greater solidarity* (pp. 1–7). 1 UN Plaza 10017, New York, NY USA. https://hs.hdr.undp.org/pdf/srhs2022.pdf

UNHCR. (2022). *Global Trends Report 2021.* https://www.unhcr.org/media/global-trends-report-2021

WB. (2021). *Gender dynamics of disaster risk and resilience (WWW Document).* World Bank (WB) Group. URL https://www.worldbank.org/en/topic/disasterriskmanagement/publication/gender-dynamics-of-disaster-risk-and-resilience. Accessed 8 September 24

Yumarni, T., Amaratunga, D., & Haigh, R. (2014). Assessing gender vulnerability within post-earthquake reconstruction: Case study from Indonesia. *Procedia Economics and Finance, 18*, 763–771. https://doi.org/10.1016/s2212-5671(14)01000-4

Zafar, S., Rahman, I. U., & Ammara, S. (2023). Disasters and corruption: An empirical analysis of 16 countries from Asia and the Middle East. *International Journal of Disaster Risk Reduction, 90*, Article 103678. https://doi.org/10.1016/j.ijdrr.2023.103678

Chapter 4
Gender-Responsive Drought Resilience and DRR Approaches in Bangladesh

Md. Shamsuzzoha, Md. Fuad Hasan, Md. Mazharul Anwar, Md. Sagirul Islam Majumder, A. B. M. Mahbub Morshed Khan, Debashish Sarker, Md. Anwarul Abedin, and Rajib Shaw

Abstract Bangladesh is increasingly affected by climate-induced droughts, which have the potential to exacerbate the risks to its agricultural systems, water resources, and vulnerable populations, especially in the northwest. Despite the evolution of national disaster risk reduction (DRR) policies, gender inequality persists as a substantial impediment to the development of effective and equitable drought resilience measures. This study explores the gender dimensions of drought risk in Bangladesh, emphasizing the disproportionate burdens faced by women and their underrepresentation in decision-making processes. It is evident that women, despite their constrained access to resources, play a pivotal role in adaptation through their local knowledge, agricultural practices, and community mobilization. The paper underscores the importance of integrating gender-responsive strategies into drought

---

Md. Shamsuzzoha (✉) · Md. F. Hasan · Md. M. Anwar
Department of Emergency Management, Faculty of Environmental Science and Disaster Management, Patuakhali Science and Technology University, Patuakhali, Bangladesh
e-mail: zohageo@pstu.ac.bd

Md. F. Hasan
Department of Environmental Science and Technology, German University Bangladesh, Gazipur, Bangladesh

Md. S. I. Majumder
Department of Soil Science, Faculty of Agriculture, Patuakhali Science and Technology University, Dumki, Patuakhali, Bangladesh

A. B. M. M. M. Khan
Department of Agricultural Botany Faculty of Agriculture, Faculty of Agriculture, Patuakhali Science and Technology University, Dumki, Patuakhali, Bangladesh

D. Sarker
Geodetic Detachment, Survey of Bangladesh (SoB), Tejgaon, Dhaka, Bangladesh

Md. A. Abedin
Department of Soil Science, Faculty of Agriculture, Bangladesh Agricultural University, Mymensingh, Bangladesh

R. Shaw
Graduate School of Media and Governance, Keio University, Endo, Fujisawa, Japan

Md. A. Abedin et al. (eds.), *Climate Crisis in South and East Asia*, Sustainability Solutions, https://doi.org/10.1007/978-3-032-01012-4_4

resilience and DRR planning by examining existing institutional frameworks, policy gaps, and grassroots innovations. The incorporation of women's leadership and voices within DRR governance has been demonstrated to enhance the relevance and reach of interventions, thereby ensuring that resilience efforts are sustainable and inclusive. The document calls for transformative approaches, centered on gender equity, capacity building, and inclusive planning to address structural vulnerabilities and build long-term community resilience in the face of escalating climate threats.

Keywords Bangladesh · Drought resilience · DRR · Gender · GIS · Satellite remote sensing

4.1 Introduction

Bangladesh is situated in the low-lying delta of the Ganges–Brahmaputra–Meghna (GBM) river system (Shamsuzzoha & Ahamed, 2023), is globally recognized as one of the most climate-vulnerable countries (Freihardt, 2024). The country frequently faces various natural hazards (Shefa et al., 2024), including floods, cyclones, riverbank erosion, salinity intrusion, and droughts. Among these hazards, drought has gained increased attention in recent years due to its growing frequency, intensity, and socio-economic impacts, especially in the country's northwestern regions (Shamsuzzoha et al., 2022, 2025a). Drought is illustrated as prolonged periods of deficient rainfall, leading to water scarcity, agricultural failures, food insecurity, and degradation of livelihoods (Shamsuzzoha et al., 2018a).

The Landsat 8/9 operational land imager (OLI) satellite-derived soil adjusted vegetation index (SAVI) is employed to rectify the normalized difference vegetation index (NDVI) for the effect of soil brightness in Bangladesh exhibiting minimal vegetative cover (Huete, 1988) (Fig. 4.1). The vegetation condition index (VCI) is a key indicator used to assess vegetation health by comparing current conditions to historical trends (Kogan, 1995). It is particularly useful in monitoring drought and agricultural conditions (Fig. 4.2).

The impact of drought is frequently underestimated in comparison to more visible natural disasters, such as cyclones and floods (Frame et al., 2020; Shamsuzzoha & Ahamed, 2024), its consequences are far-reaching and long-term (Gebremeskel et al., 2019; Meir et al., 2018), especially when it intersects existing vulnerabilities such as poverty, gender inequality, and marginalization. Drought resilience and disaster risk reduction (DRR) approaches are critical components of sustainable development and climate adaptation strategies (Shamsuzzoha et al., 2021; Talukder et al., 2018). In this context, resilience is defined as the ability of individuals, communities, institutions, and ecosystems to absorb, adapt, and recover from drought-related shocks and stress without compromising long-term development (Cai et al., 2018; Tariq et al., 2021). The DRR framework for building community resilience involves a systematic approach to analyzing and mitigating the underlying causes of disasters (Alam et al., 2024). This process encompasses various strategies, including risk assessment,

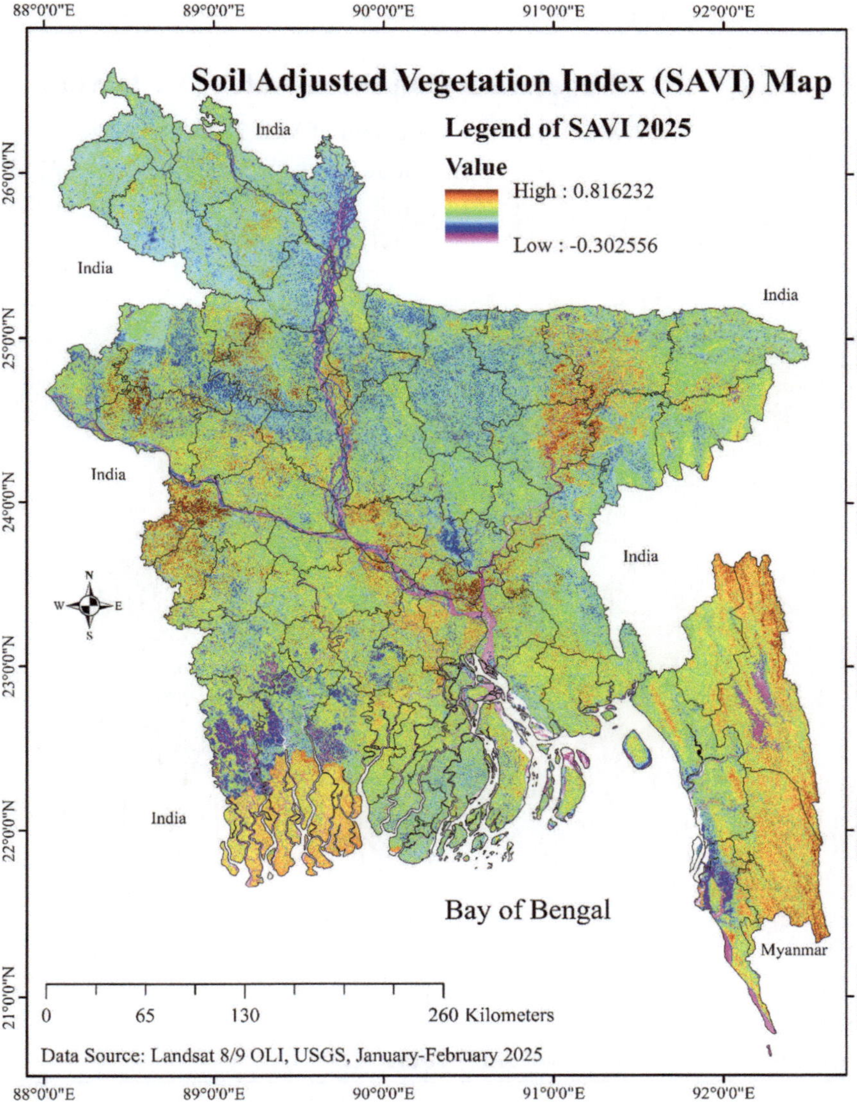

Fig. 4.1 Soil adjusted vegetation index in Bangladesh (Author generated, 2025)

the establishment of early warning systems, sustainable natural resource management, and empowerment (Vasileiou et al., 2022). These frameworks are imperative in shifting the focus from reactive responses to proactive and preventative measures.

In Bangladesh, both drought resilience and DRR strategies have been traditionally gender-blind (Alston, 2015; Shamsuzzoha et al., 2025b), often overlooking the specific needs, roles, and capacities of women, girls, and other vulnerable gender

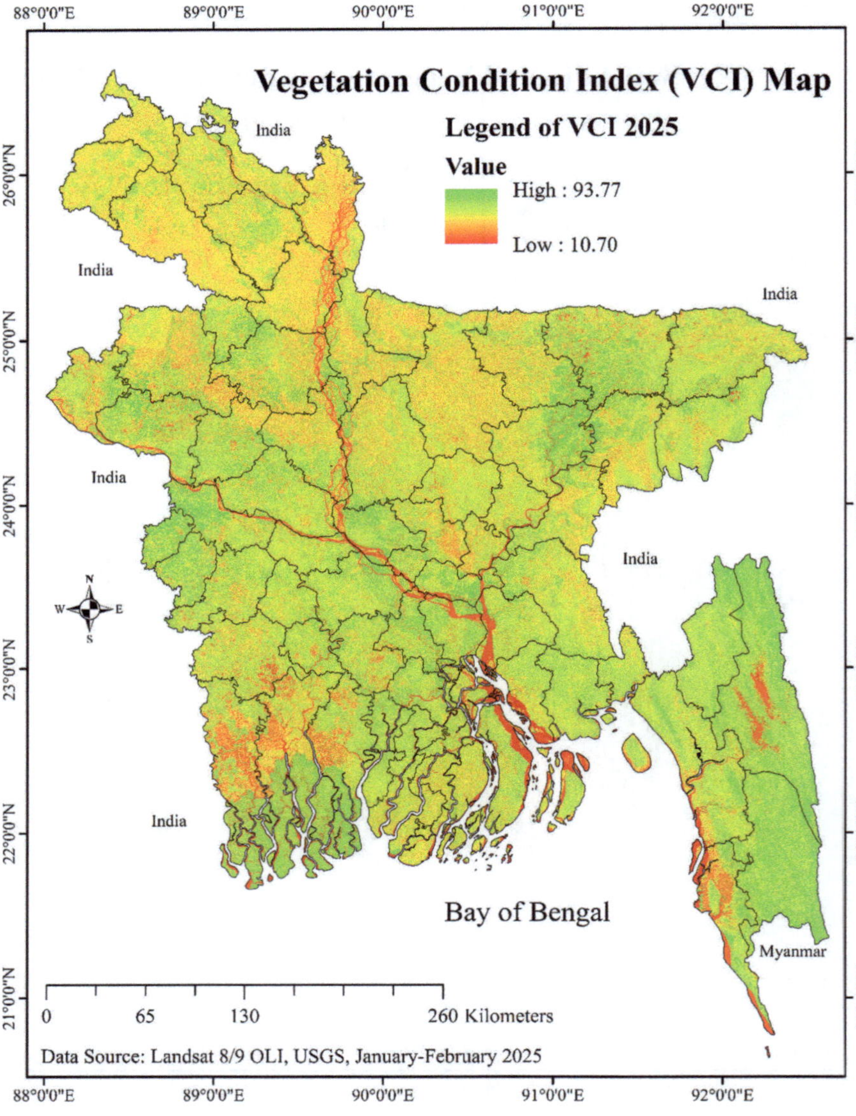

Fig. 4.2 Vegetation condition index in Bangladesh (Author generated, 2025)

groups (Ashraf & Azad, 2015; Fatema et al., 2023). Nevertheless, an increasing number of scholars have come to acknowledge that gender responsiveness is not merely an ethical imperative; it is also a strategic necessity (Chineka et al., 2019). Women and girls, particularly in rural areas with a high risk of drought, bear an unequal share of the burden during drought events. This disparity is primarily attributable to their roles in activities such as water collection, food production, caregiving, and household management (Algur et al., 2021; Shamsuzzoha et al., 2018b, 2018c). It has been demonstrated that those who are affected by droughts often have restricted access to information, decision-making processes, and resources. This restricted access renders them more vulnerable to the deleterious effects of drought (Abid et al., 2018; Chineka et al., 2019). Concomitantly, women must not be regarded as mere passive victims. Rather, they are active agents of change who possess valuable knowledge and coping strategies. These assets can meaningfully contribute to resilience-building and risk reduction (Aziz & Anjum, 2024; Mulyasari & Shaw, 2013).

Gender-responsive approaches to drought resilience and DRR aim to ensure that the diverse experiences, needs, and contributions of all gender groups are recognized and integrated at all stages of policy, planning, implementation, and evaluation. The promotion of gender equity in resource allocation is of paramount importance. Furthermore, it is essential to enhance women's participation and leadership in disaster management. Finally, addressing systemic inequalities that exacerbate vulnerability is necessary (Shamsuzzoha et al., 2025a; Zaidi & Fordham, 2021). In the context of Bangladesh, such approaches are particularly relevant given the prevailing patriarchal socio-cultural norms, high levels of gender inequality in rural areas land use and land cover, and the increasing feminization of agriculture due to male migration and livelihood shifts (Arab et al., 2022; FAO, 2024; Hasan et al., 2021; Shanta et al., 2025). This chapter explores the importance, challenges, and opportunities of integrating gender into drought resilience and disaster risk reduction in Bangladesh. The document meticulously delineates the historical and geographical context of drought in the country, underscoring the gender dimensions of drought vulnerability and resilience (Dankelman & Naidu, 2020; Tewari et al., 2023). It methodically examines the policy and institutional frameworks and engages in a discursive analysis of innovative approaches and community-based practices that have demonstrated effectiveness. The objective of this study is to establish a comprehensive foundation for understanding how gender-responsive strategies can bolster the resilience of communities vulnerable to drought in Bangladesh and contribute to sustainable development and climate justice.

4.2 Drought in Bangladesh: A Hidden Crisis

4.2.1 Geographical and Climatic Overview

Despite its frequent association with floods and cyclones (Kulatunga et al., 2013; Shamsuzzoha et al., 2021, 2024), Bangladesh is also susceptible to drought (Fig. 4.3). The northwestern region of the country, specifically the Rajshahi, Rangpur, and parts of the Khulna divisions, exhibits a high degree of vulnerability to seasonal droughts (Hosen et al., 2023; Sadiq et al., 2023). These regions are distinguished by low and erratic rainfall, sandy soils with inadequate water retention capacity, and limited surface water resources. The dry season, which occurs from November to March, has been shown to exacerbate water scarcity. The pre-monsoon period, which extends from April to May, is particularly critical for agricultural activities due to the limited availability of water during this time (Anwarul Islam et al., 2019; AI et al. 2023). The ongoing climate change phenomenon is contributing to the exacerbation of drought conditions in Bangladesh. The Intergovernmental Panel on Climate Change (IPCC) has observed that South Asia, which includes Bangladesh, is projected to experience an increase in the frequency and severity of droughts (IPCC, 2023). This phenomenon is attributed to rising global temperatures, erratic rainfall patterns, and altered monsoon systems (Jihan et al., 2025). Projections indicate that by 2050, significant portions of the country may face water stress, affecting food production, health, and livelihoods.

4.2.2 Socioeconomic Impacts of Drought

Drought in Bangladesh disproportionately affects the agriculture sector (Sarkar et al., 2024), which employs over 40% of the population (Chowhan et al., 2024) and is highly dependent on rainfall (Chowhan et al., 2024). Crop failures, particularly of paddy, wheat, maize, and vegetables, lead to income losses for smallholder farmers and landless laborers. Water scarcity affects livestock health, reduces fishery yields (Richter et al., 2020), and impairs domestic and sanitation needs (Abanyie et al., 2025). These impacts reverberate across rural economies, increasing poverty, food insecurity, malnutrition, and distress migration (Cuong et al., 2024). The effects of drought are not uniform across populations. Vulnerable groups such as women, children, indigenous people, and persons with disabilities face heightened risks due to pre-existing structural disadvantages. For instance, women are often forced to travel long distances to fetch water (Sultana, 2009), reduce their own food intake to feed family members (Parvin & Ahsan, 2013), and engage in unpaid care work during crises (Bin & Sharma, 2021). These burdens exacerbate gender inequalities and limit women's ability to recover and adapt to future shocks. To illustrate the multi-dimensional socioeconomic consequences of drought in Bangladesh, (Table 4.1). Summarizes its sectoral impacts, highlighting how interconnected vulnerabilities span agriculture, livestock, fisheries, health, and socioeconomic systems. These cascading effects undermine food security, livelihoods, and rural resilience, particularly for marginalized and resource-poor populations.

Information Source (Umma et al., 2011).

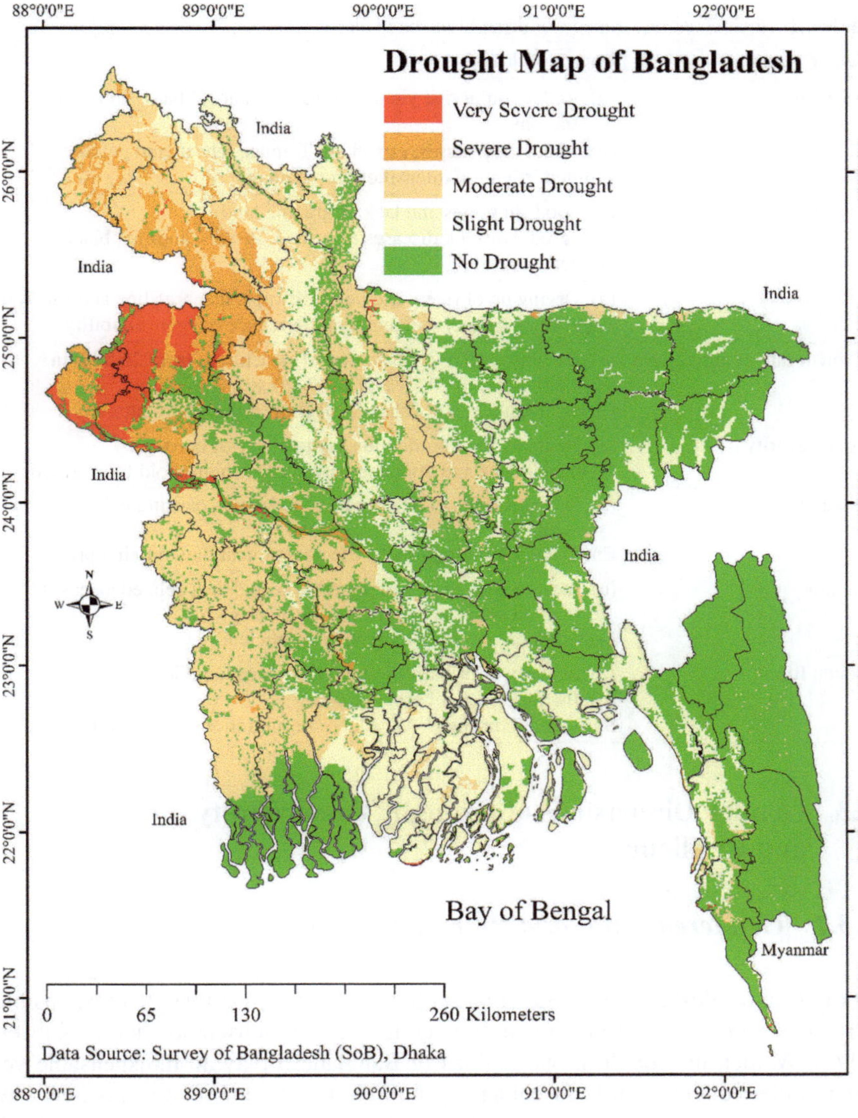

Fig. 4.3 Drought in Bangladesh (Author generated, 2025)

Table 4.1 Socioeconomic impacts of drought in Bangladesh

Sector's name	Impact of drought
Agriculture	(a) Reduction in rice yields across pre-kharif, kharif, and rabi seasons (b) Major crop losses, especially T-aman, wheat, and pulses Limited irrigation intensifies crop damage
Livestock	(a) Food shortages and lack of grazing land Increased livestock diseases (e.g., heat strokes, anthrax, black quarter)
Fisheries	(a) Drying up of rivers and ponds disrupts fish breeding and survival High water salinity and temperature reduce oxygen solubility
Growth rate of agriculture	(a) Declining GDP contribution of agriculture from 25.58% in 1999–2000 to 20.88% in 2007–2008 Crops, livestock, and fisheries all show reduced productivity
Food security	(a) Reduced food production, particularly of boro rice Food shortages and price increases reduce household food access
Price of rice	(a) Drought leads to rice production deficits, causing market prices to rise Emergency food imports by the government due to rising prices
Health	(a) Increased incidence of waterborne and heat-related diseases (e.g., diarrhea, hypertension, dysentery) Dehydration and long-term energy deficiency
Social life	(a) Scarcity of drinking water causes local conflicts (b) Increased school dropout rates Shift from farming to wage labor or migration for survival

4.3 Gender Dimensions of Drought Vulnerability and Resilience

4.3.1 Gendered Roles and Responsibilities

In rural Bangladesh, gender roles are traditionally defined, with women responsible for water management, food preparation, childcare, and household chores (Kabeer, 1991). Women in drought-prone regions of Bangladesh play an indispensable yet often underrecognized role in enhancing household and community resilience to climate-related hazards (Huq et al., 2023). As primary managers of water, food, and family welfare (Ghosh et al., 2021), their daily practices are deeply intertwined with environmental sustainability and disaster preparedness (Figueiredo & Perkins, 2013). During times of water scarcity and agricultural stress, women are frequently the first responders within households—adapting consumption patterns, reallocating scarce resources, and seeking alternative income sources to cope with adversity (Moayedi & Hayati, 2023). In the agricultural sector, women contribute significantly to crop production, post-harvest handling, seed preservation, and homestead gardening (Chakma & Ruba, 2021a). Despite limited land ownership, they possess indigenous knowledge of soil fertility, seed selection, and pest management skills that become even more critical in drought-affected contexts. For instance, many rural women have demonstrated expertise in cultivating drought-tolerant crop varieties and utilizing water-saving irrigation methods such as pitcher irrigation or mulching (Rahman et al., 2022). This locally grounded knowledge base can serve as a foundation for community-based adaptation and DRR initiatives.

Moreover, women's leadership in informal institutions, such as self-help groups, microfinance collectives, and savings associations, provides a crucial platform for community mobilization, information dissemination, and emergency response coordination (Huq et al., 2023). In several drought-prone upazilas, women-led groups have actively participated in forming water user associations (Islam et al., 2011), managing small-scale irrigation schemes (Hussain, 2007), and promoting hygiene awareness during dry spells (Alam & Rahman, 2017). Their collective agency not only fosters solidarity and social capital but also bridges gaps left by formal institutions. Importantly, empirical evidence (Akter, 2025; Thapa & Pathranarakul, 2019) suggests that gender-inclusive DRR interventions yield more effective and sustainable outcomes. When women are involved in early warning systems, contingency planning, and disaster committees, they bring distinct perspectives and prioritize community needs such as food security, safe shelter, sanitation, and protection from gender-based violence. However, their participation is often constrained by patriarchal norms, mobility restrictions, and time poverty caused by unpaid care burdens (Azad & Pritchard, 2023; Kreutzer et al., 2023; Md et al., 2022; Zehba et al., 2025). To unlock the full potential of women in drought resilience and DRR, it is essential to integrate gender-sensitive tools and indicators in project design, build their capacity through targeted training, and ensure representation in all levels of decision-making. Institutionalizing women's voices in local disaster governance, investing in gender-transformative infrastructure (e.g., community water points, childcare facilities), and recognizing women's contributions in climate financing mechanisms can catalyze long-term resilience building as like Fig. 4.4.

4.3.2 Gendered Access to Resources and Decision-Making

One of the major barriers to gender-responsive resilience is the unequal access to resources and decision-making platforms (Chidinma Favour Chikwe et al., 2024). Women in Bangladesh generally have limited control over land, water, credit, and agricultural inputs (Chakma & Ruba, 2021b; de Brauw et al., 2021). This restricts their capacity to invest in climate-resilient practices or participate in formal adaptation programs. Additionally, women are underrepresented in community-based organizations, local government bodies, and disaster management committees (Humayra et al., 2024). Social norms and mobility constraints further inhibit women's participation in training programs, public meetings, and extension services (Islam et al., 2023; Medendorp et al., 2022). Male-centric planning often overlooks women's needs and fails to address issues like safe sanitation, menstrual hygiene, or gender-based

Fig. 4.4 Women's role in drought resilience cycle *(*Akter, 2025; Chakma & Ruba, 2021a; Ghosh et al., 2021; Huq et al., 2023; Moayedi & Hayati, 2023*)*

violence during water stress and displacement (Ferdous & Uddin, 2020; Pervin & Mokhtar, 2023). Without intentional efforts to dismantle these barriers, resilience strategies risk reinforcing gender inequities and failing to reach those most in need as presented in Fig. 4.5.

4.4 Evolving Policy and Institutional Frameworks

The concise overview of the major institutional and community-level initiatives employed for drought risk management in Bangladesh (WB, 2021). It emphasizes the pivotal roles of government agencies, non-governmental organizations (NGOs), and local communities in these efforts. The document presents a range of strategies, including the implementation of early warning systems, the development of climate-resilient agricultural practices, and the establishment of infrastructure to support long-term resilience (Brady et al., 2022; Ramalho, 2019; UNFCCC, 2024) (Table 4.2). At the institutional level, the Government of Bangladesh addresses drought management through the Ministry of Disaster Management and Relief (MoDMR) and the Department of Disaster Management (DDM), which oversee policy and coordination efforts. The Climate Change and Environment Cell (CCEC) under the Ministry of Environment ensures climate adaptation is integrated into disaster management. Programs like the Comprehensive Disaster Management Programme (CDMP) help strengthen institutional resilience. Early warning systems

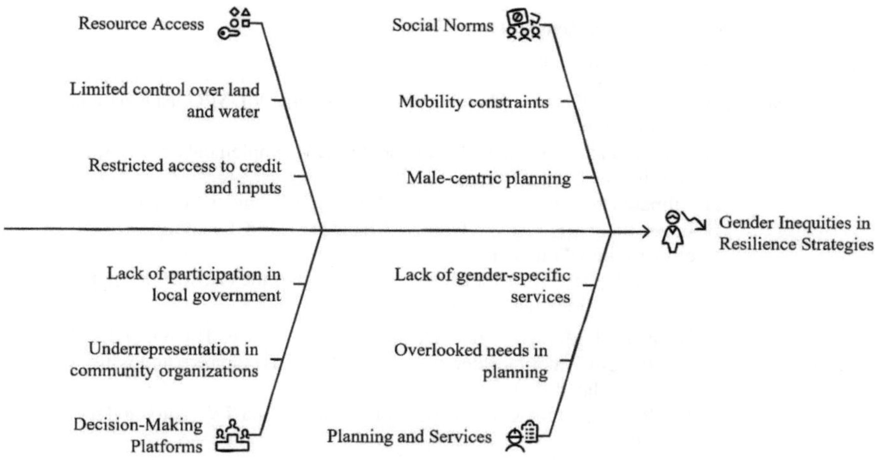

Fig. 4.5 Barriers to gender-responsive resilience in Bangladesh (Islam et al., 2017;Ferdous & Uddin, 2020; Chakma & Ruba, 2021a; Chidinma Favour Chikwe et al., 2024; Humayra et al., 2024)

are crucial for drought risk management (Akhand, 2003; Islam et al., 2024). Agencies such as the Bangladesh Meteorological Department (BMD), SPARRSO, and the Bangladesh Water Development Board (BWDB) monitor climate variables and issue forecasts via mass media. Despite challenges in predicting complex events like El Niño, community preparedness has improved.

The government and NGOs implement both short- and long-term strategies. Relief during droughts includes food, water, and work-for-food programs. Long-term efforts promote agricultural resilience through drought-tolerant crops developed by BRRI and BARI, with the Department of Agricultural Extension supporting climate-resilient practices. NGOs like BRAC and CARITAS offer microcredit, rainwater harvesting, and small-scale farming support (Akhand, 2003; Matin & Taher, 2001; Rahman & Tasnim, 2023). Infrastructure development is vital, especially in the northwest. Institutions like BMDA, BADC, and RDA expand irrigation via tube wells, canal re-excavation, and water reservoirs. At the community level, farmers adjust by planting drought-tolerant crops, using local water sources, or purchasing irrigation services. Women contribute through income-generating activities, and in severe cases, households reduce consumption, access loans, or migrate temporarily for work.

Table 4.2 Drought risk management practices in Bangladesh

Level	Key practices and initiatives
Institutional level	(a) Ministry of Disaster Management and Relief (MoDMR) leads national coordination (b) Department of Disaster Management (DDM) implements policies and coordinates local activities (c) Climate Change and Environment Cell (CCEC) under Ministry of Environment, Forest and Climate Change (MoEFCC) addresses climate-linked disasters (d) Comprehensive Disaster Management Programme (CDMP) supports capacity building and risk reduction
Early warning systems	(a) Bangladesh Meteorological Department (BMD), SPARRSO, and Bangladesh Water Development Board (BWDB) monitor and issue drought-related forecasts (b) Weather and climate information disseminated via TV, radio, and local channels
Government and NGO actions	(a) Emergency relief (drinking water, food, food-for-work, subsidies) (b) Promotion of drought-tolerant crop varieties developed by BRRI and BARI (c) Training by Department of Agricultural Extension (DAE) on AWD, seed management, and irrigation (d) NGOs (e.g., BRAC, CARITAS) support microcredit, rainwater harvesting, kitchen gardening, and small-scale irrigation
Physical/ infrastructure level	(a) Installation of deep/shallow tube wells and overhead tanks by Barind Multipurpose Development Authority (BMDA), Bangladesh Agricultural Development Corporation (BADC), and Rural Development Academy (RDA) (b) Re-excavation of canals, cross-dams, afforestation, and construction of irrigation facilities
Community and household level	(a) Cropping diversification (vegetables, pulses, spices, fruit trees) (b) Use of shallow/deep tube wells, hand pumps, and rainwater harvesting for irrigation (c) Women participate in income-generating activities (e.g., handicrafts, poultry) (d) Households reduce consumption, use savings, take microloans, or engage in seasonal migration

Source (GoB, 2020; Habiba et al., 2011, 2013)

4.5 Gender-Responsive Approaches: Principles and Practices

Gender-responsive drought resilience and DRR approaches are guided by several core principles: (i) equity and inclusion for recognizing and addressing unequal power relations, ensuring that marginalized voices are heard, (ii) participation and leadership for promoting women's and marginalized groups' involvement in decision-making and leadership roles, (iii) empowerment for building capacities through education, training, and access to resources, (iv) intersectionality for acknowledging how gender intersects with age, ethnicity, disability, and socioeconomic status, and (v) accountability for stablishing mechanisms to monitor, evaluate, and enforce gender equity commitments. (Table 4.3) outlines key urgent calls for establishing women's leadership in disaster risk management. Drawn from practical insights and policy analysis, these calls emphasize the need for meaningful participation, institutional reform, and social change to empower women as agents of resilience. Addressing these areas is essential for building inclusive, effective, and sustainable disaster preparedness and response systems.

Table 4.3 Strategic priorities for advancing women's leadership in disaster risk reduction

Theme	Urgent call to action	Details/Explanation
Recognition of local knowledge	Recognize and institutionalize women's indigenous knowledge	Acknowledge women's roles in food preservation, caregiving, water and resource management as critical to disaster risk reduction
Meaningful participation	Shift from symbolic to substantive leadership	Ensure women have real decision-making power in disaster committees, not just token roles
Access to power	Ensure access to decision-making spaces	Support voluntary and meaningful participation of women in planning and consultations
Institutional integration	Integrate women at all levels of DRM	Include women not only as volunteers but also as planners, experts, and managers within institutional structures
Ground-level planning	Include women in risk analysis and emergency planning	Engage women in village-level risk assessments, shelter design, relief planning, and emergency response
Public awareness & norm change	Challenge social norms through education and campaigns	Promote the value of women's leadership and contributions through public campaigns and school-based education
Policy & data reform	Reform data and policy frameworks	Introduce gender-disaggregated data and policies that address women's specific vulnerabilities and losses
Pilot initiatives	Promote demonstrative gender-transformative projects	Develop programs that train girls in life-saving skills, support female entrepreneurship, and include women in paid recovery work
Multi-level collaboration	Coordinating institutional change across sectors	Establish partnerships among government, NGOs, and communities to maintain women's leadership in DRM

4.6 Conclusion

The ongoing drought conditions in Bangladesh are escalating, posing significant challenges to the region's already vulnerable agricultural sector. This phenomenon can be attributed to the effects of climate change. It is imperative that we take swift and substantial measures to address this issue. These steps should be equitable and enduring. This is not merely an optional extra; rather, it is an indispensable component. Consequently, initiatives aimed at facilitating communities' recovery must be designed with a gender-responsive approach. It is imperative that they reach the most affected populations and leverage the potential of communities. A comprehensive consideration of the impact on both men and women is imperative for Bangladesh to enhance its management of droughts, promote social justice, stimulate economic opportunity, and uphold human rights. Therefore, achieving progress necessitates a shift in our cognitive frameworks, the institutions we endorse, and the financial decisions we make. The call for such measures is predicated on the necessity of robust data systems, augmented capacity, coherent policy frameworks, and active engagement of communities. Of particular significance is the assertion that women should be regarded as pivotal contributors to the creation of a robust and equitable future. In the context of the nation's ongoing challenges concerning climate change adaptation and disaster risk management, it is imperative to deliberate on the potential long-term benefits these endeavors could yield for all segments of society.

Acknowledgements The authors express their profound gratitude to the respondents and relevant officials of the Department of Agricultural Extension (DAE), the Ministry of Agriculture (MoA), the Survey of Bangladesh (SoB), the Ministry of Defence (MoD), and Patuakhali Science and Technology University (PSTU), respectively, in Bangladesh, for their invaluable assistance in facilitating the research team's collection of critical data for the study. The authors would also like to acknowledge the Environmental Systems Research Institute (ESRI) for providing the ArcGIS® 10.8.2 software utilized to create the study area map. The authors are very thankful to the United States Geological Survey (USGS) for providing the necessary open-source Landsat 8 and 9 OLI satellite data.

Author Contributions Md. Shamsuzzoha, Md. Fuad Hasan, & Md. Mazharul Anwar: Conceived and designed the research, literature review, methodology development, geospatial maps preparation, wrote and finalized the manuscript, revised the manuscript based on reviewers' comments. Md. Sagirul Islam Majumder, A.B.M. Mahbub Morshed Khan, & Debashish Sarker: Contributed materials, analyzed and interpreted the data, and edited the paper. Md. Anwarul Abedin, & Rajib Shaw: Supervised and reviewed the manuscript. Md. Shamsuzzoha & Md. Fuad Hasan: Equal contributor(s) as the first author of this review article. Following a thorough review of the final manuscript, all authors consented to its publication.

Funding The authors did not receive any internal or external funding for the research.

Data Availability Statement Upon receipt of a formal request, the corresponding author is obligated to make the dataset created and analyzed for this study available to the public. However, there are limitations to the reusability of the data, including restrictions on its reproducibility and the disclosure of commercially sensitive information.

Conflicts of Interest The authors of the research say that they have no conflicts of interest.

References

Abanyie, S. K., Amuah, E. E. Y., & Nang, D. B. (2025). Water scarcity and its implications on sanitation: A perspective study in an emerging city in Northern Ghana. *Green Technologies and Sustainability*. https://doi.org/10.1016/j.grets.2024.100138

Abid, Z., Abid, M., Zafar, Q., & Mehmood, S. (2018). Detrimental effects of climate change on women. *Earth Systems and Environment, 2*, 537–551.

Akhand, M. H. (2003). Disaster Management and Cyclone Warning System in Bangladesh. *Early warning systems for natural disaster reduction* (pp. 49–64). Springer.

Akter, S. (2025). Climate resilient development for agriculture and pathways for gender inclusivity. *Agricultural Economics*. https://doi.org/10.1111/agec.70040

Al, S. B. A., Khan, M. Z., & Islam, M. S. (2023). Spatio-temporal characteristics of meteorological and agricultural drought indices and their dynamic relationships during the pre-monsoon season in drought-prone region of Bangladesh. *Environmental Challenges*. https://doi.org/10.1016/j.envc.2023.100695

Alam, K., & Rahman, M. H. (2018) The role of women in disaster resilience. In *Handbook of Disaster Risk Reduction & Management 2018* pp. 697–719.

Alam, K. F., Shamsuzzoha, M., Arab, S. T., et al. (2024) Application of remote sensing in the analysis of climate extremes due to global climate change. In Ahamed, T. (ed.) Remote Sensing Application II. New Frontiers in Regional Science: Asian Perspectives, pp. 1–21. Singapore: Springer.

Algur, K. D., Patel, S. K., & Chauhan, S. (2021). The impact of drought on the health and livelihoods of women and children in India: A systematic review. *Children and Youth Services Review*. https://doi.org/10.1016/j.childyouth.2020.105909

Alston, M. (2015). *Women and climate change in Bangladesh* (1st ed.). Routledge.

Anwarul Islam, M., Shamsuzzoha, M., Rasheduzzaman, M., et al. (2019). Assessment on climate change adaptation: a study on coastal area of Khulna District in Bangladesh. *Australian Journal of Engineering and Innovative Technology, 6*, 14–20. https://doi.org/10.34104/ajeit.019.014020

Arab, S. T., Islam, M. M., Shamsuzzoha, M., et al. (2022). A Review of remote sensing applications in agriculture and forestry to establish big data analytics. In Ahamed, T. (ed.) Remote Sensing Application. New Frontiers in Regional Science: Asian Perspectives, pp 1–24. Singapore: Springer.

Ashraf, M. A., & Azad, M. A. K. (2015). Gender issues in disaster: Understanding the relationships of vulnerability, preparedness and capacity. *Environment and Ecology Research, 3*, 136–142. https://doi.org/10.13189/eer.2015.030504

Azad, M. J., & Pritchard, B. (2023). The importance of women's roles in adaptive capacity and resilience to flooding in rural Bangladesh. *International Journal of Disaster Risk Reduction*. https://doi.org/10.1016/j.ijdrr.2023.103660

Aziz, M., & Anjum, G. (2024). Transformative strategies for enhancing women's resilience to climate change: A policy perspective for low- and middle-income countries. *Womens Health (Lond), 20*, 17455057241302032.

Bin, I. F., & Sharma, M. (2021). Gendered dimensions of unpaid activities: An empirical insight into rural Bangladesh households. *Sustainability (Switzerland)*. https://doi.org/10.3390/su13126670

Brady, B., Rosenberg, S., Newman, C. E., et al. (2022). Gender is dynamic for all people. *Discover Psychology, 2*, 41. https://doi.org/10.1007/s44202-022-00054-2

Cai, H., Lam, N. S. N., Qiang, Y., et al. (2018). A synthesis of disaster resilience measurement methods and indices. *International Journal of Disaster Risk Reduction, 31*, 844–855.

Chakma, K., & Ruba, U. B. (2021a). Role of Bangladeshi women in diverse agricultural production: A review. *European Journal of Agriculture and Food Sciences, 3*, 1–5. https://doi.org/10.24018/ejfood.2021.3.3.284

Chakma, K., & Ruba, U. B. (2021b). Role of Bangladeshi women in diverse agricultural production: A review. *European Journal of Agriculture and Food Sciences, 3*, 1–5. https://doi.org/10.24018/ejfood.2021.3.3.284

Chikwe, C. F., Kuteesa, K. N., & Ediae, A. A. (2024). Gender equality advocacy and socio-economic inclusion: A comparative study of community-based approaches in promoting women's empowerment and economic resilience (2022). *International Journal of Scientific Research Updates, 8*, 110–121. https://doi.org/10.53430/ijsru.2024.8.2.0066

Chineka, J., Musyoki, A., Kori, E., & Chikoore, H. (2019). Gender mainstreaming: A lasting solution to disaster risk reduction. *Jàmbá: Journal of Disaster Risk Studies, 3*, 723. https://doi.org/10.4102/jamba.v11i3.723

Chowhan, S., Moshiur Rahman, M., Sultana, R., et al. (2024). Agriculture policy and major areas for research and development in Bangladesh. *Sarhad Journal of Agriculture, 40*, 819–831. https://doi.org/10.17582/journal.sja/2024/40.3.819.831

Cuong, O. Q., Suza, M., Savelli, A., & Nelson, K. M. (2024). Exploring the relationship between climate change events and migration decisions: Evidence from a choice experiment in Bangladesh. *International Journal of Disaster Risk Reduction, 113*, 104831. https://doi.org/10.1016/j.ijdrr.2024.104831

Dankelman, I., & Naidu, K. (2020). Introduction: Gender, development, and the climate crisis. *Gender and Development, 28*, 447–457. https://doi.org/10.1080/13552074.2020.1843830

de Brauw, A., Kramer, B., & Murphy, M. (2021). Migration, labor and women's empowerment: Evidence from an agricultural value chain in Bangladesh. *World Development*. https://doi.org/10.1016/j.worlddev.2021.105445

FAO, (2024). Better life: Gender equality and women's empowerment at the food and agriculture organization of the United Nations. Rome

Fatema, S. R., East, L., Islam, S., & Usher, K. (2023). Gender-based vulnerabilities for women during natural disasters in Bangladesh. *Front Commun (Lausanne)*. https://doi.org/10.3389/fcomm.2023.1180406

Ferdous, J., & Uddin, N. (2020). Women in decision-making positions: Developments and stumbling blocks, pp. 1–13.

Figueiredo, P., & Perkins, P. E. (2013). Women and water management in times of climate change: Participatory and inclusive processes. *Journal of Cleaner Production, 60*, 188–194. https://doi.org/10.1016/j.jclepro.2012.02.025

Frame, D. J., Rosier, S. M., Noy, I., et al. (2020). Climate change attribution and the economic costs of extreme weather events: A study on damages from extreme rainfall and drought. *Climate Change, 162*, 781–797. https://doi.org/10.1007/s10584-020-02729-y

Freihardt, J. (2024). Perceptions of environmental changes among a climate-vulnerable population from Bangladesh. *Climate Change*. https://doi.org/10.1007/s10584-024-03678-6

Gebremeskel, G., Tang, Q., Sun, S., et al. (2019). Droughts in East Africa: Causes, impacts and resilience. *Earth-Science Reviews, 193*, 146–161.

Ghosh, S., Sen, L. C., Mali, S. K., et al. (2021). The role of rural women in household food security and nutrition management in Bangladesh. *Asian Journal of Women's Studies, 27*, 441–459. https://doi.org/10.1080/12259276.2021.1970351

GoB, (2020). Standing orders on disaster 2019. Dhaka

Habiba, U., Shaw, R., Hassan, A. W. R. (2013). Drought risk and reduction approaches in Bangladesh, pp 131–164.

Habiba, U., Shaw, R., & Takeuchi, Y. (2011). Drought risk reduction through a socio-economic, institutional and physical approach in the Northwestern region of Bangladesh. *Environmental Hazards, 10*, 121–138. https://doi.org/10.1080/17477891.2011.582311

Hasan, M. M., Islam, R., Rahman, M. S., et al. (2021). Analysis of land use and land cover changing patterns of Bangladesh using remote sensing technology. *American Journal of Environmental Sciences, 17*, 64–74. https://doi.org/10.3844/ajessp.2021.64.74

Hosen, M. K., Alam, M. S., Chakraborty, T., & Golder, M. R. (2023). Monitoring spatiotemporal and seasonal variation of agricultural drought in Bangladesh using MODIS-derived vegetation health index. *Journal of Earth System Science*. https://doi.org/10.1007/s12040-023-02200-3

Huete, A. R. (1988). A soil-adjusted vegetation index (SAVI). *Remote Sensing of Environment, 25*, 295–309. https://doi.org/10.1016/0034-4257(88)90106-X

Humayra, J., Uddin, M. K., & Pushpo, N. Y. (2024). Deficiencies of women's participation in climate governance and sustainable development challenges in Bangladesh. *Sustainable Development*. https://doi.org/10.1002/sd.3075

Huq, H., Rahman, M. M., & Hossen, M. A. (2023). Women-led climate change adaptation: A Study in Northwestern Bangladesh. *Sustainability (Switzerland)*. https://doi.org/10.3390/su152015041

Hussain, I. (2007). Understanding gender and diversity dimensions of irrigation management for pro-poor interventions. *Irrigation and Drainage, 56*, 299–305. https://doi.org/10.1002/ird.295

IPCC, (2023). Climate change 2022: Impacts, adaptation and vulnerability. intergovernmental panel on climate change (IPCC), contribution of working group II to the sixth assessment report of the intergovernmental panel on climate change, Cambridge University Press., New York.

Islam, M. A., Islam, S. L., Hassan, A. (2017). Impact of climate change on water with reference to the Ganges-Brahmaputra-Meghna River Basin. In *Chemistry and Water: The Science Behind Sustaining the World's Most Crucial Resource*. Elsevier Inc., pp. 121–160.

Islam, A. R., Al Mamun, A., Rahman, M. N., et al. (2024). A cost-benefit analysis of flood early warning system: Evidence from lower Brahmaputra River Basin, Bangladesh. *International Journal of Disaster Risk Reduction, 104*, 104380. https://doi.org/10.1016/j.ijdrr.2024.104380

Islam, M. M., Gray, D., Reid, J., & Kemp, P. (2011). Developing sustainable farmer-led extension groups: Lessons from a Bangladeshi case study. *Journal of Agricultural Education and Extension, 17*, 425–443. https://doi.org/10.1080/1389224X.2011.596658

Islam, M. A., Hack-Polay, D., Rahman, M., et al. (2023). Gender and leadership in public higher education in South Asia: Examining the individual, socio-cultural and organizational barriers to female inclusion. *Studies in Higher Education, 48*, 1197–1215. https://doi.org/10.1080/03075079.2023.2187771

Jihan, M. A. T., Popy, S., Kayes, S., et al. (2025). Climate change scenario in Bangladesh: Historical data analysis and future projection based on CMIP6 model. *Science and Reports*. https://doi.org/10.1038/s41598-024-81250-z

Kabeer, N. (1991). Gender dimensions of rural poverty: Analysis from bangladesh. *Journal of Peasant Studies, 18*, 241–262. https://doi.org/10.1080/03066159108438451

Kogan, F. N. (1995). Application of vegetation index and brightness temperature for drought detection. *Advances in Space Research, 15*, 91–100. https://doi.org/10.1016/0273-1177(95)00079-T

Kreutzer, W., Millerd, C., & Timbs, N. (2023). Disasters and the diminishing of women's economic empowerment. *Disasters, 47*, 891–912. https://doi.org/10.1111/disa.12582

Kulatunga, U., Wedawatta, G., Amaratunga, D. G., Shamsuzzoha, M. (2013). Disaster risk and community vulnerability: The case of Patuakhali, Bangladesh.

Matin, N., & Taher, M. (2001). The changing emphasis of disasters in Bangladesh NGOs. *Disasters, 25*, 227–239. https://doi.org/10.1111/1467-7717.00174

Md, A., Gomes, C., Dias, J. M., & Cerdà, A. (2022). Exploring gender and climate change nexus, and empowering women in the South Western Coastal Region of Bangladesh for adaptation and mitigation. *Climate*. https://doi.org/10.3390/cli10110172

Medendorp, J. W., Reeves, N. P., Sal y Rosas Celi, V. G., et al. (2022). Large-scale rollout of extension training in Bangladesh: Challenges and opportunities for gender-inclusive participation. *PLoS ONE*. https://doi.org/10.1371/journal.pone.0270662

Meir, P., Mencuccini, M., Binks, O., et al. (2018). Short-term effects of drought on tropical forest do not fully predict impacts of repeated or long-term drought: Gas exchange versus growth. *Philosophical Transactions of the Royal Society b: Biological Sciences, 373*(1760), 20170311.

Moayedi, M., & Hayati, D. (2023). Identifying strategies for adaptation of rural women to climate variability in water scarce areas. *Frontiers in Water*. https://doi.org/10.3389/frwa.2023.1177684

Mulyasari, F., & Shaw, R. (2013). Role of women as risk communicators to enhance disaster resilience of Bandung, Indonesia. *Natural Hazards, 69*, 2137–2160. https://doi.org/10.1007/s11069-013-0798-4

Parvin, G. A., & Ahsan, R. M. R. (2013). Impacts of climate change on food security of rural poor women in Bangladesh. *Management of Environmental Quality: An International Journal, 24*, 802–814. https://doi.org/10.1108/MEQ-04-2013-0033

Pervin, N., & Mokhtar, M. (2023). Lifelong learning opportunities for professional women in Bangladesh: To what extent does patriarchy act as a barrier? *International Journal of Lifelong Education, 42*, 451–469. https://doi.org/10.1080/02601370.2023.2238903

Rahman, M. S., Sujan, M. H. K., Acharjee, D. C., et al. (2022). Intensity of adoption and welfare impacts of drought-tolerant rice varieties cultivation in Bangladesh. *Heliyon*. https://doi.org/10.1016/j.heliyon.2022.e09490

Rahman, S., & Tasnim, F. (2023). The role of NGOs in ensuring local governance in Bangladesh: From the perception of other actors of governance. *Asia-Pacific Journal of Regional Science*. https://doi.org/10.1007/s41685-023-00283-w

Ramalho, J. (2019). Empowerment in the era of resilience-building: Gendered participation in community-based (disaster) risk management in the Philippines. *International Development Planning Review, 41*, 129–148. https://doi.org/10.3828/idpr.2018.25

Richter, B. D., Bartak, D., Caldwell, P., et al. (2020). Water scarcity and fish imperilment driven by beef production. *Nature Sustainability, 3*, 319–328. https://doi.org/10.1038/s41893-020-0483-z

Sadiq, M. A., Sarkar, S. K., & Raisa, S. S. (2023). Meteorological drought assessment in northern Bangladesh: A machine learning-based approach considering remote sensing indices. *Ecological Indicators*. https://doi.org/10.1016/j.ecolind.2023.111233

Sarkar, S. K., Das, S., Rudra, R. R., et al. (2024). Delineating the drought vulnerability zones in Bangladesh. *Science and Reports, 14*, 25564. https://doi.org/10.1038/s41598-024-75690-w

Shamsuzzoha, M., Ghosh, R., Al-Maruf, A., et al. (2018a). Determination of economic security and social capital for climate change adaptation to cyclone risk at household level in coastal Region of Bangladesh Delta.

Shamsuzzoha, M., Ahamed, T. (2024). Application of introducing damaged area index for cultivated agricultural lands affected by recurrent tropical cyclones Bulbul and Amphan using satellite remote sensing. In Ahamed, T. (ed.) Remote Sensing Application II. New Frontiers in Regional Science: Asian Perspectives. Singapore: Springer, pp. 165–178.

Shamsuzzoha, M., Hasan, M. F., Rasheduzzaman, M., et al. (2025b). Bangladesh's progress on disaster risk reduction over the past 20 years. In: Shaw, R., Izumi, T., Djalante, R., Imamura, F. (Eds.) Two Decades from the Indian Ocean Tsunami: Key Challenges and Advancements. Singapore: Springer Nature Singapore, pp. 317–358

Shamsuzzoha, M., Biswas, A. K. M. A. A., Hasan, M. M., et al. (2025a). Determination of constraints in existing DRR coordination mechanisms among different humanitarian actors of coastal Bangladesh. In Shaw, R., Izumi, T., Djalante, R., Imamura, F. (Eds.) Two Decades from the Indian Ocean Tsunami: Key Challenges and Advancements. Singapore: Springer Nature Singapore, pp 379–409.

Shamsuzzoha, M., & Ahamed, T. (2023). Shoreline change assessment in the coastal region of Bangladesh delta using Tasseled cap transformation from satellite remote sensing dataset. *Remote Sens (Basel)*. https://doi.org/10.3390/rs15020295

Shamsuzzoha, M., Kormoker, T., & Ghosh, R. C. (2018b). Implementation of water safety plan considering climatic disaster risk reduction in Bangladesh: A study on Patuakhali Pourashava water supply system. *Procedia Engineering, 212*, 583–590. https://doi.org/10.1016/j.proeng.2018.01.075

Shamsuzzoha, M., Noguchi, R., & Ahamed, T. (2021). Damaged area assessment of cultivated agricultural lands affected by cyclone bulbul in coastal Region of Bangladesh using Landsat

8 OLI and TIRS datasets. *Remote Sensing Applications*. https://doi.org/10.1016/j.rsase.2021.100523

Shamsuzzoha, M., Noguchi, R., & Ahamed, T. (2022). Rice yield loss area assessment from satellite-derived NDVI after extreme climatic events using a fuzzy approach. *Agricultural Information Research, 31*, 32–46. https://doi.org/10.3173/air.31.32

Shamsuzzoha, M., Rasheduzzaman, M., & Ghosh, R. C. (2018c). Building resilience for drinking water shortages through reverse osmosis technology in coastal areas of Bangladesh. *Procedia Engineering, 212*, 559–566. https://doi.org/10.1016/j.proeng.2018.01.072

Shamsuzzoha, M., Shaw, R., & Ahamed, T. (2024). Machine learning system to assess rice crop change detection from satellite-derived RGVI due to tropical cyclones using remote sensing dataset. *Remote Sensing Applications, 35*, Article 101201. https://doi.org/10.1016/j.rsase.2024.101201

Shanta, M. A., Suman, S. U., & Islam, M. M. (2025). Role of formal aquaculture industry for empowering women: A study in the southern coastal region of Bangladesh. *Women's Studies International Forum*. https://doi.org/10.1016/j.wsif.2024.103046

Shefa, M. S., Hasan, M. F., Ovi, M. H., et al. (2024). Navigating the ravages of riverbank erosion: Socio-economic and environmental impacts in the Nalua Union of Bakerganj Upazila. *Asian Journal of Social Sciences and Legal Studies* 154–161. https://doi.org/10.34104/ajssls.024.015400161

Sultana, F. (2009). Fluid lives: Subjectivities, gender and water in rural Bangladesh. *Gender Place and Culture, 16*, 427–444. https://doi.org/10.1080/09663690903003942

Talukder MF, Shamsuzzoha M, Hasan I (2018) Damage and agricultural rehabilitation scenario of post cyclone Mahasen in coastal zone of Bangladesh. *Journal of Sociology and Anthropology*. https://doi.org/10.12691/jsa-2-1-7

Tariq, H., Pathirage, C., & Fernando, T. (2021). Measuring community disaster resilience at local levels: An adaptable resilience framework. *International Journal of Disaster Risk Reduction, 62*, 102358.

Tewari, N., Bush, A., Butt, M. N., et al. (2023). Gendered Dimensions of Loss and Damage in Asia.

Thapa, V., & Pathranarakul, P. (2019). Gender inclusiveness in disaster risk governance for sustainable recovery of 2015 Gorkha Earthquake Nepal. *International Journal of Disaster Risk Reduction, 34*, 209–219. https://doi.org/10.1016/j.ijdrr.2018.11.019

Umma, H., Rajib, S., & Yukiko, T. (2011). Socioeconomic impact of droughts in Bangladesh. *Community Environment and Disaster Risk Management, 8*, 25–48. https://doi.org/10.1108/S2040-7262(2011)0000008008

UNFCCC, (2024). Decision-CP. 29: Gender and climate change.

Vasileiou, K., Barnett, J., & Fraser, D. S. (2022). Integrating local and scientific knowledge in disaster risk reduction: A systematic review of motivations, processes, and outcomes. *International Journal of Disaster Risk Reduction, 81*, 103255.

WB, (2021). Gender dynamics of disaster risk and resilience. In *World Bank (WB) Group*. https://www.worldbank.org/en/topic/disasterriskmanagement/publication/gender-dynamics-of-disaster-risk-and-resilience. Accessed 8 Sep 2024.

Zaidi, R. Z., & Fordham, M. (2021). The missing half of the Sendai framework: Gender and women in the implementation of global disaster risk reduction policy. *Progress in Disaster Science*. https://doi.org/10.1016/j.pdisas.2021.100170

Zehba, M. P. F., Priya Rajendran, L., Firoz, C. M. (2025). Where is the time? Time poverty and women's urban mobility Narratives in Kochi, India. Urban Plan. https://doi.org/10.17645/up.9272

Chapter 5
Climate Crisis and Its Differential Impacts: Multi-sectoral Gender Perspective from Uttarakhand, India

Ambika Dabral, Harshit Sosan Lakra, Monu Kumari, Surabhi Mehrotra, Sonajharia Minz, Malavika Kanathoor, Shobha Poudel, Deepthi Wickramasinghe, Suvendrini Kakuchi, Ranit Chatterjee, and Rajib Shaw

Abstract Climate risks are contextual. Various systemic structures and processes, visible in physiological, cultural and socio-economic aspects of day-to-day life, make the women more vulnerable and prone to adverse effects of climate change in comparison to men. These differential impacts are not limited to mere climate-related mortality and physical damage and loss but often are long-term and intangible in nature. The chapter examines these differential impacts of climate change on women. It studies the systemic interactions among the underlying drivers of risks and underscores how the existing social, economic, and cultural inequalities are further reinforced during the climate crisis. Using the case study of Udham Singh Nagar, Uttarakhand, India, the chapter highlights how gendered roles across sectors such as agriculture, water management, and livestock management expose women to greater climate vulnerabilities and shocks, such as food and nutrition insecurity and livelihood loss. It delves into how limited access to resources and limited opportunities to meaningfully participate in decision-making processes hinder women's ability to

A. Dabral · M. Kumari · S. Kakuchi · R. Chatterjee · R. Shaw
Resilience Innovation Knowledge Academy, Delhi, India

H. S. Lakra (✉) · M. Kumari · M. Kanathoor
Indian Institute of Technology Roorkee, Roorkee, India
e-mail: harshit.lakra@ar.iitr.ac.in

S. Mehrotra
Maulana Azad National Institute of Technology, Bhopal, India

S. Minz
Jawaharlal Nehru University, New Delhi, India

S. Poudel
Science Hub, Kathmandu, Nepal

D. Wickramasinghe
University of Colombo, Colombo, Sri Lanka

Md. A. Abedin et al. (eds.), *Climate Crisis in South and East Asia*, Sustainability Solutions, https://doi.org/10.1007/978-3-032-01012-4_5

adapt to climate change. It highlights the need for gender-responsive policies that empower women as key actors in risk communication and climate resilience.

Keywords Women · Gender · Climate change · Adaptation · Resilience

5.1 Introduction

Global Risk Report 2025 ranks extreme weather events as the second most severe global risk over the short term (2 years) and as the most severe global risk over the long term (10 years) (WEF, 2025). In the long term, the risk of extreme weather events is followed by the risks of biodiversity loss and collapse of ecosystem, critical change to earth systems and shortages of natural resources (ibid). Studies project an increase of 3 °C in the global temperature by the end of the century, even if nations fulfil respective commitments made under the Paris Agreement 2015 (MoES, 2020). Such a rise will in turn aggravate the frequency and severity of extreme climatic events, though differentially, across different parts of the world.

India has already witnessed an increase of 0.7 °C in its average temperature during the period 1901–2018 (MoES, 2020). The country has experienced a decrease of approximately 6% in rainfall during the monsoon season between 1951 and 2015 (ibid). This decline has been significant over the Indo-Gangetic Plains and the Western Ghats. Further, there have been increasing cases of dry spells and severe cases of wet spells during the monsoon period along with a heightened frequency of localised events of cloud bursts and flash floods. During the period 1951–2016, not only the frequency of droughts in the country has increased but also the extent of drought-affected areas has increased by 1.3% per 10 years (ibid).

The Hindu Kush Himalayas (HKH) region spanning over 8 countries (including India) is home to sources of 10 major river systems in Asia; thus, nurturing a human population of over 1.9 billion (25% of the world's population) across the basins of these river basins (ICIMOD, 2025). The region is witnessing retreating glaciers and decreasing snowfall. It experienced an increase of around 1.3 °C between 1951 and 2014. Projections under the RCP 8.5 scenario indicate that as the century ends, there will be an increase of about 5.2 °C in the annual surface temperature in the region (MoES, 2020). A large population of the HKH region reside in remote, ecologically fragile and harsh weather conditions with many practising livelihoods which are prone to climatic uncertainties. Besides, limited access to basic services and infrastructure and degrading natural resources and ecosystem services, further expose the population and their socio-economic wellbeing to increasing disaster and climatic risks.

Among the poor and marginalised groups exposed to adverse impacts of climate change, women are particularly vulnerable, primarily because of the inherent "reproductive, productive and community roles" attributed to them (Khandelwal et al., 2025; Sardar & Bhaduri, 2022). In many traditional rural communities, women have been leading the roles linked with fetching water, farming, collecting fodder for livestock, collecting forest produce, household chores of cooking and cleaning, nursing and nurturing the household members including young, sick and elderly, along with sole responsibility of childbearing. As many of these roles are linked to the existing natural resources, any damage, loss or long-term degradation of these resources due to anthropogenic activities or induced by changing climate will directly and/ or indirectly aggravate the challenges already faced by the women in undertaking these roles. The chapter further dives into these gendered roles to capture the disproportionate impacts of climate change on women. It further studies the systemic interactions among the underlying drivers of risks and underscores how the existing social, economic, and cultural inequalities are exacerbated by the climate crisis.

5.2 Literature Review

The impacts of climate change are visible across multiple sectors and dimensions. The altered weather and climate patterns are triggering more number and severe extreme events across the globe; thus, adding more uncertainties to the disaster risk landscape. The South Asia region is facing increasing and more devastating disaster events such as floods, cyclones, landslides, droughts, forest fires, and heatwaves, among others. Between 2000 and 2025 (as of February 2025), the South Asia Region has braved a total of 1209 natural hazard-induced disaster events (EM-DAT, 2025). Out of these, 1000 disaster events were linked to hydrological hazards (659) such as flash floods, riverine floods, coastal floods, avalanches (wet), landslides (wet) and mudslides; meteorological hazards (311) such as storms, cyclones, lightning, heat wave, cold wave, hail; and climatological hazards (30) such as droughts, glacial lake outburst flood and wildfires (ibid). A total of 379 of these events occurred and affected different parts of India. Not only do these events result in high mortality of women and children but they are also found to adversely impact their health conditions differentially. Thus, women are considered among the vulnerable groups to these increasing disaster and climate risks. There is a myriad of physiological, cultural and socio-economic reasons for the same (Wright et al, 2023).

Physical and mental health: Researchers like Tiwari et al. (2022) have closely analysed around 42 studies on extreme events and have found a positive correlation with health impacts, particularly that of women and children. Three such studies brought forth strong linkages between mortality and meteorological conditions, while one study underscored how exposure of pregnant women or newborns to droughts could make the newborns more prone to being underweight (Tiwari et al., 2022). Similarly, studies suggest that in India, there has been an increase in cases of mortality among children who were born in the year following a drought year (Kumar et al., 2016). Further, while women suffer significantly from mental health impacts stemming from climate and disaster risks, researchers highlight that the aspect remains understudied in the South Asia region (Tiwari et al., 2022). A study on mental health aspects of women prone to disaster and climate risks in Bangladesh brings forth some crucial concerns that were giving rise to cases of depression among these women (Monya et al., 2024). These included among others, factors associated with the caregiving roles of women, inadequate social and financial support they have and incidences of domestic violence they experience (ibid).

Apart from the mortality and health aspects, women, particularly those from ecologically fragile geographies, often find themselves at a disadvantage to the complex nexus of climate, food & nutrition security and natural resources (Tiwari et al., 2022). Particularly in rural communities, their close association with natural resources and climate-sensitive and time-consuming livelihood options and daily chores, further make them differentially (and more) vulnerable to adverse impacts of escalating ecological degradation and climate risks. The impacts are more severe for women because of their lack of ownership of productive assets, absence of or limited access to resources, and poor representation in decision making, among others (Moon, 2024).

Some of the key multi-sectoral gendered impacts of climate change are discussed below.

Agriculture: Women in rural areas are often tied to the management of household chores along with sustaining household consumption and income through farming and allied activities. "The Unjust Climate: Measuring the impacts of climate change on rural poor, women and youth" by the Food and Agriculture Organisation (FAO) brings out glaring facts in this regard. It states that in comparison to households headed by men, women-headed households face higher losses in annual average incomes due to heat stress (8%) and floods (3%) (FAO, 2024). There is also a stark gap in the access to and uptake of appropriate climate adaptation practices in the agricultural fields managed by women and men; indicating systemic bottlenecks in the sectoral policy and programmes (ibid). The report highlights how households headed by women are spending more on agriculture compared to those headed by men but are also losing more due to climate-induced extreme events. Despite the additional burden that women undergo, during climatic shocks, they are more likely to lose their livelihood opportunities, particularly those not linked to agriculture (Mahajan, 2017 as cited in FAO, 2024).

With a majority of crops in India being rainfed, the wellbeing of Indian farmers is all the more sensitive to climate uncertainties. Though over the decades, there has been a decline in the workforce engaged in agriculture in India, still women constitute a higher percentage of this workforce. They are majorly involved in roles such as seed selection, propagation, preparation of fields, sowing, cultivation and even harvesting, among others. Many of them also work as agricultural labourers on farms not owned by them. During extreme weather conditions, the prevailing socio-cultural fabric of the country allows fewer opportunities for women to opt for migration or alternate sources of livelihood in comparison to men (Agarwal et al., 2021). Such scenarios involving the out-migration of men further aggravate the burden on women for shouldering both agricultural and other household responsibilities. In the absence of adequate and appropriate tools, resources, extension services, and knowledge of climate resilient agriculture along with decision making power, the degradation of farms and/or their lower production and productivity would only continue to burden these women without yielding the required results.

Household food and nutrition security: Various literature highlights the direct correlation between climatic uncertainties and threats to food security. Rise in temperature by one degree is likely to lead to a decrease of around 1.2% and 1.7% in food security over the short and long-term (Bhuyan et al., 2023). Similarly, a shortage of rainfall can adversely impact food security in the subsequent years for a short period of time (ibid).

During times of food shortage, varied systemic socio-cultural and economic reasons often make the women and children among the first ones to be affected while limited food resources get directed towards men so that they can continue working (Agarwal et al., 2023). Further, the nutritional intake of women is more likely to be compromised during such times with priority being given to the nutritional needs of men and children (Sorensen et al., 2018). As the nutritional needs of women, particularly those mensurating, pregnant or lactating are higher, the impact of its deficiency can be significant and long-lasting in terms of not only their own health and well-being but also their neonatal health (ibid). It is witnessed that with aggravating conditions of food insecurity in a household, there is also an increase in cases of gender-based violence; further widening the underlying gender-based disparities (Agarwal et al., 2023).

Natural resources management: Climate change significantly alters the water cycle, with widespread consequences for water resources and their management (Madhusoodhanan et al., 2016). It is likely to significantly impact water availability and natural water resources in India. Climate change and various other human-induced factors have already led significant loss and degradation of wetlands, lakes, ponds, and other such ecosystems in the country (Chakrabarty, 2016). Women being associated with time and effort intensive roles of fetching water, water management within households and for livestock, food cultivation, etc. face higher challenges due to the impact/loss/degradation of these water bodies. To better gauge the risk and differential impacts on women due to degrading natural water resources, it is crucial that hydrological cycle be rather studied as hydrosocial cycle, a concept used by Linton and Budds (2014) to better understand the social dimensions with respect to water.

With regard to forests as natural resources, climate change is providing an enabling environment for different disruptions with the potential to adversely impact the forests. These include among others, increased incidences of forest fires, pest infestation, and new/emerging species (alien or invasive) in the local ecosystem (Hussain et al., 2024). Similarly, in the Indian Himalayan region, the impacts include a reduction of alpine plantations and receding tree-line due to increasing temperature while in the coastal areas, there has been a degradation of mangrove plantations due to increasing intrusion of saline water and decreasing productivity (ibid). This resultant loss of terrestrial, riverine and marine resources in turn is leading to loss of biodiversity. With the reduction in production and productivity of forest produce, the burden on women who are often closely associated with the collection and management of these is increasing.

Livestock management: Livestock sector is also likely to be affected by climate change negatively because of increasing degradation of natural resources and associated increase in the competition for the existing ones, deterioration in the quality and also the quantum of the forage, increases in incidences and the emergence of new livestock diseases along with their increasing exposure to extreme climatic events such as heat stress, among others (Rojas-Downing et al., 2017). About 70% of the workforce in the livestock sector in India are women (Patel et al., 2016). They mostly undertake activities such as collecting fodder, feeding and watering the livestock, accompanying them for grazing, milking them, managing their health and caring for newborns (ibid). With the degrading sources of water and fodder and increasing incidence of diseases among the livestock and also in some cases, their transmission to humans, challenges are becoming increasingly complex, time- and effort-consuming for these women. Disaster events such as droughts, floods, and extreme temperatures, induced by climate change, further aggravate these challenges for women amidst the other caregiving and household responsibilities they undertake during such times.

The above literature review helps map the following critical and systemic linkages among impacts of climate change and underlying vulnerabilities of women that reinforces these impacts into cascading ones (Fig. 5.1).

5.3 Conceptual Framework

The analysis uses an intersectional framework to understand the systemic interplay of underlying drivers of climate vulnerability and risk instead of mere consideration of climatic hazards. The framework strives to unravel how the vulnerable groups such as women and young girls face the compounding and cascading vulnerabilities due to historic and prevailing socio-cultural and economic contexts and associated inequalities as triggered by intersection of determinants such as gender, age, religion, caste, and ability, among others (Daalen et al., 2024).

The framework helps understand how gender roles and intersecting inequalities give rise to multi-dimensional vulnerabilities of women to climate risks (Minjung, 2025). It helps in better comprehension of nuanced contexts in which women face the risks of climate change. It further enables one to recognise that despite belonging to a larger social group of women, how the manifestation of risk varies from one woman to another depending on their power and resource dynamics within their contexts, vulnerabilities and adaptive capacities.

An intersectional analysis helps identify the root causes along with others bottlenecks and gaps in policy and actions that fail to address or even continue to perpetuate

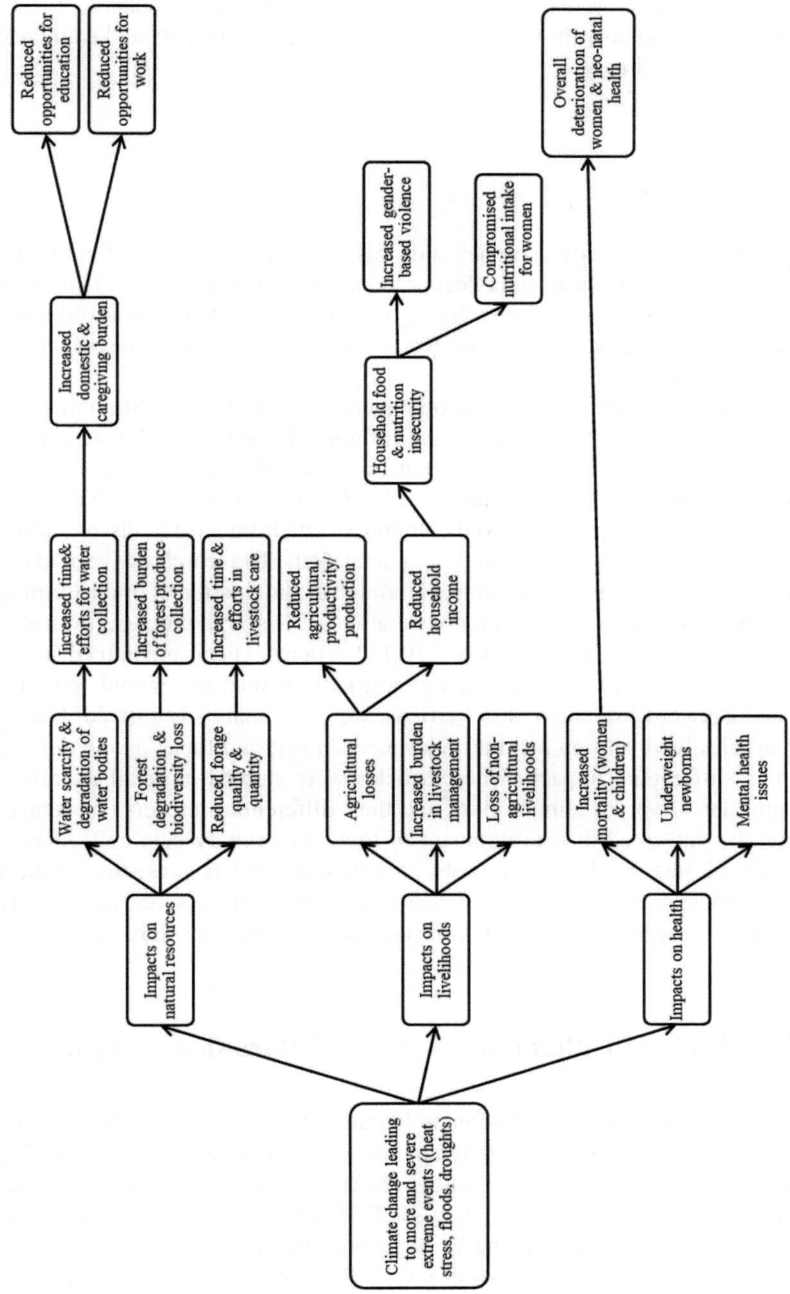

Fig. 5.1 Systemic linkages among impacts of climate change and underlying gendered-vulnerabilities that reinforces these impacts

the prevailing disparities. Such identification of socio-cultural and economic dimension of climate change is crucial in designing more equitable and sustainable institutional mechanisms and instruments for building the adaptive capacity and resilience of women to climate change.

5.4 Methodology

The chapter uses a case study approach. For this, literature review and field-based tools in the study area were used to collect primary and secondary data and information. The key field-based tools used in this regard include stakeholder consultations, semi-structured interviews, focus group discussions, story-telling, transect walks, and participatory observation.

The case study is developed using Systematic Literature Review (SLR) to study review and synthesise existing literature on differential impacts of climate change using a gender lens. To achieve this, a systematic search of relevant research papers was conducted using the Scopus database. This platform was selected because it offers a broad spectrum of peer-reviewed literature in a wide range of fields, providing a large and relevant collection of research for our analysis. The search was carried out using certain keywords which were selected through a combination of brainstorming sessions, group discussions, and expert consultations. The search process, illustrated in Fig. 5.2, initially identifies a pool of 210,182 articles. Using predefined inclusion and exclusion criteria, such as language restrictions, year, review-only articles, citation, and keyword limitation, 1253 articles were automatically filtered. Out of these, 51 articles were chosen for their relevance, clarity, and focus on the climate crisis and its differential impacts. These articles were studied in detail to inform the emerging landscape of climate risks and their differential impacts on women. This study also informed the development of tools for primary data collection in the study area. Using the developed tools, various stakeholders consisting of local government officials, community representatives, youth volunteers, and community members were engaged in different field-based activities and consultations.

5.5 Case Study of Udham Singh Nagar, Uttarakhand, India

Uttarakhand is among the 11 mountainous Indian states lying in the HKH region. It is characterised by its distinctive geographical and ecological features, including high mountain ranges, dense forests, glaciers, and river systems. It lies at the foothills of the Himalayas between the 28°43′ N & 31°27′ N and 77°34′ E & 81°02′ E. The two main river systems, the Ganga and Yamuna, originate and flow through the state which is rich in water resources, mostly fed by the Himalayan glaciers (Singh & Pandey, 2023). The state occupies 53,483 km^2 and is primarily mountainous (93%) and has forests on 65% of its terrain (Nandargi et al, 2016). The elevation of the

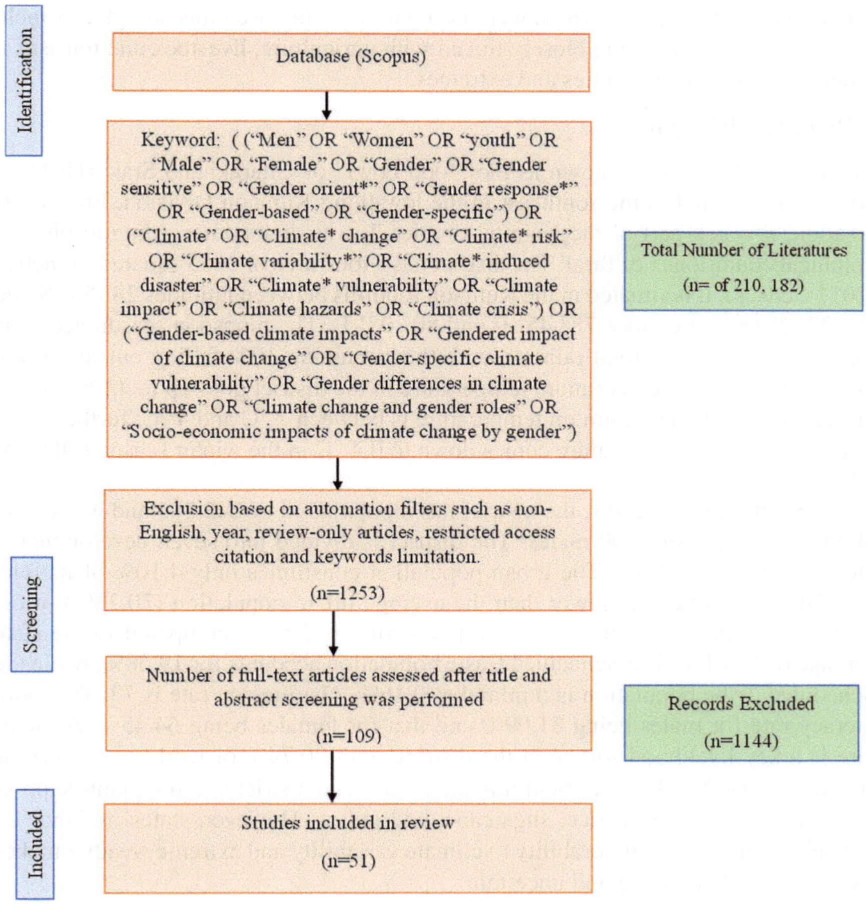

Fig. 5.2 Flowchart illustrating the SLR steps undertaken

state varies from 210 m to 7817 m, which is a considerable variation. Summers in the state are extremely hot, with temperatures varies between 40 °C and 45 °C. The state can experience very cold winters with sometimes temperatures dropping below 5 °C (Department of Health & Family Welfare, Uttarakhand, 2022). The southwest monsoon (June to September) is the major source of rainfall in the region; accounting for around 90% of the annual rainfall (Nandargi et al., 2016). Majority of the population (around 75%) of the state resides in rural parts and primarily engaged in agriculture-based economy (Government of Uttarakhand, 2014).

Uttarakhand is susceptible to a variety of climate-related hazards because of its geographical location and environment. These include riverine floods, cloudbursts, flash floods, droughts, landslides, forest fires, hailstorms, cold waves, etc. Examples of such occurrences include the June 2013 flash floods in Uttarakhand and the

Chamoli tragedy in 2021, which were brought on by intense rains and glacier melt. The state's economy being closely linked with agriculture, livestock and tourism is prone to climate uncertainties and extremes.

Udham Singh Nagar

Udham Singh Nagar is known as the "food bowl" of Uttarakhand State (UCOST, 2020) It is situated in the southeast in the low-lying Kumaon Division. Prior to its creation, it was a part of the Nainital district. It was divided according to physiographical conditions i.e. Tarai. The district has a total area of 3055 square kilometres (2011 Census). It is situated in the Kumaon foothills between latitudes 28° 53' N and 29° 23' N and longitudes 78° 45' E and 80° 08' E. The monsoon season accounts for about 90% of the total rainfall, with the remaining 10% falling outside of the monsoon season. The maximum temperature in the district goes up to 42 °C during the summers and the minimum temperature is between 1 °C and 4 °C, further north of the district, the temperature comes down to 0.4 °C in the winter season (DDMA, 2024).

As per the Census 2011, the district has a population of 242,285 and a sex ratio of 1114 females per 1000 males. The district is divided into seven developmental blocks and seven tehsils. The urban population constitutes only 4.10% of its total population, significantly lower than the average urban population (30.23%) in the state. The decadal growth rate of the population is 6.53%, compared to the state average of 18.81%. The Scheduled Caste population accounts for 19.68%, while the Scheduled Tribe population is minimal at 0.16%. The literacy rate is 73.10%, with literacy rate for males being 81.09% and that for females being 64.45%. Agriculture is a key livelihood source in the district, with 20.74% of workers engaged as cultivators and 27.94% as agricultural labourers. The district's fertile plains support the cultivation of wheat, rice, sugarcane, and pulses. However, states' reliance on agriculture amidst its vulnerability to climate variability and extreme weather makes the situation challenging and uncertain.

Based on the literature review and primary engagement with different stakeholders, the following sectoral insights have been captured.

Agriculture: Climate change has significantly impacted agricultural practices, with rising temperatures contributing to heat stress in crops, altering growing seasons, and reducing yields in the area. Women farmers, who play a critical role in agricultural labour and subsistence farming, are disproportionately affected by these changes. Inconsistent rainfall and temperature anomalies can lead to crop failures, affecting the livelihoods of farmers and the region's food security (Tyagi, 2022). *"Our summers are much warmer now, making cultivation more challenging than it was in the past"*, a female farmer stated during consultations. 11.481 ha of agricultural land were damaged in 2015–16 in the district due to excess rainfall. Excess rainfall is the primary cause of disasters in the district (UCOST, 2020).

Another female farmer said that "*heavy rainfall causes us to lose our crops, which causes great distress throughout the family*". During a focus group discussion, a community trainer stated, "*In our village, the men only prepare the land and give water to the field; the rest of the work of farming lies on the shoulders of women."* Women perform more physical work with the household responsibility which often also compromise their health status; making them more vulnerable to adverse impacts. These observations are in congruence with a study by the FAO which found that women in agriculture bear a disproportionate share of the burden during climatic disruptions due to their dual responsibilities in the field and at home (FAO, 2011).

Water resources: Water resources are being adversely impacted by climate change, which changes the hydrological cycle and affects water availability and quality. Rural women, often tasked with fetching water, must travel greater distances to collect it, increasing their already substantial workload (FAO, 2024). Similar experiences came forth in the study area. A woman in her 30 s pointed out, *"I have to walk extra miles to fetch water when there are disruptions in the supply of tap water. My health is impacted because the route doesn't have many trees or shaded spots where I can take rest"*.

Besides, during the transect walk across one of the villages in the district, it was observed that one of the river channels has completely dried up over the years and is now being used as a landfill. The channel's drying and changes in land use highlight how climate change combined with adverse anthropogenic actions have impacted the local resources which were earlier used as source of clean water for the community. The community informed this change has taken place over the last decade and a half or so and negligence by the community in taking appropriate actions timely further worsened the current situation of this natural resource.

According to a 2023 WHO/UNICEF Joint Monitoring Programme (JMP) study, 1.8 billion people worldwide reside in homes without on-site water services. In seven out of ten of these homes, females above 15 years of age are largely responsible for gathering water, compared to three out of ten households for their male counterparts. Additionally, girls under 15 are more likely to collect water (7%) than boys under 15 (4%). Women and girls typically go farther to pick it up, which means they miss out on time for work, school, and leisure and run the risk of sustaining injuries or encountering other hazards along the route (UNICEF & WHO, 2023). It was noted during focused group discussion in the study area that women take up to 1.5–2 hours to fetch water during flood/ waterlogging situations. This time-consuming chore of water collection, compounded by water shortages during the summers, adds a significant load to their daily tasks. In addition, during period of heatwave, long distance walks to fetch water expose them to conditions of heat stress and heat stroke; and also, in some cases comprise their safety by exposing them to gender-based violence/crime (Jalees & Shiva, 2005).

It is pointed out that women as the main caregivers and home managers, already handle a variety of responsibilities such as cooking, cleaning, childcare, and agricultural labour. Their physical and mental health are further strained by the extra time and effort needed to obtain water, which reduces their time for other important activities and detrimentally affect their health.

Livestock management: Literature suggests that climate change impacts livestock production and leads to livestock loss primarily due to changing pattern of production and deteriorating quality of feed crops and forage (IFAD, 2010), decreasing availability of water, affected growth of the livestock, impacted production of milk, diseases, reproduction (Nardone et al., 2010), and biodiversity. Udham Singh Nagar is an agrarian society, and most of the household depends on livestock rearing. Here also, women and young girls share the major responsibilities of livestock management in the study area. However, according to the UCOST Report 2020, livestock suffered a significant loss in the years 2011, 2014, and 2024 as a result of climate-induced events such as heat waves and extreme floodings during monsoon.

Extreme flooding during monsoon often results in loss of livestock and other assets, which affects the rural subsistence, food security, and income of rural communities. *"This monsoon flooding was painful since our cattle, cows and goats were drowned and we couldn't save them,"* said a community trainer on their experience during a recent monsoon flood. This first-hand story demonstrates the devastating emotional and financial toll that climate-induced hazards have on communities.

5.6 Results and Discussion

Climate change impacts are not neutral to gender but are rather gender-differentiated, as it is experienced by men and women differently. These are often felt more severely by the women due to underlying socio-economic challenges within the society, which also limit their coping capacity. The case study has highlighted how Udham Singh Nagar is facing interrelated climate change challenges, especially pertaining to agriculture, water security, and livestock management. Interim findings show that women, who are primarily responsible for household tasks and family caring roles, are also overburdened by farming and livestock management activities. This is because, in the majority of cases, traditionally male counterparts have not been sharing these responsibilities and in some cases, this is also due to the outmigration of the male population for livelihood opportunities. Women are disproportionately impacted by these changes due to their socioeconomic status, dual responsibility of work, and limited opportunity for adaptation. Participants identified that recent changes in climate variability, such as monsoon floods, had increased hardship through increased workloads for women fetching water and firewood for the household.

Climate change also had severe impacts on livestock management, impacting livelihoods and increasing food insecurity. The economic vulnerabilities of rural communities are further exacerbated by the loss of livestock due to disease, feed shortages, and flooding. These losses have a greater impact on women, as they are the primary caregivers. The findings reveal a complex network of climate-related risks that has worsened the already-existing gender disparities and highlight the importance of inclusivity and representation of women in community-based initiatives and policies to promote long-term sustainability and resilience.

However, studies suggest various existing barriers for designing and executing effective adaptation and resilience strategies. Some of these are discussed below.

Social and cultural inequalities: Discrimination based on various socio-cultural parameters significantly impedes access of women to critical resources, exacerbating climate vulnerability. Mobility restrictions, particularly for women, further curtail their ability to seek alternative livelihoods or access support during climate shocks. Compounding these challenges, women often experience increased workloads and time poverty, as they shoulder additional burdens related to water collection, food preparation, and caregiving in the face of environmental stress. These intersecting factors highlight how social hierarchies and gendered norms create distinct and heightened vulnerabilities to climate change, demanding targeted interventions that address these complex realities.

During a focus group discussion, a mother said, *"Due to the family's financial position, we can afford only one child's education, and we want our son to study so that he can get work and take care of us when we are older. My daughter will assist me with domestic tasks because she will be married off."* Gender gaps in education persist in the area, especially in rural households where economic restrictions frequently lead families to prefer sons' education over daughters. Girls are more likely to be kept out of school to help with household chores, which reinforces traditional gender norms. The narrative reflects the patriarchal setup of the society, which shows men's preference for socio-economic opportunities and limits the choices for the women's economic stability.

Limited access and ownership of resources: Access to fundamental assets like land, financial resources, and information technology significantly shapes the adaptive capacity and resilience of women to climate change. Limited land ownership and restrictive inheritance rights curtail their control over productive resources and their influence to implement sustainable practices. Similarly, restricted access to financial resources and credit hampers their ability to invest in climate-resilient technologies or diversify their livelihoods.

Limited access to resources, financial restrictions, and inequality in ownership of property and agricultural resource management in Udham Singh Nagar, Uttarakhand, indicate the larger socioeconomic challenges that women in the rural households face. During an FGD, a married woman in her 30s stated that "*my husband makes all key decision-making in the household, including money, land, and children, because he is the head of the household.*" Similar gender-based inequalities prevailing in the management of agricultural assets and resources are also highlighted in many studies. These indicate how the women engage in tasks and duties that are very intensive of labour and time such as weeding and produce sorting while their male counterparts continue to own and manage major farm and financial resources and taking key decisions, such as purchasing land and agricultural tools (Pandey et al., 2024).

Moreover, imbalances in access to information and technology, such as those on early warning systems and agricultural extension services, disable women from performing critical emergency roles and making informed decisions to mitigate and address climate risks. These systemic inequalities underscore the urgent need for interventions that promote their equitable access to these critical resources, empowering them to build their socio-economic resilience while imbibing the principles of ecological sustenance.

Lack of robust sex-disaggregated datasets: Global frameworks and agreements such as Paris Agreement and Sendai Framework for Disaster Risk Reduction have given a clarion call for having in place sex-disaggregated datasets and baselines pertaining to disaster and climate risks and impacts. While progress is being made in this direction, still more needs to be done as the same serves as a critical foundation for better comprehension of gender-differentiated systemic factors and impacts and hence laying down of the gender-responsive policies and programmes (UNFCCC, 2022).

Constraints on decision-making power: Women, though, playing dual role of managing household chores and that of agricultural and livestock management, often gets restricted by socio-cultural norms and practices to participate on platforms and mechanisms that take critical planning, financial and governance decisions, including those on risk management and planning. With limited or no engagement in these decision-making processes and risk planning discourses or ability to influence them in anyway, even at a household level, women find themselves at disadvantage to effectively prepare, mitigate, adapt or respond to climate risk and associated challenges. This calls for interventions that are not only informed by the deep-rooted power and resource disparities but are also able to challenge or navigate them by promoting meaningful engagement and empowerment of women. Laying down and promoting such gender-responsive climate policies, planning, and programmes necessitates that these are duly informed by risk and resilience assessments which are guided by local evidence on "historical, social, and political realities" (Sidloski et al., 2025).

Studies highlight the instrumental role women can play in the uptake of policies and programmes aimed at ecological restoration and climate change adaptation because of their exposure and sensitivity to such issues. Women at all levels are required to be brought on board for ensuring effective climate actions. If provided with same resources that are available with men, studies suggest, women can help significantly enhance agricultural productivity and output (UNFCCC, 2023). This can trigger positive spiral loop in terms of increase food security, reduced poverty and hunger and when further supported with appropriate capacity building and access to better technology and resources, can result in making the management of agriculture, livestock and natural resources more sustainable (ibid). With improved engagement with women, one can enrich the decision making and program designing with more nuance and specific needs and priorities; thus, moving beyond designing and executing "one-size-fits-all" interventions to tailormade and targeted planning and interventions.

Table 5.1 enlists the underlying factors determining or aggravating different dimensions of vulnerabilities and potential way forward as captured by study of literature and case study discussion.

Table 5.1 Key gender-determinant of climate vulnerability and risks

Dimensions	Underlying factors determining/ aggravating vulnerability to climate risks	Way forward
Social	• Burden due to gendered roles-caregiving, food preparation, fetching of water and firewood for household • Reduced food and nutritional intake during period of distress and extreme events • Reduced opportunities for education/increased drop out of girls during extreme events • Limited participation in household level decision making processes	• Effective documentation and communication of adverse impacts on women health and well-being • Improved awareness of community and family members on above mentioned impacts • Long-term advocacy on rights of women related to ownership of assets, inheritance, participation in decision-making processes
Economic	• Participation in farming and livestock management activities—time and labour-intensive activities • Supporting livelihood of the household in case of migrating male counterparts • Mobility restriction limiting access to alternative/better livelihood options • Limited or no access to productive assets and resources such land, and house • Restrictive inheritance rights • Limited opportunities for credit	• Enabling women with scientific tools and technologies for enhancing productivity and output of agricultural and livestock management activities • Upskilling women for uptake of alternative and climate-smart livelihood options • Improving the access of women to financial services including credit, and insurance
Institutional	• Poor representation in decision-making processes • Limited participation in risk planning discourse • Limited opportunity for adaptation • Imbalance in access to information and technology • Limited access to agricultural extension services • Inadequate targeted interventions on gender-inclusive adaptation and resilience building • No or limited access to mental health and psycho-social services	• Ensuring improved access to climate information and services • Formulating targeted schemes and programmes on adaptive social protection • Improved participation of women in adaptation and resilience building planning and interventions • Adequate gender-disaggregated datasets on climate risk drivers and impacts for formulating gender-responsive plans and policies

(continued)

Table 5.1 (continued)

Dimensions	Underlying factors determining/aggravating vulnerability to climate risks	Way forward
Ecological	• Degradation of natural resources • Extreme heat and weather conditions making fetching of water and firewood challenges	• Enabling effective participation of women in decision making processes and discourses on climate actions • Leveraging indigenous and/or ecosystem approaches for conservation of natural resources and climate actions • Sensitising women and building their capacities for being prepared against potential adverse impact of climate change on their health and daily lifestyle

5.7 Conclusion

Informed by the study of "adopters" and "non-adopters" of Climate Smart Agriculture (CSA) practices in the state of Bihar, Agarwal et al. (2021) bring forth evidence of how the adopters were able to better mitigate the risks to food security during periods of climatic stress. It states that almost across all categories of food (pulses, vegetables, legumes, cereals, fruits, eggs, meat), climate stress reduced the intake of nutrition more in non-adopters of CSA than in the adopters (Agarwal et al., 2021). Thus, to support the continued food and nutrition security among these households, particularly among the women, it is crucial that they are increasingly engaged and up-skilled on such coping strategies and practices. Further, the need for accessible and affordable services on mental health and psycho-social support to women, particularly those belonging to ecologically fragile geographies is crucial. Researchers such as Moyna et al. (2024) further uphold the integration of these considerations in the risk management policies and governing tools.

At the same time, it is crucial that women have improved access to climate information and services such as that on early warning, weather forecasting, tools & technologies supporting and building resilience of their livelihood options, insurance and other financial services and products, among others. While making the technologies available to them, care should also be taken in ensuring that their indigenous and traditional knowledge of the local contexts and associated skills do not get replaced but are carefully integrated into climate adaptation practices, striking a fine balance between the traditional and modern practices as it suits the local needs and priorities.

Above all, it is the need of the hour to be sensitive of and recognising the differential vulnerabilities of women to climate risks but at the same time considering and engaging them not just as a vulnerable group but rather also as champions of climate change adaptation amidst the evolving landscape of climate risks.

Acknowledgements The case study under the chapter is based on the work supported by the Asia-Pacific Network for Global Change Research (APN) under Grant No. CBA2023-06MY-harshit.

References

Agarwal, T., et al. (2021). Gendered impacts of climate-smart agriculture on household food security and labor migration: Insights from Bihar, India. *International Journal of Climate Change Strategies and Management, 14*(1), 1–19. https://doi.org/10.1108/IJCCSM-01-2020-0004

Agarwal, P., et al. (2023). The interrelationship between food security, climate change, and gender-based violence: A scoping review with system dynamics modeling. *PLOS Glob Public Health, 3(2).* https://doi.org/10.1371/journal.pgph.0000300

Bhuyan, B., Mohanty, R. K., & Patra, S. (2025). Impact of climate change on food security in India: An evidence from autoregressive distributed lag model. *Environment, Development and Sustainability, 27*, 6349–6369. https://doi.org/10.1007/s10668-023-04139-3

Chakrabarty M. (2016). *Climate change and food security in India.* ORF Issue Brief. https://www.orfonline.org/public/uploads/posts/pdf/20230523125345.pdf

Daalen, K. R., et al. (2024). Bridging the gender, climate, and health gap: The road to COP29. *The Lancet Planetary Health, 8*(12), e1088–e1105. https://doi.org/10.1016/S2542-5196(24)00270-5

District Disaster Management Authority (DDMA). (2024). District Disaster Management Plan, Udham Singh Nagar.

EM-DAT (2025). *The international disaster database.* Centre for Research on the Epidemiology of Disasters. https://www.emdat.be/

FAO. (2011). The state of food and agriculture. Rome: Food and Agriculture Organization of United Nations.

FAO. (2024). *Climate smart agriculture sourcebook.* Retrieved from Food and Agriculture Organisation of United Nations. https://www.fao.org/climate-smart-agriculture-sourcebook/enabling-frameworks/module-c6-gender/chapter-c6-1/en/

Government of Uttarakhand. (2014). UAPCC. Government of Uttarakhand.

Hussain, S., et al. (2024). Navigating the impact of climate change in India: A perspective on climate action (SDG13) and sustainable cities and communities (SDG11). *Frontiers in Sustainable Cities 5*, 1308684.

International Centre for Integrated Mountain Development (ICIMOD). (2025). *Mountain livelihoods lens.* https://www.icimod.org/our-themes/livelihoods/

International Fund for Agricultural Development (IFAD). (2010). *Livestock and climate change.*

Jalees, K., & Shiva, V. (2005). *Water & women: A report by research foundation for science, technology, and ecology for national commission for women.* National Commission for Women (India) & Research Foundation for Science.

Khandelwal, S., et al. (2025). Assessment of climate change over the Indian region. In Pachauri (Ed.) *Health and climate change*, (pp 289–311). https://doi.org/10.1016/B978-0-443-29240-8.00013-4

Kumar S et al (2016). Drought and early child health in rural India. *Population and Development Review, 42*(1), 53–68. https://onlinelibrary.wiley.com/doi/abs/10.1111/j.1728-4457.2016.00107.x?msockid=1523ae386c90642f3a36ba256d966505

Linton, J., & Budds, J. (2014). The hydrosocial cycle: Defining and mobilizing a relational-dialectical approach to water. *Geoforum, 57*, 170–180. https://doi.org/10.1016/j.geoforum.2013.10.008

Madhusoodhanan, C. G., Sreeja, K. G., & Eldho, T. I. (2016). Climate change impact assessments on the water resources of India under extensive human interventions. *Ambio, 45*, 725–741. https://doi.org/10.1007/s13280-016-0784-7

Mahajan, K. (2017). Rainfall shocks and the gender wage gap: Evidence from Indian agriculture. *World Development, 91*, 156–172. https://doi.org/10.1016/j.worlddev.2016.11.004

Ministry of Earth Sciences (MoES), Government of India. (2020). Assessment of climate change over the Indian Region. https://doi.org/10.1007/978-981-15-4327-2

Ministry of Health and Family Welfare, Government of India. (2022). SAPCCHH-Uttarakhand. Roorkee: Ministry of Health and Family Welfare, Government of India.

Minjung, J. (2025). Shedding Light on the Hidden Struggles: Intersectional Vulnerabilities of Urban Women in Deyang during the COVID-19. *International Journal of Disaster Risk Reduction, 121*, 105363. https://doi.org/10.1016/j.ijdrr.2025.105363

Moon, M. P. (2024). How does climate change affect the food security and vulnerability of women? A systematic review of gender perspectives. *Frontiers in Climate 6*, 1374469. https://doi.org/10.3389/fclim.2024.1374469

Moyna, S. A., et al. (2024). Depressive symptoms among women in disaster-prone region in Bangladesh. *Journal of Affective Disorders Reports, 16*, 100762. https://doi.org/10.1016/j.jadr.2024.100762

Nandargi, S., Gaur, A., & Mulye, S. S. (2016). Hydrological analysis of extreme rainfall events and severe rainstorms over Uttarakhand, India. *Hydrological Sciences Journal, 61*(12), 2145–2163. https://doi.org/10.1080/02626667.2015.1085990

Nardone, A., et al. (2010). Effects of climate changes on animal production and sustainability of livestock systems. *Livestock Science, 130*, 57–69. https://doi.org/10.1016/j.livsci.2010.02.011

Pandey, P., et al. (2024). A study on gender disparity in agricultural resource management in the state of Uttarakhand, India. *Archives of Current Research International, 24*(11), 486–493.https://doi.org/10.9734/acri/2024/v24i11987

Patel, S. J., et al. (2016). Role of women gender in livestock sector: A review. *Journal of Livestock Science, 7* , 92–96. https://livestockscience.in/wp-content/uploads/gender-livestock.pdf

Rojas-Downing, M. M., et al. (2017). Climate change and livestock: Impacts, adaptation, and mitigation. *Climate Risk Management, 16*, 145–163. https://doi.org/10.1016/j.crm.2017.02.001

Sardar, S., & Bhaduri, S. K. (2022). An exploratory analysis of the differential climate change impact on women in India.

Sidloski, M., Reed, M. G., & Andrews-Key, S. (2025). "Are you prepared or not?": An intersectional analysis of a community-engaged climate change vulnerability assessment and adaptation planning process with Tsáá? Ché Ne Dane. *Environmental Science and Policy, 163*, 103984. https://doi.org/10.1016/j.envsci.2024.103984

Singh, G., & Pandey, A. (2023). Climate change induced disasters and highly vulnerable infrastructure in Uttarakhand, India: Current status and way forward towards resilience and long-term sustainability. *Sustainable and Resilient Infrastructure, 9*(2), 145–167. https://doi.org/10.1080/23789689.2023.2253409

Sorensen, C. (2018). Climate change and women's health: Impacts and opportunities in India. *Geo Health, 2*, 283–297. https://doi.org/10.1029/2018GH000163

Tiwari, I., et al. (2022). Climate change impacts on the health of South Asian children and women subpopulations—A scoping review. *Heliyon*. https://doi.org/10.1016/j.heliyon.2022.e10811

Tyagi, N., et al. (2022). Evaluation of observed and future climate change projection for Uttarakhand, India, using CORDEX-SA. *Atmosphere, 13*(6), 947. https://doi.org/10.3390/atmos13060947

UNICEF & WHO. (2023). *Progress on household drinking water, sanitation and hygiene 2000–2022: Special focus on gender*. New York: United Nations Children's Fund (UNICEF) and World Health Organization (WHO).

United Nations Framework Convention of Climate Change (UNFCCC). (2022). *Dimensions and examples of the gender-differentiated impacts of climate change, the role of women as agents of change and opportunities for women.* https://unfccc.int/sites/default/files/resource/sbi2022_07.pdf

United Nations Framework Convention of Climate Change (UNFCCC) (2023). *Five reasons why climate action needs women.* https://unfccc.int/news/five-reasons-why-climate-action-needs-women

Uttarakhand State Council for Science and Technology (UCOST) (2020). *Resource Atlas of Udham Singh Nagar District.*

World Economic Forum (2025) *Global Risk Report 2025.* https://reports.weforum.org/docs/WEF_Global_Risks_Report_2025.pdf

Wright, M. L., et al. (2023). Climate change and the adverse impact on the health and well-being of women and girls from the women's health expert panel of the American academy of nursing. *Nursing Outlook, 71*, 101919. https://doi.org/10.1016/j.outlook.2023.101919

Chapter 6
Gender Perspectives on Women's Role in Landslide Risk Reduction and Early Warning Systems in the Nepal Himalaya

Ranjan Dhungel, Puja Maharjan, Suraj Gautam, Sanjaya Devkota, and Basanta Raj Adhikari

Abstract Landslides in the Himalayas pose a major challenge due to active seismotectonic activities, steep terrains, fragile geology, intense rainfall, and increasing human activities. The daily interaction and mobility of women and marginalized people with natural resources is high, making them more vulnerable to landslide risks. Their livelihood is dependent on firewood collection in forests and practicing agriculture on hill terraces, placing them at the forefront of both experiencing challenges and implementing Indigenous landslide risk mitigation techniques on hill slopes. Traditionally, these communities have mitigated landslide risks, practiced settlement planning, and managed agricultural farms by observing trees and their inclination with respect to hill slopes, managing drainage, and planting trees and shrubs. They know the hill slopes where they live and have been practicing slope-stabilizing techniques for a long time. The gender-based initiatives to manage natural resources, particularly the community forest in Nepal, is a successful model which is a participatory and informed decision-making approach. These initiatives have further strengthened the role of women and marginalized communities in natural resource management, making their contributions more structured and impactful. The federal constitution promulgated in 2015 provides concurrent power to the municipal government to practice legislative, judiciary, and implementation rights in a three-tier governance system: federal, province, and local. This is a major shift in power from central to local government, providing plenty of opportunity to act locally on disaster risk reduction and management. However, due to climate change and changes in seasonal rainfall patterns, they are now facing more frequent surprises and challenges in predicting landslides. The shifting rainfall patterns have made it increasingly difficult to anticipate hazardous events, highlighting the need for improved risk reduction strategies.

R. Dhungel · P. Maharjan · B. R. Adhikari (✉)
Department of Civil Engineering, Pulchowk Campus, Institute of Engineering, Tribhuvan University, Lalitpur, Nepal
e-mail: bradhikari@ioe.edu.np

S. Gautam · S. Devkota
Institute of Himalayan Risk Reduction (IHRR), Patan, Nepal

Md. A. Abedin et al. (eds.), *Climate Crisis in South and East Asia*, Sustainability Solutions, https://doi.org/10.1007/978-3-032-01012-4_6

Now, they realize the importance of Early Warning Systems (EWS) and integrating EWS into landslide risk reduction.

Keywords Landslide · Gender · Indigenous knowledge · EWS

6.1 Introduction

People in the Himalayas have unique living habits due to limited access to modern facilities. Their livelihood is integrated with natural resources such as forests, rivers, and rugged hilly terrains, etc. The steep slope, abrupt change in elevation, unplanned settlement, infrastructure particularly rural roads and changing climate added hardship in the lives and livelihood of people. The monsoon pattern is unpredictable to locals and challenges livelihoods, particularly for agriculture. The hill slope is unpredictable when intense monsoons threaten the slope stability. The reported cases of landslides, debris flow and flash floods are increasing in recent years. The behavior, extent and scale of landslides are complex to anticipate. The damage and loss induced by small to moderate landslides increases when frequency increases. The severity of landslides is further amplified by anthropogenic activities, which result in loss of life and interrupt lifeline services such as water supply, telecommunications, electricity, and roads. The local infrastructure is usually impacted and adds hardship to daily functions. The resilience infrastructure is one aspect of disaster management that demands early warning systems (EWS) as an essential component of infrastructure resilience. The social cohesion and the strength of local governance are disturbed due to such events. Therefore, infrastructure resilience, social cohesion and disaster governance need to be in place to enhance the disaster resilience threefold disaster resilience strategy.

In Nepal, the role of women in landslide resilience is increasingly recognized as a vital role. The unique role of women at the household level has positioned them as a driver of change for disaster preparedness and response. The societal role assigned to women in most South Asian countries as a primary caregiver, users of natural resources and influencers for decision-making at the household level. The social role that has been performed by the women if boosted with skills, knowledge, awareness and advocacy, the overall community resilience is sustainably enhanced. There is a local saying that if ?women are empowered the whole family will be empowered and then the entire society is empowered?. The social roles performed by the Nepalese women equip with indigenous knowledge and practical experience of working in the Himalayas slope where the occurrence of landslide hazards frequently disturbs daily life. Real-life experiences of women living in the hills are assets for landslide risk reduction.

In Nepal, the involvement of women in the political system and governance has significantly increased due to the change in election policy after the promulgation of Nepal?s new Constitution in 2015. According to the Election Commission of Nepal, the representation of women leaders in elections held in 2022 for local municipalities is 41.2% across 753 local government units. Elected 14,465 women leaders are now delivering their service as mayors, deputy mayors in most cases, ward presidents and members. The representation of women in governance is a favorable environment despite the disparity that exists in executive roles at the federal and provincial tiers of governance.

On the contrary, women?s involvement in disaster risk reduction and management, mostly in Early Warning Systems (EWS) is limited. Social, cultural, and family responsibilities are often discouraging women for their innovative and full engagement in accessing resources for EWS and acquiring training and capacity-building efforts. Like many training and awareness packages implemented in Nepal, early warning systems adopt gender-specific needs for women, in principle. This is important to demonstrate success in women?s involvement as a change agent for landslide risk management. The access to information about the capacity building initiatives, the training spots, time, and duration are adopted women-friendly in case of EWS interventions. The women?s requirement for learning and growing environment is different than the males since they must deal with caring for children, elderly family members and household responsibilities. For instance, a successful example of EWS initiatives highlights the intentional selection of training venues within the participants? communities. The course schedule should be designed to accommodate women?s needs, allowing them to balance household responsibilities.

Nevertheless, there are increasing initiatives to minimize the shortcomings in integrating women in disaster management and climate issues. The example of the community forest management system in Nepal sets standards and demonstrates the success of women?s involvement in managing and decision-making. Their close relationship to the forest resources for fuelwood, fodder, medicinal plant harvesting and other day-to-day needs gives them a living experience with forest interaction. National statistic highlights that 33% of women are in leadership roles in forest groups. They can be a change agent for landslide risk reduction and promoting EWS in landslide risk management to strengthen community resilience by narrowing the gap between the formal local institutions and the community which also aligns with the spirit of sustainable development goals (SDGs).

This chapter investigates the role of women in understanding physical science for landslide risk reduction, promoting social and cultural harmony to enhance preparedness and enhancing the capacity of local governance for landslide risk mitigation and disaster preparedness promoting early warning system in the Himalayan region. The case study discussed in this chapter highlights the challenges and opportunities in managing and making decisions for better governance through the involvement of women. The analysis of existing practices, successful case study and promotion of gender-inclusive policy in practice discussed in this chapter gives insights into how landslide risk can be promoted effectively and sustainably in the face of an early warning system.

6.2 Traditional Knowledge and Role of Women in Landslide Risk Reduction

Traditional knowledge evolves from generation to generation, adapting to changing contexts and incorporating innovative local solutions. In Nepal, traditional knowledge is used by local communities to mitigate soil erosion and landslides through practices such as planting trees on steep slopes, employing agroforestry techniques, building terraces to reduce water runoff, and reinforcing soil with bamboo and other vegetation (Luintel, 2018). In addition to these mitigation measures, communities rely on traditional knowledge to recognize the early signs of landslides. These signs are identified through observation of environmental parameters such as soil movement, surface cracks, rainwater flow path, plant and animal behaviors, as well as traditional beliefs, myths, and trust in knowledgeable individuals within the community.

Building on this traditional knowledge, natural resource management, such as forest management plays a crucial role in reducing landslide risks (Dolidon et al., 2009). In Nepal, the community-based forest management program was established to address environmental degradation, improve rural livelihoods, and empower women and marginalized groups while rehabilitating degraded land (Ghimire & Lamichhane, 2020). As primary caretakers of household resources, women rely on forest-based ecosystem services such as food, fodder for livestock, and fuel, and they have gained the capacity to manage forests sustainably through community-based forest management programs (Sedhain et al., 2022). Their deep interaction with forest ecosystems enables them to recognize ecological changes, including variations in vegetation, slope stability, and water flow. Similarly, women have a strong understanding of landslide risks and slope stability in local terrains, coupled with their traditional knowledge of identifying and cultivating plants with armoring properties. This awareness and understanding influence their forest management practices and empower them to take proactive roles. Thus, women are recognized not only as a vulnerable group but also as crucial contributors to natural resource management (Joshi & Bhardwaj, 2015). Their deep connection to the environment equips them with valuable knowledge, allowing them to engage in decision-making and inspire others to join the process. Furthermore, women are key communicators of risk, using their social networks to disseminate information about hazards and assist vulnerable members of the community, and they are more receptive to emergency warnings, showing heightened safety awareness (Mehta, 2007). This is critical in ensuring the effectiveness of early warning systems (EWS) for landslides. Their involvement ensures that information reaches diverse groups, including those who may be marginalized, such as elderly individuals, children, and those with disabilities.

In addition, women?s leadership roles in local networks and organizations, along with their ability to mobilize resources and monitor programs (Roy & Mukherjee, 2024), contribute to strengthening the Early Warning System. Women took leadership in evacuation before a landslide in the Nuwakot district of Nepal and served as first responders, alerting their families and community during landslide events

(Pradhan et al., 2014), highlighting how their integration into landslide early warning systems enhances communication, resource mobilization, and preparedness, thereby reducing overall vulnerability to disasters from landslides. Therefore, women?s involvement extends beyond household responsibilities and environmental management, reaching into disaster risk reduction where they along with other marginalized groups, play direct or indirect roles in mitigating potential future landslide risks through community-based initiatives, such as community forest groups (Basyal, 2021), and Early warning systems (Kafle, 2017).

6.3 Landslide Early Warning System in the Nepal Himalayas

Early warning systems in landslide risk reduction are important for saving lives and property in the Himalayas. The spatiotemporal pattern of landslides informed that Nepal lost 111 people annually from 1971 to 2019. It is important to note that 97.7% of annual landslide events happen during monsoons between June and September (Adhikari et al., 2022). The restructure of the Nepal governance structure allows plenty of opportunity to act for landslide risk reduction and preparedness. The local government can enhance preparedness for landslides by establishing early warning systems. Landslide Early Warning Systems consist of major four components; rainfall monitoring, soil moisture sensing, real-time data and ground movement evidence for its efficient functions.

The EWS can be learned easily and is possible to integrate with the emergency preparedness for the community at risk. This system is generally established by combining sensors on the ground, satellite data and weather forecasts to raise alert. The system set-up in susceptible locations sometimes raises alerts based on rainfall threshold value (Pradhan et al., 2014). For this, rainfall monitoring is one of the crucial parameters for anticipating landslides. The system is linked to the threshold value of rainfall; after reaching the threshold the likelihood of landslides is higher and raises alert when rainfall exceeds the threshold value set up for the area. The use of devices for soil moisture information and ground movement sensors are additional devices that are placed on vulnerable hill slopes to determine slope stability. These sensors sense the water content in soil mass which is correlated to interpret landslide risk. Ground motion sensors trace shifting in the ground during the start-up of a landslide (Thapa et al., 2023; Thapa & Adhikari, 2019). These are most practices in local level landslide anticipation which can be linked with advanced techniques such as remote sensing and satellite technology. The advanced techniques of remote sensing advise for changes happening in the landscape and local environment conditions.

6.3.1 *Landslide Monitoring Through Moving-Peg Method and Citizen Science*

A moving-peg method for landslide monitoring is very effective in the rural part of Nepal. One of the successful examples is from Bosimpa, Bhimeshwor Municipality, Dolakha district of Nepal where a 15 membered LEWS User Committee has been formed from the marginalized community for the effective implementation of LEWS. The LEWS User Committee had 1 Male Coordinator and 2 Female Coordinators with 60% of the members being Female. The LEWS User Committee were sensitized about the four pillars of early warning system and the importances of local monitoring in LEWS. In landslide monitoring, the Moving Peg Method is a technique used to evaluate the stability and movement of hillsides and slopes over time. Using this technique, markers or pegs are placed at certain points down a slope, and their positions are periodically measured by the female members to identify any displacement or movement. It?s an easy yet efficient method of keeping an eye on changes in the landscape that could point to the possibility of a slope failure or landslide. Besides, through this User Committee, the community also communicates the risk information among the households in Bosimpa. Women members have been involved during the installation of peg in the vulnerable areas. The initial measurements were taken. Every month, the women members meet and measure the displacements and share it with the municipality (Fig. 6.1).

Besides, the municipality has also deployed more than 20 Female Citizen Scientist to monitor daily accumulated rainfall in the nine wards of the Municipality. Through

Fig. 6.1 Women leading landslide early warning system in Dolakha, Nepal. *Source* IHRR

Fig. 6.2 Women citizen scientist reading rain gauge at community, Nepal. *Source* IHRR

the low-cost rain gauges, every day at 8:45 AM the citizen scientists are reporting the amount of 24 h accumulated rainfall. Besides, with the help of VICTORY mobile application, the female citizen scientists are also reporting landslides in their nearby areas. As a citizen scientist, they are also serving as a messenger to disseminate the weather and risk information to their communities (Fig. 6.2).

Communication through the mixed method: (a) using mobile technology for data reporting and information dissemination (b) oral tradition plays a crucial role in implementing effective early warning systems. By efficiently disseminating vital information about the reporting of daily rainfall accumulation and nearby landslides, citizen scientists ensured that warnings reach community members on time. Besides, the female citizen scientists were also provided with the information related to Forecast and Impact based Weather Forecasting Bulletin (IBF). With the available information, the female citizen scientists were able to make data-based decision such as skipping schools on a very rainy day, attending their children from school, harvesting paddy on time, avoiding the areas with potential debris flow, and so on.

Besides that, the simulation exercises have equipped the female with essential knowledge and skills on the significance of rain gauge stations, and extensometer focusing on rainfall-induced landslides and safety measures. They are now able to assemble in a safe place, assemble Go Bags in their houses, assess potential landslide scenarios through early signs, evacuation procedures, and community safety protocols. Women are also now able to come up with first aid solutions, preparation of stretcher for immediate evacuation, and transportation in the difficult mountainous terrains (Fig. 6.3).

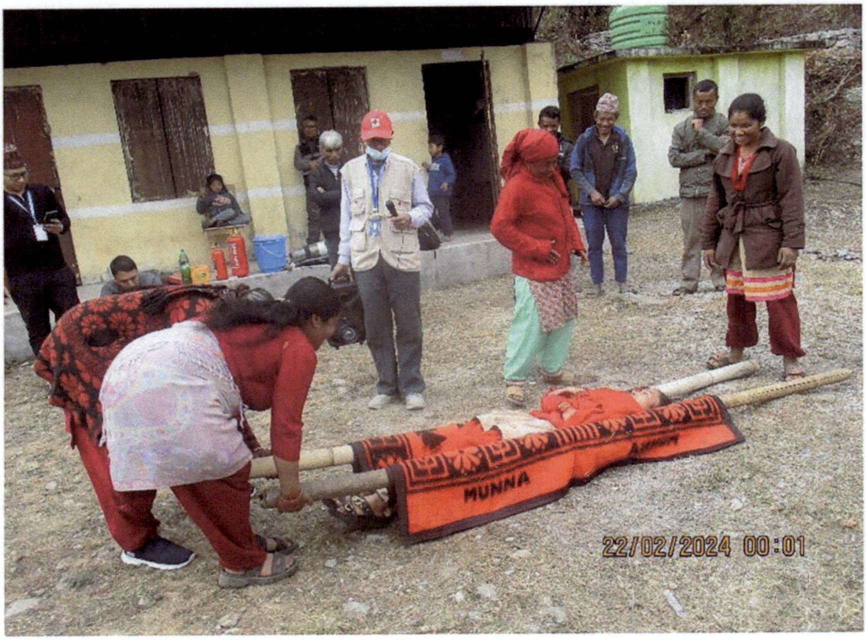

Fig. 6.3 Women participating in preparedness drill to enhance efficiency in emergency response. *Source* IHRR

The success of EWS is more on a community-led monitoring approach, community participation increases the efficiency of the systems and action after an alert. Simply, the EWS system consists of data collection and monitoring, data analysis and prediction, alert generation, information dissemination and community response actions. Data collection and data analysis require system support to collect real-time information on soil moisture, rainfall, and slope movement. The data information is processed through the predictive algorithms using models to assess the likelihood of landslides. The landslide?s favorable conditions force EWS to raise the alerts and local media SMS, radio messages, and social media is used for alert dissemination. Upon listening alert for the landslide evacuation, the community starts safe evacuation according to the preparedness plan and remains safe. This is how early warning works by allowing sufficient time to escape from the landslide risk area to the identified safe zone.

6.3.2 Challenges and Opportunities to Women in Using Their Full Potential in EWS

The Early Warning System (EWS) plays a crucial role in mitigating the impacts of disasters, and women?s participation is vital for its effectiveness. However, women from the South Asia region, including Nepal, face several socio-economic, cultural, and institutional barriers that limit their full participation in Early Warning Systems (EWS). Shrestha et al. (2014) highlight the limited role of women in Early Warning Systems (EWS) across four key components: risk knowledge, monitoring and warning services, communication and dissemination, and response capability. They found out that women have minimal involvement in risk assessments, and their traditional knowledge is often overlooked. Monitoring and warning services remain male-dominated, restricting women?s access to timely warnings. In communication and dissemination, men primarily handle disaster alerts, and women often receive warnings indirectly, with no feedback mechanisms to assess their understanding. While women?s involvement in response capability building has been increased due to mandatory government provisions and organizational strategies, social norms and barriers continue to hinder their active participation in response preparedness and capacity planning.

Several factors contribute to these challenges. One of the major factors is the lack of access and control over information and resources, as traditional communication channels may not effectively reach women, especially in rural areas where mobility and literacy rates are lower (Dhungel & Ojha, 2012). They also highlighted that social and cultural norm, including gender discrimination and heavy workloads, often restrict women?s participation in community activities and decision-making processes. It limits their ability to contribute to Early Warning Systems (EWS). Additionally, limited technical skills and training opportunities create barriers for women in engaging with disaster management systems (Roy & Mukherjee, 2024), including Early Warning Systems (EWS), by restricting their ability to participate in the design, operation, and dissemination of early warning messages. Moreover, Adhikari et al. (2014) reported that institutional challenges, such as the lack of gender-sensitive policies, result in the exclusion of women from key roles within formal disaster management structures, thereby limiting their involvement in the development and implementation of EWS.

Despite these challenges, integrating women into early warning systems (EWS) presents significant opportunities to enhance disaster resilience by utilizing their unique strengths and contributions. Women are vital communicators within their communities (Ashraf & Azad, 2015), and utilizing their social networks offers an opportunity to strengthen community connections and ensure that critical disaster information is disseminated effectively and widely. Women?s deep knowledge of local resources, traditional practices, and environmental patterns (Roy & Mukherjee, 2024) provides valuable insights that enhance the accuracy and relevance of early warning messages, allowing for better-prepared communities. Furthermore, women bring unique perspectives to decision-making processes, ensuring that disaster response strategies are more balanced and considerate of all community member needs. Thus, involving them at all stages of decision-making within disaster risk management efforts offers the opportunity to improve overall disaster adaptation and response (Mehta, 2007).

Gender-transformative approaches to early warning systems provide opportunities for women and marginalized groups to engage in leadership roles, contribute their knowledge, and shape inclusive disaster risk management strategies (Sarah et al., 2019). Nepal (2020) highlighted the importance of collaborating with government bodies, local organizations, and community leaders to empower women, ensuring their active participation in disaster risk management, from decision-making and capacity building to leading early warning systems and search and rescue efforts, thereby enhancing community resilience. Furthermore, training programs play a crucial role in enhancing women?s confidence, enabling them to take active roles in safeguarding their families during disasters such as landslides (Bajracharya et al., 2017). This is particularly important as women often bear a disproportionate burden in disaster recovery, and their active involvement in EWS significantly contributes to more effective disaster response and recovery efforts (Stern et al., 2018). Interestingly, awareness and advocacy for gender equality are increasing, and as part of the broader political-economic shifts occurring in Nepal, the social narratives and roles of women are slowly changing (Adhikari & Sharma, 2022). The shift in women?s roles from caregivers to leaders, with women being recognized as active agents of cultural change, is encouraging their involvement in disaster management efforts (Poudel, 2018). The cultural change adopted by the community is a good indication for establishing women?s identity for utilizing their full potential in disaster early warning systems.

6.4 Policy and Institutional Support for Gender Inclusion

The political awareness among women significantly increased after the promulgation of the federal constitution in 2015. The Nepalese constitution guarantees the gender equality and protection of women as a constitutional right. The National Women Commission is playing an important role in advocacy for gender equality. The Local Government Act (2017) incorporates many clauses for empowering women at the local level. However, the honest policy enforcement mechanism and robust institutional setup are essential to bring a drastic change in societal transformation for an inclusive environment. The legal and institutional mechanisms on DRR have embraced the whole-of-society approach to inclusive disaster management. In order to guarantee that disaster risk reduction and management initiatives are inclusive and equitable, addressing the various needs and vulnerabilities of everyone, including women, people with disabilities, and marginalized communities, Nepal?s NDRRMA (National Disaster Risk Reduction and Management Council) has already endorsed Strategic Action Plan for Gender Equality, Disability and Social Inclusion (GEDSI) in Disaster Risk Reduction and Management, 2024 on 3 May 2024 during the 22nd

National Disaster Risk Reduction and Management Executive Committee Meeting. This Strategic Action plan has envisioned a strategy for the GEDSI consideration into multi-hazard early warning systems and anticipatory action which includes developing inclusive and accessible multi-hazard early warning systems, making sure that DRR operations follow inclusive approaches, and integrating GEDSI into institutional frameworks across the government. Similarly, to support vulnerable groups with the social protection, Nepal has also endorsed the National Standard on Shock Responsive Social safety (SRSP).

Besides, there has been provisions of the formation of Ward-level and community-based Disaster Risk Management Committee inclusive (Adhikari & Gautam, 2022). Similarly, the Disaster Risk Reduction and Management Volunteer Bureau Formation and Mobilization Guidelines, 2022 have also envisioned the inclusive engagement of the women and vulnerable groups in the bureau formation and development of rosters of volunteers. The volunteers can be mobilized for the public awareness, early warning, preparedness and response. Similarly, the newly endorsed Local Disaster and Climate Resilience Framework (LDCRF) 2024 of NDRRMA has also prioritized the meaningful engagement of women in the community based early warning system. Nepal adopted preferential policies for supportive actions to ensure gender equality by reserving 20% of positions in government offices and mandating 33% representation in parliamentary seats. Affirmative action of state policies increases women?s participation in governance and leadership roles. National Disaster Risk Management Authority (NDRRMA) outlines specific actions to mainstream gender inclusion through disaster risk reduction national strategic plan of action (2018?2030). Local adaptation Plans of action (LAPAs) prioritize the active involvement of women and youth in identifying climate risks and acting on them locally. Nepal?s policies are progressive to ensure women?s increasing role positioned them to active change makers from ideal victims of landslides.

6.4.1 Conceptual Framework for Women?s Engagement in Early Warning System (EWS)

This conceptual framework is rooted in community centered, gender transformative approach to EWS. It integrates the daily life lessons and the ancestor?s consolidated learnings to adopt landslide resilience actions formed as a women?s indigenous knowledge, leadership role of the women?s in local governance and trustable communication roles linking with major four pillars: understanding risk, monitoring and warning, dissemination and response capacity strengthening. The conceptual framework emphasized on empowering women in informed decision making, trainings and integration of science based local technology, the framework envision trusted, timely and community system integrated warning systems. This transform women from vulnerable group to the agent of social change in landslide risk reduction and contribute on community resilience. The credibility and outreach of warning messages will be enhanced.

6.5 Conclusion

The role of women in every aspect of life is crucial since the global male-to-female ratio is almost equal. The family responsibility and societal imposed role are more pivoted to the women restricting them from opening up for leadership roles, placing them in a decision-making position and limiting them in formal education. Sustainable development goals target 5.5 strive the need to ?ensure equal access for women to affordable and quality technical vocational and tertiary education, including university?. The successful examples women?s involvement in its effectiveness are inspirational and replicable for regional and global contexts. To scale it up, the national, regional, and global networks need to be fostered and enhanced in their leadership for resilient infrastructure and community resilience in landslide risk reduction and management in the Himalayas. The conceptual women centric landslide risk reduction framework proposed in Table 6.1 above emphasized how gender inclusive approach translate to sustainable landslide reduction outcomes through EWS. The proposed framework fosters the strategic need for landslide early warning system establishment in the highly vulnerable Himalayas.

Table 6.1 Conceptual framework for Women-centric landslide risk reduction and early warning system

Core problem	Women in Himalaya are often vulnerable to landslides are in increased risk but remain less engaged in landslide risk reduction, communication and decision making		
Key concept	? Community-centered landslide risk management ? Gender-Responsive landslide risk reduction ? Leadership of women in local government ? Inclusive landslide early warning system ? Blending of local knowledge and citizen science ? Capacity building and landslide knowledge reinforcement ? Timely and effective communication and Resilience		
Assumptions	? Women?s holds indigenous knowledge to early identification of landslide risk ? Gender inclusive response results sustainable and effective outcomes for landslide risk management ? Empowered and inclusive local government play important role in communication, disaster preparedness and early actions ? Women?s involvement improved retarding social barriers to take early actions for EWS		
Conceptual framework logic			
Inputs	Activities	Outputs	Outcomes
Gender-transformative policies, Local women leadership, technical support and informed decision-making system	Capacity building trainings on EWS, Mapping Of women Leadership at local government, women lead hazard mapping, Awareness raising and information for decision making	Risk mapping, decision making women?s forum, women DRR group at community, EWS and evacuation plan	Informed decision making, increased women?s leadership, landslide risk reduction and resilience
Increasing Resources for DRR, Funding, prioritized support from local government and civil society organization	Development program linked with DRR, Livelihood program linked for risk reduction and early actions, retrofitting/ soil anchoring trainings and landslide risk mitigation trainings for women	Women-led mitigation programs for slope stabilization	Long-term local community engagement in Landslide risk reduction and management
Cross cutting theme	? Intersectionality for addressing the need of different age group women based on their background and local condition ? Aligning policy with Sendai Framework and SDG 5 (gender equality) and national acts ? Transforming community perception to accept and support women leadership at local level and community		

References

Adhikari, B. R., & Gautam, S. (2022). A review of policies and institutions for landslide risk management in Nepal. *Nepal Public Policy Review, 2*, 93?112. Department of Hydrology and Meteorology. (2019). Annual Climate Report. Government of Nepal.

Adhikari, B. R., Gautam, S., and Paudel, B. (2022). Landslide, landcover and landuse changes in its impacts in Nepal: In: R. Sarkar, R. Shaw, and B. Pradhan (Eds) *Impact of climate change, land use and land cover, and socioeconomic dynamics on landslides* (pp. 149?164). Singapore: Springer.

Adhikari, R., & Sharma, J. R. (2022). Gendered consequences of social changes in Nepal: Rich possibilities. *European Bulletin of Himalayan Research, 58*. https://doi.org/10.4000/ebhr.549

Adhikari, P., Koirala, S., & Sharma, D. (2014). Gender and disaster risk reduction in Nepal: The role of women in managing risks. *Journal of Disaster Risk Studies.*

Ashraf, M. A., & Azad, M. A. K. (2015). Gender issues in disaster: Understanding the relationships of vulnerability, preparedness and capacity. *Environment and Ecology Research, 3*(5), 136?142. https://doi.org/10.13189/eer.2015.030504

Bajracharya, B., et al. (2017). Gender, knowledge and disaster risk reduction: A case study from Nuwakot District, Nepal. *Disaster Management Journal, 25*(3), 105?117.

Basyal, G. K. (2021). *Living with landslides: Integrating knowledges for landslide risk reduction in rural Nepal*, (Vol. 299).

Bhandari, D. (2025). Challenges and opportunities in Nepal?s early warning communication. *International Journal of Disaster Risk Reduction*, 105318.

Dhungel, R., & Ojha, R. N. (2012). Women?s empowerment for disaster risk reduction and emergency response in Nepal. *Gender and Development, 20*(2), 309?321. https://doi.org/10.1080/13552074.2012.687220

Dolidon, N., Hofer, T., Jansky, L., & Sidle, R. (2009). Watershed and forest management for landslide risk reduction. In *Landslides?Disaster Risk Reduction*, (pp. 633?649). https://doi.org/10.1007/978-3-540-69970-5_33

Ghimire, P., & Lamichhane, U. (2020). Community based forest management in Nepal: Current status, successes and challenges. *Grassroots Journal of Natural Resources, 3*(2), 16?29. https://doi.org/10.33002/nr2581.6853.03022

Joshi, K., & Bhardwaj, N. (2015). Women and natural resource management: A study of ?Communities of Practice? prevailing in women farmers? community for management of water and forests of lesser Himalayan Region in India. *International Journal of Advanced Research, 3*(7), 363?374. http://www.journalijar.com

Kanta Kafle, S. (2017). Disaster early warning systems in Nepal: Institutional and operational frameworks. *Journal of Geography & Natural Disasters, 07*(02). https://doi.org/10.4172/2167-0587.1000196

Luintel, Y. R. (2018). *Indigenous knowledge for disaster risk mitigation?: Field notes from Nepal?s four* (Vol. 8, No. 1).

Mehta, M. (2007). *Gender matters: Lessons for disaster risk reduction in South Asia.* https://www.preventionweb.net/files/2406_GenderandDisasters.pdf

Nepal, U. (2020). A study on potential leadership approaches to address gender-based constraints effective preparedness for disasters in Nepal. *Journal of Disaster and Emergency Research, 3*(1), 40?57. https://doi.org/10.18502/jder.v3i1.2655

Poudel, B. (2018). The gendered impact of natural disasters in Nepal: Challenges and opportunities for women. *Journal of Gender and Development.*

Pradhan, P., et al. (2014). Community-based landslide early warning systems in Nepal. In *Proceedings of the International Conference on Early Warning Systems*, Kathmandu, Nepal.

Roy, D., & Mukherjee, M. (2024). Challenges and opportunities for women in disaster risk management. *Journal of Integrated Disaster Risk Management, 14*(2), 228?240. https://doi.org/10.5595/001c.123771

Sarah, B., Mirianna, B., Alison, S., David, L., Puja, S., & Sujana, U. C. (2019). Gender transformative early warning systems: Experiences from Nepal and Peru. March. https://floodresilience.net/resources/item/gender-transformative-early-warning-systems-experiences-from-nepal-and-peru/

Sedhain, J., Ian, E., & Ebreo, N. (2022). Human-nature interactions. *Human-Nature Interactions*, January. https://doi.org/10.1007/978-3-031-01980-7

Shrestha, M. (2014). *Flood early warning systems in Nepal: a gendered perspective*. October, 52.

Stern, R., Piers, P., & Green, L. (2018). Women?s role in disaster risk reduction and early warning systems in South Asia. *Journal of International Development*.

The Local Government Act. (2017). Local Government Act, 2017 [Law]. Government of Nepal. https://www.lawcommission.gov.np/en/archives/11674

Thapa, P. S., & Adhikari, B. R. (2019). Development of community-based landslide early warning system in the earthquake-affected areas of Nepal. *Journal of Mountain Science, 16*(12), 2701?2713.

Thapa, P. S., Adhikari, B. R., Shaw, R., Bhattarai, D., & Yanai, S. (2023). Geomorphological analysis and early warning systems for landslide risk mitigation in Nepalese mid-hills. *Natural Hazards, 117*(2), 1793?1812.

Chapter 7
Gender-Focused Social Capital and Community Actions to Mitigate Flood Risks in Pakistan

Shehla Gul, Atta-ur Rahman, and Rajib Shaw

Abstract This chapter examines the role of gender-focused social capital in mitigating flood risks in Pakistan, emphasizing how women's networks and community engagement strengthen flood disaster resilience. Pakistan has experienced mega devastating flood events during the past two decades including riverine floods, flash floods, and pluvial flood events. It has been observed that the flood characteristics in upper reaches are to largely different then the middle and lower reaches of the river basin. In Pakistan, where large share of the population is residing in rural areas, therefore, the role of women in mitigating flood risk is of key significance. Utilizing case studies from flood-prone regions, the research explores how women leverage social connections to mobilize resources, share knowledge, and implement adaptive practices during flood events. These efforts, often overlooked in formal disaster management, are critical to building local resilience. The chapter also discusses the involvement of NGOs, local institutions, and government agencies in fostering gender-sensitive disaster risk reduction strategies. By integrating qualitative and quantitative data from recent floods, the study offers insights into the importance of gender perspectives in climate resilience and disaster mitigation, with implications for policy in Pakistan and other South Asian countries.

Keywords Social capital · Flood mitigation · Risk · Gender · Community action

S. Gul (✉) · A. Rahman
Department of Geography and Geomatics, University of Peshawar, Peshawar, Pakistan
e-mail: sgul@uop.edu.pk

A. Rahman
e-mail: atta-ur-rehman@uop.edu.pk

R. Shaw
Keio University, Tokyo, Japan
e-mail: shaw@sfc.keio.ac.jp

Md. A. Abedin et al. (eds.), *Climate Crisis in South and East Asia*, Sustainability Solutions, https://doi.org/10.1007/978-3-032-01012-4_7

7.1 Introduction

Globally, the frequency and trends of hydro-meteorological disasters are increasing. A disaster occurs when a hazard interacts with conditions of exposure, vulnerability, and capacity and disrupts the functioning of a community or a society at any scale leading to human, economic, material and environmental impacts and losses (UNDRR, 2022). Disasters, both natural and man-made, pose significant challenges to societies worldwide. They have far-reaching impacts, affecting not only the immediate survivors but also the long-term resilience of communities. Disaster management is a process or strategy implemented in the pre, during, and post-disaster phases to prepare for, prevent, forecast, respond to and recover from the disaster event (Oloruntoba et al., 2018, p. 544). It aims at reducing the potential losses caused by hazards, ensuring the provision of immediate and appropriate assistance to those affected, and facilitating a rapid, effective recovery (Warfield, 2013). Disaster management is traditionally divided into four phases or stages: preparedness, mitigation, response, and recovery (Alrehaili et al., 2022). Preparedness and mitigation and are proactive phases, taking place before a disaster (Gul & McGee, 2024). Mitigation measures focus on preventing disasters or minimizing damage, while preparedness activities include long-term protection measures to ensure readiness for any upcoming disasters in future (Oloruntoba et al., 2018). The recovery phase encompasses measures taken after a disaster aiming to reduce both the immediate as well as long-term impacts of the disaster and enable a swift recovery process (Gul & McGee, 2022, 2024; Oloruntoba et al., 2018).

According to The United Nations Office for Disaster Risk Reduction (DRR) is aimed at preventing new and reducing existing disaster risk and managing residual risk, all of which contribute to strengthening resilience and therefore to the achievement of sustainable development. Disaster risk reduction (DRR) is the main goal of disaster risk management, and it is outlined through various strategies and plans that define its specific objectives. On a global scale, the United Nations has endorsed the Sendai Framework for Disaster Risk Reduction 2015–2030, which was adopted in March 2015. This framework sets clear goals for the next 15 years, aiming to significantly reduce disaster risks and the associated losses. These losses include not only lives and livelihoods but also the damage to health, businesses, communities, and the environment. The overarching goal is to protect the physical, social, cultural, and economic assets of individuals and countries from the impact of disasters.

In Pakistan, the frequency of flood disasters is on rise (Manglore et al., 2024). It is evident from the past few decades that riverine floods, pluvial floods in urban areas and flash floods are occurring with much more severity and intensity (Malik et al., 2023; Rahman et al., 2024). The analysis revealed that riverine flood has seriously affected the communities in upper and middle reaches, whereas the urban floods disasters are mostly reported from major cities of Pakistan including Rawalpindi, Islamabad, Peshawar, Lahore, Multan, etc. However, under the changing climate scenario, during past couple of years, the flash floods mostly hit the arid and semi-arid regions of Baluchistan, Sindh and Khyber Pakhtunkhwa (Khan & Rahman,

2024). It is therefore, this chapter focuses on the gender focused social capital and community actions in flood mitigation.

7.2 Gender and Disasters

Disasters affect men and women differently. A multitude of factors—social, economic, cultural, and political—determine an individual's or community's vulnerability to disaster risks. Gender, is a critical dimension of the social structure that becomes even more visible during disasters (Enarson, 2009). Women, girls, and gender minorities often experience increased vulnerability during and after disasters due to limited or no access to resources and gendered responsibilities (Gaillard et al., 2017). While disaster risk reduction (DRR) strategies have been developed to mitigate and manage the risks of such events, these approaches often overlook the social dynamics that exacerbate vulnerabilities. Understanding the gendered dimensions of disaster vulnerability is key to creating inclusive, sustainable, and effective disaster risk reduction strategies.

Women and girls are disproportionately affected by disasters (Cutter, 2017; Gul & McGee, 2022). In many parts of the world, the gendered roles confine women to domestic spheres, limiting their access to education, economic opportunities, and decision-making processes. In times of disasters, the gendered division of labor becomes more pronounced, exacerbating their vulnerabilities. Pregnant women or those with children often face increased Physical Vulnerabilities due to health and safety risks (Gul et al., 2024a, 2024b, 2024c, 2024d). In the post-disaster phase, women can experience increased susceptibility to disease and injuries due to restricted access to healthcare services (Akter et al., 2023; Fatema et al., 2021; Tearne et al., 2021). Women can also face increased economic vulnerability during disasters. Women are often employed in informal or low-wage sectors, leaving them with limited financial resources and security. In the post-disaster recovery phase, they can face further economic challenges due to loss of income generating activities and trainings as well as lack of compensation. Disasters may also lead to increased social vulnerability for women as cultural norms often restrict their mobility and ability to evacuate or access relief and compensation. Women are rarely involved in the decision-making process which might lead to the absence of women's priorities in the disaster response and mitigation strategies. This, in turn might lead to ineffective and unsustainable disaster risk reduction strategies. Another important issue women face during and after disasters is the increased risk of Gender-Based Violence (GBV) which is usually caused by population displacement, loss of livelihood opportunities, the breakdown of social structures and lack of appropriate security in emergency shelters. Women and girls, especially those from marginalized communities, are at a heightened risk of physical and sexual violence in post-disaster settings. These risks can be intensified by the lack of adequate facilities and the absence of law enforcement or support structures in crisis situations.

7.3 Understanding Social Capital in Disaster Contexts

The gendered vulnerability of women in disasters highlights the need for a more inclusive approach to disaster risk management. While women often face heightened risks due to societal norms and limited access to resources, it is essential to recognize that resilience in disaster contexts is not just about survival—it's about the ability to recover and thrive as a community. In this context, the concept of community resilience has been introduced in disaster risk reduction and management process which focuses on the collective ability of a community to adapt, withstand, and recover from adverse events (Berkes & Ross, 2013; Saul, 2022). A resilient community is one that comes together, leveraging its strengths, resources, and networks to face challenges. In this context, community participation and the development of social capital are key components (Gul & McGee, 2024). When communities actively engage in decision-making and collaborate, they build trust and shared resources, which are crucial for both immediate recovery and long-term sustainability. This collaborative approach ensures that disaster risk reduction strategies are inclusive and cater to the needs of all members, including the most vulnerable, like women and gender minorities.

Social capital is an important component of resilience which is considered to enhance community resilience (Gul, 2019). Social capital is defined as the "features of social organization such as networks, norms and social trust that facilitate coordination and cooperation for mutual benefit" (Putnam, 1995, p. 67). Social capital includes a variety of key elements that help strengthen communities. These include the size and strength of social networks, active civic participation, trust between individuals, organizations, and institutions, a shared sense of community, a strong attachment to place, and the involvement of citizens in decision-making processes (Gul & McGee, 2024; Kasemsap, 2017).

Szreter and Woolcock (2004) proposed a framework of social capital; according to this framework, social capital is generally divided into three main types: bonding, bridging, and linking. Bonding social capital is defined as relations between "family members, close friends, and neighbors" and it connects similar individuals often living in close proximity to each other and it grows from organizations and activities among these individuals (Chawa, 2017; Gul & McGee, 2024). Due to these strong connections, bonding social capital is good to provide social support as well as personal assistance, especially in times of need such as disaster (Panday et al., 2021).

Bridging social capital refers to the connections between individuals who may not be closely linked but come from different social groups, such as those from different classes or races (Aldrich & Meyers, 2015; Chawa, 2017). Bridging social capital creates opportunities for people to access new information and resources, which can help them progress in life. This type of social capital is often built through participation of community members in community organizations including civic or political groups, sports clubs, parent-teacher associations as well as religious and educational institutions (Gul, 2019).

Linking social capital comprises "norms of respect and networks of trusting relationships between people who are interacting across explicit, formal or institutionalized power or authority gradients in society" (Szreter & Woolcock, 2004. p. 655). This type of social capital brings together people of unequal status this providing a connection between citizens and the decision-makers and leaders holding positions of authority and can often distribute scarce resources (Aldrich & Meyer, 2015).

7.4 Gendered Dimensions of Social Capital

In the context of disasters, gender plays a significant role in shaping how individuals can access and contribute to social capital. Women, particularly in vulnerable communities, often rely on strong social networks to access resources, share information, and provide mutual support (Gul & McGee, 2024). These networks, which form the foundation of social capital, can be crucial for women's survival and recovery during and after a disaster (Table 7.1). However, gender inequalities can affect women's ability to fully participate in these networks, limiting their access to critical resources and decision-making processes. When women are empowered to actively engage in social capital-building activities, they not only enhance their own resilience but also contribute to strengthening the community's overall capacity to respond and recover. Despite this intricate connection between gender, social capital, and disasters, there has been limited focus on research related to the role of gender-focused social capital in disaster risk reduction measures including mitigation and preparedness.

7.5 The Context of Pakistan, Floods, and Social Capital

Flooding is referred to as the overflow of water onto typically dry land, occurs due to heavy rainfall, river overflows, glacial melts, and coastal surges (Gul et al., 2024a, 2024b, 2024c, 2024d). It threatens nearly 2 billion people worldwide and has severe economic, social, and infrastructural consequences (Islam & Wang, 2024). Developing countries, particularly those with inadequate infrastructure, are highly vulnerable (Otto et al., 2023).

Natural disaster such as floods devastate human use system, killing thousands of people and destroying infrastructure which results in billion dollars' economic losses (Bazai et al., 2024). Depending the socioeconomic resilience and the geomorphology of the lands, the effects of the catastrophic floods vary from area to area. Developing nations record more economic losses and human mortality than developed countries. In contrast to the 2081 natural disasters that occurred between 1900 and 2000, the CRED (2015) statistics indicates that frequency of 50 natural disaster increased from 100 natural disasters each decade between 1900 and 1940. Water, energy, infrastructure, food and human lives are constantly under serious risk from hydro meteorological events, which have become more frequent and intense due to

Table 7.1 Comparative roles of men and women in flood risk mitigation through social capital

Role/activity	Male participation (description and level)	Female participation (description and level)	Impact on flood risk mitigation
Early warning communication	Often receive official warnings; act as communicators in formal networks; moderate to high involvement	Use informal networks for information sharing within households and communities; high involvement especially in rural areas	Women's informal networks critical in timely spread of warnings; men's formal role supports coordination
Resource mobilization	Mobilize physical resources such as manpower, infrastructure repair; moderate involvement	Mobilize food, medical supplies, shelter support through kinship and neighborhood networks; high involvement	Women's networks fill crucial gaps when external aid is delayed; men provide large-scale physical support
Community leadership (formal and informal)	Predominantly hold formal leadership roles in disaster management committees; high visibility	Informal leadership in household and neighborhood-level relief coordination; moderate but growing due to NGO support	Informal female leadership vital in grassroots mobilization and resilience; male formal leadership important for policy
Decision-making in disaster preparedness	Participate in community decision-making bodies; moderate involvement	Limited participation in formal bodies due to social norms; active in informal, family-level decisions	Limited female participation in formal planning can weaken gender-sensitive policies; informal female influence strong
Risk communication and education	Lead public awareness campaigns and community meetings; moderate to high involvement	Disseminate information via social networks and care roles, especially among women and children; high involvement	Women's role ensures reach to vulnerable household members; men focus on public messaging
Emotional and psychosocial support	Less visible but contribute through community groups; low to moderate involvement	Primary providers of emotional support, care, and trauma mitigation within families and communities; very high involvement	Women's role critical for community recovery and psychosocial resilience

climate change (Soomro et al., 2022). Significant contributors to disasters in Pakistan include variations in temperature and much higher precipitation, particularly during the monsoon season.

From 1994 to 2013, floods accounted for 43% of natural disasters in South Asia, impacting billions and resulting in significant fatalities (Chaudhary & Piracha, 2021; Nanditha et al., 2023). The increased frequency of severe flooding events in South Asian nations, such as India, Pakistan, Bangladesh, and Nepal, is largely attributed to climate change and rapid urbanization, which compound the impact of monsoon-related rainfall (Rahman et al., 2023; World Bank, 2022a, 2022b).

Pakistan is highly vulnerable to extreme weather events, ranking among the top five countries at risk due to climate change impacts like floods, droughts, and landslides (Shehzad, 2023). Floods are the most significant hazards, severely affecting lives, property, agriculture, and health, and exacerbating social and economic issues (Ashraf et al., 2013; Otto et al., 2023). The country has witnessed a surge in climatic events, such as volatile temperatures, glacier melting, and droughts, which threaten food security (Shehzad, 2023). Pakistan's average temperature has increased by 0.5 °C since the early twentieth century, with projections of 1.3–4.9 °C by the century's end, driving glacier melt and erratic monsoon rainfall (Khan et al., 2021; Pakistan Meteorological Department, 2019). Major floods, including those in 1950 and 2010, caused significant human and economic losses, with 2010 alone resulting in US $10 billion in damages (Federal Flood Commission, 2010; Khan & Iqbal, 2013). The Global Climate Risk Index highlights Pakistan's growing vulnerability, placing it among the top eight most affected nations (Eckstein et al., 2021).

The 2011 flood in Pakistan affected 9.6 million people, resulting in 520 deaths and over 1180 injuries, causing extensive damage to villages, cities, and agricultural areas, and worsening food insecurity (National Disaster Management Authority, 2010). The following year, heavy rains and floods damaged over 636,000 buildings, impacted 1.172 million acres of agricultural land, and affected 14,159 villages, with 571 fatalities and 4.85 million people impacted, particularly in Punjab, Sindh, and Baluchistan (Federal Flood Commission, 2010).

The 2022 flood were even more severe than the 2010 and 2011 floods as the 2022 floods affected a larger area, more people, and caused more damage than the previous floods (World Bank, 2022a, 2022b). These floods were attributed to global warming and climate change due to high temperatures, melting glaciers, and heavy monsoon rains (Nanditha et al., 2023; Otto et al., 2023; Shehzad, 2023). All four provinces were affected, with 15% of Pakistan's population impacted, and millions of people are losing lives and properties (Zaman et al., 2024). Sindh and Khyber Pakhtunkhwa suffered the most, with significant damage to health, education, and infrastructure. Health facilities were destroyed, leading to a rise in waterborne diseases like cholera and malaria (World Health Organization, 2022). The damage to infrastructure, including roads and bridges, was extensive, costing billions and hampering recovery (Government of Pakistan, 2023). Over 33 million people were affected, with 1700 deaths, 2 million homes damaged, and 4.4 million acres of crops lost, threatening food security. International aid fell short despite appeals (UNICEF, 2022).

7.6 Gender Roles and Vulnerabilities in Flood-Prone Areas

Pakistan has faced increasingly extreme weather events, with devastating floods being a major consequence of climate change (Malik et al., 2023; Otto et al., 2023). The 2022 floods, among the worst in the country's history, affected one-third of the nation, surpassing the destruction of the 2010 and 2011 floods. These floods resulted in over 1700 deaths, impacted 33 million people, and caused economic losses exceeding $39 billion, with Khyber Pakhtunkhwa (KP) and Sindh being the most affected (World Bank, 2022a, 2022b). In KP, rapid glacial melting has further heightened flood risks, leading to over 3000 glacial lakes in northern Pakistan (Ashraf et al., 2021). The 2022 floods caused significant economic and social losses in KP, amounting to over $15 billion and severely impacting homes, agriculture, and infrastructure (Hina et al., 2024). Vulnerable groups, particularly women and children, faced the greatest hardships (Gul et al., 2024a, 2024b, 2024c, 2024d).

The National Disaster Management Authority (NDMA, 2024) reported that over 33 million people were impacted by the 2022 floods, with 1 million homes being damaged. Among those affected, women and children were the most vulnerable. During the disaster, women faced numerous challenges, including limited access to healthcare, cultural constraints, and significant physical and mental health issues, such as post-traumatic stress disorder (PTSD) and gender-based violence. Pregnant women and lactating mothers, in particular, experienced heightened risks, including malnutrition, food shortages, and sexual violence in relief camps, compounded by poor sanitation conditions (Eckstein et al. 2021). The United Nations reported that approximately 130,000 pregnant women required immediate medical attention in Pakistan following the 2022 flood (UNICEF, 2022). Historical data has consistently shown that women tend to suffer more during disasters, with higher mortality rates observed in events such as the 2004 tsunami and the 2010 floods in Pakistan (Hussain, 2023). As climate change increases the intensity of natural disasters, women's vulnerabilities deepen, driven by limited access to resources, education, and disaster preparedness (Nosheen et al., 2023). Furthermore, in countries like Pakistan, gender norms severely limit women's mobility and their access to emergency aid, leaving them even more exposed to risk and violence (Soomar et al., 2023).

7.7 Women as an Agents of Resilience Through Social Capital: Context of Floods in Pakistan

Floods are recurring and devastating natural hazards in Pakistan (Malik et al., 2023). During floods, Pakistani women face heightened vulnerability due to existing social, economic, and cultural challenges (Gul et al., 2024a, 2024b, 2024c, 2024d). However, despite their vulnerabilities and sufferings, women also play pivotal roles in rescue and relief efforts during and after the floods. Women can also serve as powerful agents of resilience during such disasters, particularly through the lens of social capital.

Table 7.2 Types of social capital, gender focus, and community flood resilience outcomes

Type of social capital	Description	Gender focus (male/female/both)	Examples of community actions	Outcome on flood risk reduction and recovery
Bonding	Close-knit relationships between family, friends, neighbors; provides emotional support and immediate assistance	Female-focused (strong kinship networks), both genders in family roles	Sharing resources within families; emotional support during floods	Enhances household resilience and immediate recovery support
Bridging	Connections across different social groups, facilitating access to new resources and information	Both genders, female participation growing via community groups	Participation in NGOs, religious groups, civic associations	Broadens resource access; facilitates collective community action
Linking	Relationships between individuals and institutions or authorities; enables resource allocation and policy influence	Predominantly male formal leaders; emerging female leadership	Engagement with disaster management authorities; advocacy for aid	Enables access to formal aid, policy support, and institutional resources

Women's strong kinship networks can help their families to access resources, social and emotional support, security and comfort leading to the self-sufficiency of victim families (Table 7.2; Gul & McGee, 2024; Hamilton & Halvorson, 2007). Women's access to social networks, their social relationships and access to resources through these relationships can significantly enhance community resilience.

7.8 Informal Leadership and Decision-Making During Disasters

During flood disasters, formal leadership structures may be disrupted, leaving gaps in effective response and recovery and mitigation activities. In such instances, women often step into informal leadership roles, using their social networks and knowledge of local conditions to guide and mobilize their communities. These leadership actions play a key role in organizing evacuation efforts, providing essential resources, and making decisions that directly impact their families and neighborhoods. For example, during the 2010 floods in Pakistan, women were instrumental in coordinating relief efforts, ensuring the safety of children and elderly members, and managing household resources (Ullah et al., 2024). This informal leadership often comes from women's roles as caregivers and their deep involvement in the community.

NGOs, local institutions, and government agencies also play a vital role in supporting women's leadership during disasters (Gul, 2019). These organizations provide training, resources, and support, helping women build the skills and networks needed for effective disaster management (Gul & McGee, 2024). Through their partnerships with community leaders, these agencies can strengthen the capacities of women to take on these leadership roles, ensuring that disaster preparedness and recovery efforts are inclusive.

7.9 Women's Networks for Resource Mobilization and Risk Communication

One of the most significant contributions women make to community resilience in flood-prone regions is through their social networks, which serve as vital mechanisms for resource mobilization and risk communication (Table 7.3) (Gul & McGee, 2024). These networks, often community-based and deeply rooted in everyday interactions, allow women to share resources, exchange information, and coordinate collective action (Gul, 2019).

a. ***Resource Mobilization in Flood Prone Areas***

Women's networks are often critical in mobilizing resources such as food, shelter, medical supplies, and financial assistance during times of crisis (Aldrich et al., 2021; Gul & McGee, 2024). For instance, women in flood-affected communities frequently engage in informal bartering or pooling of resources to support one another. In the aftermath of the 2010 floods in Pakistan, women's groups were instrumental in organizing the distribution of aid when external resources were delayed (Table 7.3; Muzamil, 2021).

b. ***Flood Risk Communication***

Women often play a key role in disseminating flood warnings, preparedness information, and emergency evacuation plans (Shah et al., 2022). Their networks enable them to communicate critical information to households, especially in rural areas where official channels may not reach. In this way, women become crucial players in risk communication, ensuring that the community is well-prepared and informed about disaster events.

The involvement of NGOs and local institutions further strengthens these networks (Gul & McGee, 2024). Organizations focused on disaster management and climate resilience often provide women with the training and tools needed to communicate risks effectively. The collaboration between these organizations and women's groups ensures that information about flood risks reaches those who need it most.

Table 7.3 Challenges and opportunities in gender-focused social capital for flood mitigation

Challenge/ opportunity	Description	Gender impact (men/ women/both)	Potential solutions/ recommendations
Limited female mobility	Cultural norms restrict women's movement, limiting access to emergency aid and participation	Women	Promote community awareness and gender-sensitive mobility policies; safe transport options during disasters
Underrepresentation in formal decision-making	Women have limited roles in formal disaster committees and planning	Women	Encourage inclusive governance; quotas for women's representation in disaster risk reduction bodies
Access to resources	Unequal access to resources like health care, food, and relief materials during floods	Women	Establish women-led resource distribution networks; gender-sensitive aid programs
Psychosocial vulnerabilities	Women face higher risks of PTSD, gender-based violence, and stress post-disaster	Women	Integrate mental health and gender-based violence prevention in disaster response and relief
Informal female leadership	Women stepping into informal leadership roles during disasters but lack formal support	Women	Provide training and support to formalize women's leadership in community disaster management
Strong female social networks	Women's kinship and neighborhood networks effectively mobilize resources and information	Women	Strengthen and support women's networks through NGOs and community institutions
Male-dominated institutional linkages	Formal connections to government and aid agencies dominated by men, limiting women's influence	Both, but male dominated	Foster gender-balanced institutional linkages; create platforms for women to engage with authorities

7.10 Implications of Gender-Sensitive Approaches to Policy

Incorporating gender-sensitive disaster risk reduction (DRR) strategies is essential to address the unique vulnerabilities women face during floods (Abdalla et al., 2024). The integration of qualitative and quantitative data from recent flood events in

Pakistan underscores the importance of considering gender perspectives when developing climate resilience and disaster mitigation strategies (Gul et al., 2024a, 2024b, 2024c, 2024d; Shah et al., 2022). Gender-specific data, such as the disproportionate impact of floods on women's health, safety, and economic well-being, can provide valuable insights for improving flood preparedness and response plans.

The active involvement of NGOs, government agencies, and local institutions in fostering gender-sensitive DRR strategies is crucial for creating inclusive disaster policies (Gul, 2019; Gul & McGee, 2024). These stakeholders can work together to design programs that not only address the physical impacts of floods but also support the social and psychological needs of women. For instance, ensuring access to reproductive health services, safe spaces for women in relief camps, and gender-based violence prevention programs can significantly improve women's resilience during disasters.

This approach has implications for disaster risk reduction policies not only in Pakistan but across other South Asian countries, where gender inequalities and flood risks intersect. By recognizing and addressing these unique challenges, governments and organizations can develop more effective, gender-sensitive policies that enhance overall community resilience.

7.11 Conclusion

In Pakistan, flood is the most devastating and recurrently occurring extreme event. The country is exposed to riverine, flash floods, urban floods, and coastal floods. However, the community exposure to extreme floods and its adverse consequences are gradually multiplying day by day. The role of women plays an important role in mitigating the flood hazards and most often, in rural areas, the first responders are women. Women's involvement in informal leadership roles and their active participation in social networks are critical factors in building disaster resilience in flood-prone areas. By empowering women through gender-sensitive DRR strategies, we can strengthen the resilience of entire communities. The collaboration between NGOs, local institutions, and government agencies is essential to support women in these roles, ensuring that disaster response and recovery efforts are inclusive and effective. By integrating gender-specific data and promoting women's leadership, we can develop disaster policies that not only respond to the immediate needs of women but also contribute to long-term, sustainable resilience.

References

Abdalla, S., Ramadan, E., & Al Mamari, W. (2024). Enhancing gender-responsive resilience: The critical role of women in disaster risk reduction in Oman. *Progress in Disaster Science, 24*, 100376.

Akter, R., Roy, T., & Aktar, R. (2023). The challenges of women in post-disaster health management: A study in Khulna District. *International Journal of Disaster Risk Management, 5*(1), 51–66.

Aldrich, D. P., Kolade, O., McMahon, K., & Smith, R. (2021). Social capital's role in humanitarian crises. *Journal of Refugee Studies, 34*(2), 1787–1809.

Aldrich, D. P., & Meyer, M. A. (2015). Social capital and community resilience. *American Behavioral Scientist, 59*(2), 254–269. https://doi.org/10.1177/0002764214550299

Alrehaili, N. R., Almutairi, Y. N., Alghamdi, H. M., & Almuthaybiri, M. S. (2022). A structural review on disaster management models and their contributions. *International Journal of Disaster Management, 5*(2), 93–108. https://doi.org/10.24815/ijdm.v5i2.27087

Ashraf, A., Iqbal, M. B., Mustafa, N., Naz, R., & Ahmad, B. (2021). Prevalent risk of glacial lake outburst flood hazard in the Hindu Kush–Karakoram–Himalaya region of Pakistan. *Environmental Earth Sciences, 80*(12), 451.

Ashraf, S., Iftikhar, M., Shahbaz, B., Khan, G. A., & Luqman, M. (2013). Impacts of flood on livelihoods and food security of rural communities: A case study of southern Punjab, Pakistan. *Pakistan Journal of Agricultural Sciences,* 50–57.

Bazai, N. A., Alam, M., Cui, P., Hao, W., Khan, A. P., Waseem, M., et al. (2024). Dynamics and impacts of monsoon-induced geological hazards: A 2022 flood study along the Swat River in Pakistan. *Natural Hazards and Earth System Sciences Discussions, 2024*, 1–27.

Berkes, F., & Ross, H. (2013). Community resilience: Toward an integrated approach. *Society & Natural Resources, 26*(1), 5–20.

Centre for Research on the Epidemiology of Disasters (CRED). (2015). *Annual disaster statistical review 2014: The numbers and trends*. ReliefWeb. https://reliefweb.int/report/world/annual-disaster-statistical-review-2014-numbers-and-trends

Chaudhary, M. T., & Piracha, A. (2021). Natural disasters—Origins, impacts, management. *Encyclopedia, 1*(4), 1101–1131.

Chawa, A. F. (2017). The formation of bridging and bonding social capital to empower mothers from low-income families. *International Journal of Social Science Studies, 5*, 13.

Cutter, S. L. (2017). The forgotten casualties redux: Women, children, and disaster risk. *Global Environmental Change, 42*, 117–121.

Eckstein, D., Künzel, V., & Schäfer, L. (2021). Global climate risk index 2021: Who suffers most from extreme weather events? In *Weather-related loss events in 2019 and 2000 to 2019*. Germanwatch e.V. https://www.germanwatch.org/en/19777

Enarson, E. (2009). *Prairie women prepared for disaster*. Retrieved June 3, 2011, from http://www.pwhce.ca/pdf/disasterI/MainReport.pdf

Fatema, S. R., East, L., Islam, M. S., & Usher, K. (2021). Health impact and risk factors affecting south and southeast Asian women following natural disasters: A systematic review. *International Journal of Environmental Research and Public Health, 18*(21), 11068.

Federal Flood Commission Islamabad. (2010). In Flood Report, Office of the Chief Engineering Advisor & Chairman (Ed.), *Annual flood report 2010* (1st ed.).

Gaillard, J. C., Sanz, K., Balgos, B. C., Dalisay, S. N. M., Gorman-Murray, A., Smith, F., & Toelupe, V. A. (2017). Beyond men and women: A critical perspective on gender and disaster. *Disasters, 41*(3), 429–447.

Government of Pakistan. (2023). *Post-flood infrastructure assessment report*. Ministry of Planning, Development, and Special Initiatives.

Gul, S., Ali, Z., & Ullah, U. (2024a). Integrating GIS and remote sensing for comprehensive flood risk zonation in Tehsil Shah Alam (Peshawar). *Journal of Asian Development Studies, 13*(2), 1422–1431.

Gul, S., Khalil, I., & Hidayat, K. (2024b). Impact of 2022 flood on girls' education: A case study of District Nowshera, Khyber Pakhtunkhwa, Pakistan. *International Journal of Disaster Risk Reduction, 114*, 104988.

Gul, S., & McGee, T. (2022). Women's participation in disaster recovery after the Kashmir, Pakistan earthquake. *Disasters, 46*(4), 1007–1026.

Gul, S., Khan, N., Nisar, S., Ali, Z., & Ullah, U. (2024c). Impact of climate change-induced flood on women's life: A case study of the 2022 flood in District Nowshera, Khyber Pakhtunkhwa, Pakistan. *Journal of Asian Development Studies, 13*(2), 1468–1482. https://doi.org/10.62345/jads.2024.13.2.117

Gul, S., Khan, S. N., Malik, F., Zahid, S. M. H., Khan, R. A., & Fazl-ur-Rahman. (2024d). Post-GLOF recovery analysis: A case study of the 2019 event in Golain Gol, Lower Chitral, Pakistan. *International Journal of Contemporary Issues in Social Sciences, 3*(3), 1303–1322. https://ijciss.org/index.php/ijciss/article/view/1289

Gul, S., & McGee, T. (2024). Factors influencing women's participation in disaster recovery after the 2005 earthquake in Kashmir, Pakistan. *Environmental Hazards,* 1–18. https://doi.org/10.1080/17477891.2024.2422353

Gul, S. (2019). *Impact of women's participation in disaster recovery on their lives: A case study from Pakistan* (Doctoral dissertation, University of Alberta). University of Alberta Education & Research Archive. https://ualberta.scholaris.ca/items/ebd7a1cf-3841-415d-ad01-5a96e43b221c

Hamilton, J. P., & Halvorson, S. J. (2007). The 2005 Kashmir earthquake: A perspective on women's experiences. *Mountain Research and Development, 27*(4), 296–301.

Hina, Gul, S., Ali, Z., & Ullah, U. (2024). Integrating GIS and remote sensing for comprehensive flood risk zonation in Tehsil Shah Alam (Peshawar). *Journal of Asian Development Studies, 13*(2), 1422–1431. https://doi.org/10.62345/jads.2024.13.2.113

Hussain, S. (2023). The intersection of patriarchal chains and the natural disasters: Exploring the impact of flood on women in the rural areas of Dera Ghazi Khan, Pakistan. *Contemporary Issues in Social Sciences and Management Practices, 2*(3), 276–291.

Islam, M. Z., & Wang, C. (2024). Cost of high-level flooding as a consequence of climate change driver?: A case study of China's flood-prone regions. *Ecological Indicators, 160*, 111944.

Kasemsap, K. (2017). The fundamentals of social capital. In *Cultural influences on architecture* (pp. 259–292). IGI Global.

Khan, B., & Iqbal, M. J. (2013). Forecasting flood risk in the Indus River system using hydrological parameters and its damage assessment. *Arabian Journal of Geosciences, 6*, 4069–4078.

Khan, A., & Rahman, A. (2024). Spatial analysis and extent of soil erosion risk using the RUSLE approach in the Swat River Basin, Eastern Hindukush. *Applied Geomatics, 16*(2), 1–16.

Khan, A. A., Rana, I. A., Nawaz, A., & Waheed, A. (2021). Gender-based emergency preparedness and awareness: empirical evidences from high-school students of Gilgit, Pakistan. *Environmental Hazards, 20*(4), 416–431.

Khandelwal, S., Mehra, M., & Parwez, S. (2025). Climate change and its impact on women in South Asia: evidence from a gender lens. In *Health and cimate change* (pp. 289–311). Academic Press.

Malik, F., Khan, A. N., Rahman, A., Iqbal, S., & Nawaz, M. A. (2023). Extent and evaluation of factors causing flash floods in Watalai Khwar, District Bajaur, Pakistan. *CARC Research in Social Sciences, 2*(3), 141–147.

Manglore, U., Rahman, A., Marwat, F., Naz, T., Dilbar, S., & Khan, F. (2024). Geospatial analysis of flood causes and extent of flood damages in Swat Valley, North Pakistan. *Natural and Applied Sciences International Journal (NASIJ), 5*(1), 130–153.

Muzamil, M. R. (2021). Climate-related disasters, conflict and development: Reflections about the past and insights into the future from the Khyber-Pakhtunkhwa Province of Pakistan.

Nanditha, J. S., Kushwaha, A. P., Singh, R., Malik, I., Solanki, H., Chuphal, D. S., et al. (2023). The Pakistan flood of August 2022: Causes and implications. *Earth's Future, 11*(3), e2022EF003230.

National Disaster Management Authority Flood Report, Office of the Chief Engineering Advisor & Chairman, Federal Flood Commission Islamabad (Ed.). (2010). *Annual flood report 2010* (1st ed.).

National Disaster Management Authority. (2024). *Comprehensive review of Monsoon 2024*. https://www.ndma.gov.pk/storage/publications/October2024/GyCDL3UasaOBmN4NVkB5.pdf

Nosheen, M., Iqbal, J., & Ahmad, S. (2023). Economic empowerment of women through climate change mitigation. *Journal of Cleaner Production, 421*, 138480.

Oloruntoba, R., Sridharan, R., & Davison, G. (2018). A proposed framework of key activities and processes in the preparedness and recovery phases of disaster management. *Disasters, 42*(3), 541–570. https://doi.org/10.1111/disa.12268

Otto, F. E., Zachariah, M., Saeed, F., Siddiqi, A., Kamil, S., Mushtaq, H., et al. (2023). Climate change increased extreme monsoon rainfall, flooding highly vulnerable communities in Pakistan. *Environmental Research: Climate, 2*(2), 025001.

Pakistan Meteorological Department. (2019). *Climate change in Pakistan*. PMD. http://www.pmd.gov.pk/

Panday, S., Rushton, S., Karki, J., Balen, J., & Barnes, A. (2021). The role of social capital in disaster resilience in remote communities after the 2015 Nepal earthquake. *International Journal of Disaster Risk Reduction, 55*, 102112.

Putnam, R. D. (1995). Bowling alone: America's declining social capital. In *Culture and politics* (pp. 223–234). Palgrave Macmillan.

Rahman, H., Khan, A., Rahman, A., & Shaw, R. (2024). Spatio-temporal dynamics of land use land cover and its impact on flood-prone drainage basin of River Swat, Eastern Hindukush. In R. Mazumder & R. Shaw (Eds.), *Surface environments and human interactions*. Springer Singapore.

Rahman, M. M., Hossain, M. S., & Ahmed, M. (2023). Flooding and tropical river systems in South Asia: Climate change and urbanization impacts. In *Climate change and water security in South Asia* (pp. xx–xx). Springer. https://link.springer.com

Saul, J. (2022). *Collective trauma, collective healing: Promoting community resilience in the aftermath of disaster*. Routledge.

Shah, A. A., Ullah, A., Khan, N. A., Pal, I., Alotaibi, B. A., & Traore, A. (2022). Gender perspective of flood early warning systems: People-centered approach. *Water, 14*(14), 2261.

Soomar, S. M., Arefin, A., & Soomar, S. M. (2023). "Women are again unsafe": Preventing violence and poor maternal outcomes during current floods in Pakistan. *Journal of Global Health, 13*, 03005.

Soomro, S. E. H., Hu, C., Boota, M. W., Ahmed, Z., Chengshuai, L., Zhenyue, H., et al. (2022). River flood susceptibility and basin maturity analyzed using a coupled approach of geo-morphometric parameters and SWAT model. *Water Resources Management, 36*(7), 2131–2160.

Szreter, S., & Woolcock, M. (2004). Health by association? Social capital, social theory, and the political economy of public health. *International Journal of Epidemiology, 33*(4), 650–667.

Tearne, J. E., Guragain, B., Ghimire, L., Leaning, J., & Newnham, E. A. (2021). The health and security of women and girls following disaster: A qualitative investigation in post-earthquake Nepal. *International Journal of Disaster Risk Reduction, 66*, 102622.

UN-Habitat. (2023). *Pakistan country report 2023*. United Nations Human Settlements Programme. https://unhabitat.org/sites/default/files/2023/06/4._pakistan_country_report_2023_b5_final_compressed.pdf

UNDRR. (2022). *Disasters*. https://www.undrr.org/terminology/disaster

UNICEF. (2022). *Schools for more than 2 million children in Pakistan remain inaccessible due to devastating floods* [Press release]. UNICEF. https://www.unicef.org/rosa/press-releases/schools-more-2-million-children-pakistan-remain-inaccessible-due-devastating-floods

Ullah, W., Haijun, D., Shah, A. A., Alotaibi, B. A., & Rauf, A. (2024). Behind the floodwaters: Violence against women, and disaster management capacities in flood-affected areas of Pakistan. *Climate Risk Management, 46*, 100653.

United Nations Children's Fund (UNICEF). (2022). *Impact of the 2022 floods on education in Pakistan*. Retrieved from UNICEF Website.

Warfield, C. (2013). *The disaster management cycle, 2008*. http://www.gdrc.org/uem/disasters/1-dm_cycle.html

World Bank. (2022a). *PAKISTAN FLOODS 2022: Post-disaster needs assessment*. The Government of Pakistan, Asian Development Bank, European Union, United Nations Development Programme, World Bank. https://thedocs.worldbank.org/en/doc/4a0114eb7d1cecbbbf2f65c5ce0789db-0310012022/original/Pakistan-Floods-2022-PDNA-Main-Report.pdf

World Bank. (2022b). *South Asia climate change and flooding: Risks and policy responses.* World Bank Group. https://openknowledge.worldbank.org

World Health Organization (WHO). (2022). *Pakistan floods: Health emergency response.* Retrieved from WHO Website https://www.who.int/emergencies

Zaman, A., Ahmed, H., Usman, M., & Haque, M. A. (2024). Request for aid in flood-impacted regions following the 2022 flood crisis in Pakistan. *IJS Global Health, 7*(3), 04–55.

Chapter 8
Women in Sustainable Watershed Management: Building Resilience to Climate-Induced Disasters in Sri Lanka

Wittachchi Koralalage Athula Chammika Gnanapala, Ranathunga Arachchillage Danushka Chandimal Ranathunga, and Sudeshika Jayasekara

Abstract Sustainable Watershed Management (SWM) is a vital approach for addressing climate-induced disasters in vulnerable countries like Sri Lanka. To achieve the overarching aims of the SWM, the Integrated Watershed Management (IWM) offers an operational framework which uses variety of tools. Mainstreaming gender is one of the key principles of operationalizing IWM. However, despite the critical role of women's engagement, it is often limited or overlooked. This chapter explores how women can be effectively involved in IWM to strengthen climate resilience. The findings are based on a literature review, validated and supplemented by field observations and informal discussions. Although many IWM tools are in use across the country, women's participation remains low or partial. The chapter also highlights the key challenges that restrict women's involvement in IWM practices and presents strategies to enhance their participation in SWM aiming at building resilience against climate-induced disasters.

Keywords Climate resilience · Gender mainstreaming · Integrated Watershed Management · Sustainable Watershed Management · Women's participation

---

W. K. A. C. Gnanapala (✉) · R. A. D. C. Ranathunga
Faculty of Management Studies, Sabaragamuwa University of Sri Lanka, Belihuloya 70140, Sri Lanka
e-mail: gnanapala@mgt.sab.ac.lk

R. A. D. C. Ranathunga
e-mail: chandimalrane@mgt.sab.ac.lk

S. Jayasekara
Faculty of Agricultural Sciences, Sabaragamuwa University of Sri Lanka, Belihuloya 70140, Sri Lanka
e-mail: sjayasekara@std.agri.sab.ac.lk

Md. A. Abedin et al. (eds.), *Climate Crisis in South and East Asia*, Sustainability Solutions, https://doi.org/10.1007/978-3-032-01012-4_8

8.1 Introduction

Sri Lanka, with its varied topography and tropical monsoon climate, is highly vulnerable to climate-induced disasters. The country frequently experiences extreme weather events such as floods, landslides, and droughts, which disrupt agriculture, water security, ecosystem functions, and socio-economic stability (Samaraweera et al., 2024). These challenges highlight the need for sustainable and long-term approach to reduce risks and enhance community resilience against climate-induced disasters.

The watershed-based approach offers a promising avenue for building resilience by promoting the integrated and sustainable management of environmental and socio-economic resources within a watershed (Wani & Garg, 2009). A watershed—or drainage basin—is a natural unit that collects and channels rainfall and runoff to a common outlet. As a critical hydrological system, it supports water supply, soil retention, biodiversity, food production, and rural livelihoods (Khurana et al., 2024).

However, Sri Lanka's watersheds are increasingly threatened by deforestation, unsustainable land use, and poor management, particularly in upland areas (Surasinghe et al., 2019). Climate change has intensified these pressures. In response, Sustainable Watershed Management (SWM) has gained attention as a holistic solution that balances ecological integrity, economic viability, and social equity, while improving resilience to climate-related disasters (Azmi et al., 2024).

Integrated Watershed Management (IWM) provides the operational framework to implement SWM through participatory, inclusive, and adaptive tools (Wang et al., 2016). Yet, the inclusion of gender—especially women's roles—remains limited or overlooked. This chapter examines how IWM tools can enhance SWM, with a focus on increasing women's participation and leadership. It also explores barriers to their engagement and outlines strategies for more inclusive, gender-responsive watershed governance that supports climate resilience.

8.2 Literature Review

8.2.1 Sustainable Watershed Management (SWM)

Sustainable Watershed Management (SWM) is a coordinated, multidisciplinary approach to managing land and water resources within a defined hydrological boundary. It aims to achieve long-term ecological balance, drive economic development, and mainstream social equity (Azmi et al., 2024). The key goals of SWM include improving water availability and quality, enhancing agricultural productivity, conserving biodiversity, controlling soil erosion, and restoring degraded ecosystems. Additionally, SWM supports climate resilience, disaster risk reduction, food and livelihood security, community empowerment, and conflict resolution in resource management (Gonenc et al., 2014).

8.2.2 Role of SWM in Building Resilience to Climate-Induced Disasters

SWM plays a vital role in strengthening resilience to climate-induced disasters, such as floods, droughts, and landslides, which are becoming increasingly common in Sri Lanka. By promoting the sustainable use and conservation of land and water resources, SWM not only helps prevent these disasters but also enables faster and safer recovery when they occur. Moreover, SWM contributes to maintaining ecological balance and socio-economic stability during and after such events (Basuki et al., 2022).

8.2.3 Integrated Watershed Management (IWM)

Integrated Watershed Management (IWM) is a holistic and participatory approach that acknowledges the interdependence of land, water, ecosystems, and communities within a watershed (Erdogan, 2013). It serves as the practical framework for achieving the broader goals of SWM. IWM is guided by key principles including being community-based, participatory, inclusive, holistic, adaptive, data-driven, and resource-efficient. Its tools and practices are designed to address all three pillars of sustainability—environmental, social, and economic—ensuring balanced and integrated watershed governance (Van Der Zaag & Savenije, 2004) (Fig. 8.1).

8.2.4 Women and Integrated Watershed Management (IWM)

Women play a critical yet underrecognized role in IWM, which is frequently overlooked in natural resource management policies. Though they contribute significantly to watershed activities and management, men retain ownership and control. Limited decision-making power for women stems from socio-political inequalities, male dominance, and stereotypes, leaving their role undervalued despite active participation (Singh, 2018).

8.3 Methodology

This study employed a mixed-methods approach to examine women's participation in Integrated Watershed Management (IWM) tools. Initially, a comprehensive literature review was conducted to identify various IWM tools relevant to Sustainable

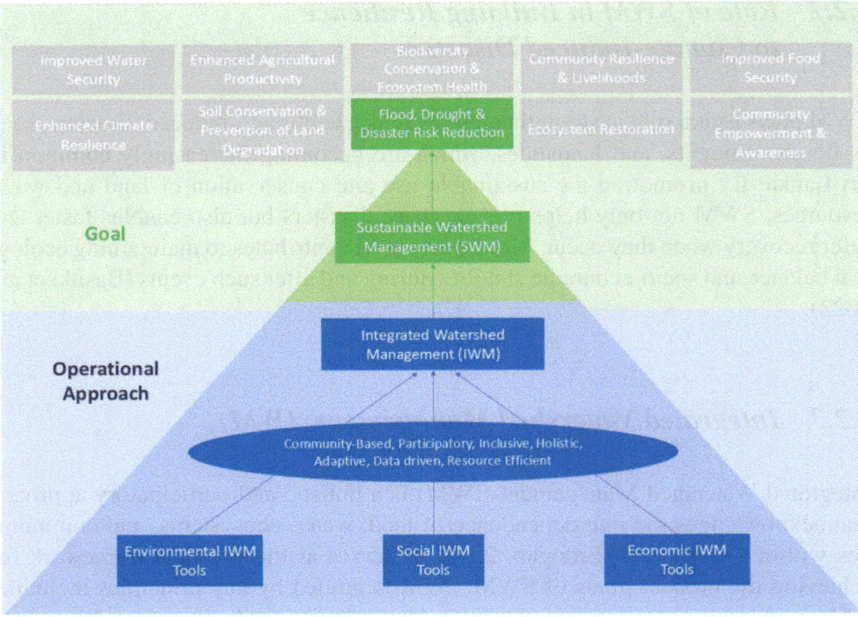

Fig. 8.1 Conceptual framework depicting how IWM tools support and enhance SWM

Watershed Management (SWM), particularly those aimed at reducing the climate-induced disaster risks. Based on this review, key tools were derived separately for three pillars of sustainability.

To validate and enrich these findings, several field observations and informal discussions were carried out. Fieldwork was primarily conducted in the Belihuloya area of the Ratnapura District, focusing on agricultural fields and home gardens. Photographic evidences were collected during these visits. Informal discussions included both in-person conversations during field visits and remote discussions via phone with selected women stakeholders involved in various IWM tools. In cases where direct contact with women stakeholders was not possible, indirect inquiries were made through known contacts in different regions of the country. These regions included Badulla, Kegalle, Ratnapura, Anuradhapura, Puttalam, and Monaragala districts. Wherever possible, photographic documentation of women's engagement in IWM activities was obtained with their consent. All collected data were synthesized and summarized into a table. The study also highlights the key challenges and barriers faced by women in IWM and concludes with a discussion on strategies to enhance women's participation in order to achieve the overarching objectives of Sustainable Watershed Management.

8.4 Results and Discussion

8.4.1 IWM Tools to Achieve Sustainable Watershed Management

A total of 20 IWM tools were identified through an extensive literature review and further validated through field observations and stakeholder discussions. These tools were categorized under the three SWM pillars: 10 environmental tools, 5 social and governance tools, and 5 economic tools (Fig. 8.2).

8.4.2 Women's Roles in IWM Tools to Achieve Sustainable Watershed Management

Table 8.1 presents a detailed overview of selected Integrated Watershed Management (IWM) tools. It outlines each tool's description, its importance in achieving Sustainable Watershed Management (SWM), scope of application, implementation methods, role played by women and the references.

8.4.3 Current Level of Engagement of Women in IWM Tools

Although the level of women's engagement in Integrated Watershed Management (IWM) tools identified through this study was not assessed through a fully scientific or quantitative method, field observations and stakeholder conversations provided key points. These interactions revealed that women's participation generally ranges from low to moderate across most regions, with only a few notable exceptions where certain IWM tools have been well established and actively promoted over time.

In several regions, some IWM tools are either poorly implemented or entirely absent, offering limited or no opportunities for women's engagement. This uneven implementation highlights significant gaps in accessibility, awareness, and institutional support for women's participation in certain watershed management practices. It underscores the need for targeted interventions to strengthen women's involvement, especially in areas where IWM tools remain underutilized or overlooked. Figure 8.3, 8.4, 8.5, 8.6, 8.7, 8.8, 8.9 and 8.10 present evidence of women's engagement in various IWM tools across the island.

Table 8.1 Role of women in IWM tools explored in the chapter

SWM pillar	IWM tool	Tool description	Importance of the IWM tool in SWM	Scope of the tool	Implementation	Role of women	Reference/s
Environment	Afforestation/ reforestation	• Afforestation is the process converting non-forested land with no previous tree cover into forested land. Reforestation is the process of replanting trees in an area that was previously forested but has been deforested or degraded	• Improving soil stability • Improving infiltration and groundwater recharge • Reducing surface runoff • Reducing soil erosion	• Regional level • Community level	• Government initiatives • Non-government programs/ initiatives • Community forestry programs	• Plant species selection • Nursery management • Planting location selection • Engage in tree planting	Ronderos (1989) and Ullah et al. (2021)
	Check dams	• Check dams are relatively small, temporary structures constructed within concentrated-flow areas such as swales, channels and ditches	• Controlling erosion • Controlling sediment transportation • Improving infiltration and groundwater recharge	• Community level	• Community water user groups • Farmer societies	• Site selection • Providing labor for construction • Engage in maintenance	Renganayaki and Elango (2013)

(continued)

Table 8.1 (continued)

SWM pillar	IWM tool	Tool description	Importance of the IWM tool in SWM	Scope of the tool	Implementation	Role of women	Reference/s
	Rainwater harvesting	• Collection, storage and use of rainwater from catchments (rooftops, land surfaces)	• Reduce surface runoff • Direct source of potable and domestic water requirement • Reduces reliance on freshwater sources • Reduce pressure on groundwater	• Community level • Household level	• Non-government programs/ initiatives	• Decision-making on the type, size, and location of the tanks • Labor for construction maintenance • Engage in water usage and quality monitoring • Engagement in community level rainwater harvesting groups	Kanyi et al. (2017)
	Irrigation management	• Irrigation management refers to the efficient planning, application, and control of water for crops to ensure healthy growth, optimize yields, and conserve water resources	• Efficient management of both groundwater and surface water sources	• Community level • Household level	• Farmer societies • Household level implementation	• Engagement in irrigation planning • Involve in agricultural field irrigation	Dasthagir (2013)

(continued)

Table 8.1 (continued)

SWM pillar	IWM tool	Tool description	Importance of the IWM tool in SWM	Scope of the tool	Implementation	Role of women	Reference/s
	Vegetative barriers	• Vegetative barriers are strips of dense vegetation, such as grasses, shrubs, or small trees, planted along contours, field boundaries, or slopes to reduce soil erosion and manage water flow in agricultural and watershed areas	• Reduces soil erosion • Reduce sediment transportation	• Community level • Farmer level	• Farmer societies • Individual farmer level implementation	• Engagement in planting vegetation • Engagement in maintenance	Khandelwal et al. (2024)
Environment	Retaining walls	• Retaining walls are engineered structures built to hold back soil, prevent landslides, and stabilize slopes, especially in hilly or uneven terrains	• Prevent soil erosion • Support steep slopes • Control water runoff	• Regional level • Community level	• Regional level implementation	• Providing labor for construction • Engage in maintenance • Engage in monitoring	Paolisso et al. (1999)

(continued)

Table 8.1 (continued)

SWM pillar	IWM tool	Tool description	Importance of the IWM tool in SWM	Scope of the tool	Implementation	Role of women	Reference/s
	Contour trenches/ bunds	• Excavating trenches or constructing earthen barriers along the contours of a slope	• Reduce soil erosion • Improve water infiltration • Improving groundwater recharge • Stabilizing slopes	• Farmland level	• Farmer level implementation • Farmer societies	• Providing labor for construction and maintenance • Deciding placement of trenches/bunds • Participating in farmer groups	Bohluli et al. (2015)
	Percolation pits	• Small, dug pits or trenches designed to collect surface runoff and allow it to slowly percolate into the sub-surface	• Groundwater recharge • Reduce surface runoff	• Community level • Household level	• Household level implementation	• Identification of suitable sites for pits • Providing labor for digging and maintaining the pits	Arun and Kulkarni (2011)

(continued)

Table 8.1 (continued)

SWM pillar	IWM tool	Tool description	Importance of the IWM tool in SWM	Scope of the tool	Implementation	Role of women	Reference/s
	Wetland restoration	• Restoring the ecological functions and biodiversity of degraded wetlands, including marshes, swamps, and rice paddies	• Regulate water flow • Filtering pollutants • Provide habitats • Water quality improvement • Soil erosion reduction • Flood control	• Regional level • Community level	• Government initiatives • Non-government programs/ initiatives • Farmer societies	• Engaging in planting wetland-appropriate vegetation • Participating in monitoring the health and biodiversity of restored wetlands • Sharing traditional ecological knowledge • Advocating for the protection of wetlands within their communities	Richardson et al. (2011)
	Riparian buffers	• Strips of vegetation (trees, shrubs, or grasses) planted or preserved along the banks of rivers, streams, lakes, or wetlands	• Stabilizes riverbanks • Filters pollutants from agricultural runoff • Provides habitat for wildlife • Improving water quality	• Community level • Farmer level	• Non-government programs/ initiatives • Farmer societies • Farmer level implementation	• Participating in planting and maintaining buffer vegetation • Engaging in awareness programs about water pollution and the role of buffers	Anbumozhi et al. (2005)

(continued)

Table 8.1 (continued)

SWM pillar	IWM tool	Tool description	Importance of the IWM tool in SWM	Scope of the tool	Implementation	Role of women	Reference/s
Social and governance	Community-based watershed committees (CBWCs)	• The local, participatory groups formed within a watershed area to manage and oversee sustainable water and land use practices	• Promote inclusive community participation and equitable decision-making • Support local knowledge, monitoring, and conflict resolution to build resilience	• Community level	• Non-government programs/ initiatives • Community level	• Engage in leadership • Participate in decision-making processes	Qianxiang et al. (2005)
	Awareness and capacity building	• A strategy to build knowledge and capacity among local community to make better decisions to manage watershed resources efficiently	• Improve community capacity • Empower community leaders for better decision making	• Community level	• Non-government programs/ initiatives • Community level	• Participate in training and capacity development programs • Participate in knowledge dissemination • Influence on decision-making	Aban (2019)

(continued)

Table 8.1 (continued)

SWM pillar	IWM tool	Tool description	Importance of the IWM tool in SWM	Scope of the tool	Implementation	Role of women	Reference/s
	Water user associations (WUAs)	• WUAs are community-based groups or organizations formed by local water users (farmers, households, industries) who collectively manage, operate, and maintain water resources, particularly for irrigation and watershed management	• Efficient use of water resources • Local stewardship and participation • Conflict resolution • Adaptive management • Linkage with government and NGOs	• Community level	• Non-government programs/ initiatives • Community level	• * Participate in water budgeting, irrigation planning, and promoting fair, efficient water use	Wang and Wu (2018)
	Early warning systems (community-based)	• The local systems that help communities detect, monitor, and respond to risks like floods, droughts, or landslides before they happen, allowing people to takc action and reduce harm	• Reduces disaster risks • Improves community preparedness • Supports adaptive planning • Strengthens resilience to climate variability and natural hazards	• Community level	• Non-government programs/ initiatives • Community level	• Support communication, monitoring, and community response through awareness, warnings, and evacuation planning	Macherera and Chimbari (2016)

(continued)

Table 8.1 (continued)

SWM pillar	IWM tool	Tool description	Importance of the IWM tool in SWM	Scope of the tool	Implementation	Role of women	Reference/s
	Formal governance structures	• Officially established government institutions or bodies at the local level that have legal authority and responsibility for governance and management of resources in watersheds	• Regulate and manage water and land resources efficiently • Coordinate watershed policies and plans • Mobilize resources and funding for watershed projects • Monitor, enforce rules, and ensure compliance to protect the watershed	• Sub-national level • Regional level • Community level	• Government initiatives	• Engagement in formal government bodies • Providing assistance in government initiatives	Imperial (2005)
Economic	Diversified cropping systems	• Involve growing a variety of crops (e.g., intercropping, crop rotation, agroforestry) instead of relying on a single crop	• Soil health improvement • Soil erosion reduction • Reduces runoff • Enhances crop productivity	• Community level • Farmer level	• Farmer societies • Farmer level implementation	• Selection of crops • Engage in planting • Engage in harvesting	Sharda and Venkateswarlu (2007)

(continued)

Table 8.1 (continued)

SWM pillar	IWM tool	Tool description	Importance of the IWM tool in SWM	Scope of the tool	Implementation	Role of women	Reference/s
	Micro-irrigation (drip/sprinkler)	• Low-pressure, low-flow irrigation methods like drip and sprinkler systems that deliver water directly to the plant root zone, minimizing wastage	• Reduces water wastage • Minimizes soil erosion and runoff • Enhances crop productivity	• Farmer level	• Non-government programs/ initiatives • Farmer level	• Involving designing and maintenance • Managing small-scale irrigation systems	Kannan et al. (2024)
	Livelihood diversification	• Process where rural households engage in a variety of income-generating activities (e.g., farming, fishing, handicrafts, small businesses, eco-tourism) rather than depending solely on agriculture or a single income source	• Reduces pressure on land and water resources • Improves community resilience • Improve community income	• Community level • Farmer level	• Non-government programs/ initiatives • Farmer level	• Engage in trainings and capacity building • Leading small-scale enterprises	Manjur et al. (2014)

(continued)

Table 8.1 (continued)

SWM pillar	IWM tool	Tool description	Importance of the IWM tool in SWM	Scope of the tool	Implementation	Role of women	Reference/s
	Payment for ecosystem services (PES)	• Voluntary agreement where beneficiaries of ecosystem services (like clean water) provide financial or other incentives to landowners or communities who manage ecosystems sustainably	• Protection of water sources • Promotes reforestation, soil conservation, and pollution control • Provides financial support for communities	• Regional level • Community level	• Private organizations • Non-government programs/initiatives	• Engage in trainings and capacity development programs • Engage in conservation efforts • Engage in local research programs	Rigonato et al. (2023)
	Watershed employment programs	• Programs create jobs for local people—especially in rural areas—through activities such as soil conservation, afforestation, water harvesting, and sustainable land management in watershed areas	• Support conservation initiatives • Builds local capacity and awareness • Creates economic incentives for local communities	• Regional level • Community level	• Private organizations • Non-government programs/initiatives • Community level	• Provide income-generating opportunities • Engage in community-based planning and decision-making • Improves women's access to skills training and financial independence	Joshi et al. (2008)

Fig. 8.2 Derived IWM tools

Fig. 8.3 Household-level rainwater harvesting project (location: Belihuloya, Ratnapura District)

8.4.4 Challenges and Barriers Hindering Women's Participation

Despite their vital contributions, women face numerous barriers to fully participating in IWM tools, which are predominantly led by men. Their limited involvement is mainly due to socio-cultural norms, institutional and policy gaps, and restricted access to resources, education, and training opportunities necessary for effective engagement.

Fig. 8.4 Female farmer engaged in irrigation in a farmland (location: Rajanganaya, Anuradhapura District)

Fig. 8.5 A female is engaging in agricultural irrigation management (location: Belihuloya, Ratnapura District)

Fig. 8.6 A female government officer representing divisional secretariat division engages with field observation (location: Nivithigala, Ratnapura District)

Fig. 8.7 A female government officer representing irrigation department engages with field supervision (location: Mahiyanganaya, Badulla District)

Fig. 8.8 Female farmer engaged in sprinkler irrigation in her farmland (location: Wellawaya, Monaragala District)

Fig. 8.9 Female farmer engages with intercropping of her farmland (location: Bandarawela, Badulla District)

8.4.4.1 Socio-cultural Constraints

Many of the Integrated Watershed Management (IWM) tools discussed in this chapter are particularly applicable to rural and remote regions of the country. However, these areas are often shaped by cultural norms and traditions that restrict women's mobility, voice, and participation in public life. Leadership roles in such communities are typically held by men. One key reason for the limited participation of women in IWM activities is their traditional household and caregiving responsibilities.

Fig. 8.10 Female workers engage with drip irrigation in community level agricultural fields (location: Norochcholai, Puttalam District)

Field observations, however, revealed that women are actively engaged in household-level IWM practices such as rainwater harvesting and micro-irrigation. Informal discussions also indicated that many women are now involved in other forms of livelihood, such as working in government institutions or factories, which limits their availability to participate in community-level IWM activities.

In many rural areas, land ownership is predominantly held by men, which further excludes women from decision-making processes related to land and water governance (Kunihira, 2014). Additionally, the engagement of young women in these initiatives is limited, as many migrate from their villages in pursuit of higher education or employment opportunities. As a result, participation in IWM initiatives tends to be concentrated among middle-aged and older women who remain in the community.

8.4.4.2 Limited Availability of Resources and Training

To effectively implement and realize the benefits of IWM tools, adequate resources must be allocated to the regions, areas, and communities most in need. Observations and discussions revealed that when financial resources are made available, women in these regions actively engage in the initial stages of such programs and are able to benefit from them.

During the early phases—such as planning and community mobilization—women often take part on behalf of their households, as men are typically occupied with income-generating activities. However, as programs move into implementation and operational stages, male participation tends to increase, often overshadowing women's roles and leadership in the process. This shift highlights the need for sustained efforts to ensure women remain actively involved throughout all phases of IWM initiatives.

8.4.4.3 Policy and Institutional Barriers

Government policies in Sri Lanka are still not strong or specific enough to promote women's participation in watershed management. Many environmental and watershed policies remain gender-blind. While gender equality is recognized in theory, implementation mechanisms, monitoring indicators, and accountability frameworks are often lacking. Institutional cultures also tend to reinforce male dominance, especially in leadership and fieldwork roles. As a result, women's involvement in on-the-ground activities and field operations remains limited.

8.4.5 The Way Forward: Gender-Inclusive Watershed Management Strategies

To mainstream women in Sustainable Watershed Management requires both policy reform and cultural change. Women are already active in resource stewardship, yet their formal participation remains marginal due to structural and socio-cultural barriers. Addressing these limitations through inclusive strategies can transform watershed governance into a more participatory and sustainable process. Without supportive interventions such as childcare services, flexible meeting schedules, or transportation allowances, women's ability to engage in formal watershed governance remains limited. Public education campaigns, community dialogues, and the inclusion of male allies are crucial to challenging restrictive norms and creating enabling environments for women's leadership in environmental governance. Global organizations such as the FAO, UNDP, and UNEP consistently emphasize that gender-mainstreamed governance leads to more durable and equitable outcomes in natural resource management. Empowering women as active agents, rather than passive beneficiaries, is the key to achieving integrated and sustainable watershed solutions.

8.4.5.1 Enhancing Women's Leadership and Representation

To enhance women's participation in Integrated Watershed Management (IWM), a phased approach can be adopted by initially focusing on household-level tools and gradually expanding their involvement to community and regional levels. In areas where these tools are actively implemented, middle-aged and elderly women are typically the most engaged. To ensure effective participation, women must be empowered through access to financial resources, training, and capacity-building programs.

Discussions revealed that women are more likely to engage in these initiatives when they receive tangible financial benefits. However, field observations also indicated that many projects tend to underperform over time due to a lack of post-implementation monitoring—for example, rainwater harvesting projects initiated by non-governmental organizations. As a solution, women can be actively involved in post-project monitoring and documentation, areas where male participation is often limited. This not only improves project sustainability but also strengthens women's roles in environmental governance.

Additionally, introducing gender quotas in watershed management committees and user associations can ensure women's representation in decision-making processes. Evidence from South Asia demonstrates that including women in resource governance leads to more sustainable and equitable environmental outcomes. Leadership development programs and peer-to-peer mentoring can further empower women to take on influential roles within community-based watershed organizations (Nimavat, 2018).

8.4.5.2 Capacity Building and Technical Training

Improving the technical capacity of women is essential, as they often lag slightly behind men in technical areas. To address this gap, training and capacity development programs should be carefully planned and implemented. After providing initial training, the same group of women should continue to receive further opportunities to build on their skills through advanced and ongoing programs.

These programs must be tailored to women's specific needs and challenges—such as limited mobility, time constraints due to caregiving duties, and restricted access to resources. Targeted capacity-building in technical knowledge, entrepreneurship, and financial literacy is crucial to empower women and enhance their participation (Blokland et al., 2019). Additionally, improving digital literacy should be a key focus area to ensure women are not left behind in an increasingly technology-driven world.

8.4.5.3 Policy Reforms and Institutional Support

Mainstreaming gender into watershed management policies is essential. This involves revising existing policies to formally recognize women's roles and actively promote their participation. Incorporating gender-disaggregated data in policy development will support evidence-based interventions that reflect on-the-ground realities. Additionally, integrating gender-specific performance indicators into program monitoring and evaluation can help ensure accountability for inclusive practices.

At the institutional level, the involvement of female staff in watershed initiatives should be strengthened. To encourage their participation—particularly in field-based activities—providing additional allowances or incentives for female workers can be an effective strategy. This support will help overcome barriers and promote greater gender equity in watershed management efforts.

8.4.5.4 Economic Empowerment and Support for Women

Many rural women involved in watershed projects are financially dependent on male family members and are primarily engaged in household and caregiving responsibilities. If the implementation of watershed tools can provide financial benefits to these women, it would not only offer them a degree of financial independence but also contribute additional income to their households. This, in turn, would motivate women to engage more actively and effectively in these initiatives.

Limited financial autonomy also restricts women's participation in income-generating activities linked to watershed projects, such as ecotourism, organic farming, and nursery development. Promoting women-led enterprises in these areas presents new opportunities that align with watershed conservation goals. Strengthening women's cooperatives and farmer organizations—through funding, access to extension services, and improved market linkages—can enhance livelihoods while supporting sustainable land-use practices.

8.5 Conclusion

This chapter explored key Integrated Watershed Management (IWM) tools aligned with the three pillars of Sustainable Watershed Management (SWM)—environmental, social, and economic-with a focus on their potential to reduce flood, drought, and disaster risks. It also examined the extent and nature of women's participation in these tools to achieve the broader objectives of SWM.

A total of 20 IWM tools were identified through an extensive literature review and further validated through field observations and stakeholder discussions. These tools were categorized under the three SWM pillars: 10 environmental tools, 5 social and governance tools, and 5 economic tools. Although the study did not scientifically quantify the current level of women's engagement, qualitative evidence suggests that many of these tools are not fully utilized, and women's participation remains relatively low. In cases where tools are actively implemented, they are often male-dominated, with women only partially involved in certain phases of implementation.

Despite the wide availability and theoretical relevance of IWM tools, women's roles in achieving SWM outcomes are frequently overlooked, resulting in a lack of inclusivity and gender sensitivity. This highlights a significant gap between theoretical frameworks and on-the-ground application, where the principles of IWM and SWM are not consistently upheld in practice. To move towards truly sustainable and inclusive watershed management, there is a pressing need to bridge this gap by strengthening the participation of women in the design, implementation, and governance of IWM tools.

8.6 Future Research Directions

Future research should aim to quantify the current level of women's engagement in Sustainable Watershed Management across different contexts. Comparative studies between various regions of the country would help identify regional disparities and context-specific challenges. There is also a need to comprehensively investigate the factors that hinder women's active participation, including socio-cultural, institutional, and economic barriers. Additionally, future studies should focus on developing and testing participatory strategies that effectively enhance women's involvement in watershed management, ensuring their voices are meaningfully integrated into planning, implementation, and decision-making processes.

Acknowledgements The authors would like to express their sincere gratitude to the local community of the Belihuloya region for their support during the field observations and discussions. Special appreciation is extended to individuals who generously provided valuable insights and evidence from various parts of the country, including Mr. Chinthaka Dissanayake from the Puttalam District; Ms. Thamburu Mihisarani and Ms. Nadeeka Samanmali from the Badulla District; Mr. Kaveesha Malshan from the Monaragala District; and Mr. Nalin Sampath from the Anuradhapura District.

References

Aban, J. L. (2019). Stakeholders' awareness and initiatives to achieve sustainable watershed resources management: A brief cross sectional case report. *EurAsian Journal of BioSciences.*

Anbumozhi, V., Radhakrishnan, J., & Yamaji, E. (2005). Impact of riparian buffer zones on water quality and associated management considerations. *Ecological Engineering, 24*(5), 517–523.

Arun, S. V., & Kulkarni, S. S. (2011). Watershed management—A means of sustanbile developement—A case study.

Azmi, N., Ahmed, S. F., & Syed, Z. A. (2024). Sustainable water management: Understanding a theoretical perspective. *Environmental Reports: An International Journal.*

Basuki, T. M., Nugroho, H. Y. S. H., Indrajaya, Y., Pramono, I. B., Nugroho, N. P., Supangat, A. B., Indrawati, D. R., Savitri, E., Wahyuningrum, N., & Purwanto. (2022). Improvement of integrated watershed management in Indonesia for mitigation and adaptation to climate change: A review. *Sustainability, 14*(16), 9997.

Blokland, M., Alaerts, G., Kaspersma, J., & Hare, M. (2019). *Capacity development for improved water management.* CRC Press.

Bohluli, M., Teh, C., Hanif, A. H. M., & Rahman, Z. A. (2015). Review on the use of silt pits (contour trenches) as a soil and water conservation. *Advances in Tropical Soil Science,* 63–80.

Dasthagir, K. G. (2013). Women inclusion in participatory irrigation management: Predilections and predicaments. *IASSI-Quarterly, 32*(4), 45–59.

Erdogan, R. (2013). Stakeholder involvement in sustainable watershed management. In *Advances in landscape architecture.* IntechOpen.

Gonenc, I. E., Wolflin, J. P., & Russo, R. C. (2014). *Sustainable watershed management.* CRC Press.

Imperial, M. T. (2005). Using collaboration as a governance strategy: Lessons from six watershed management programs. *Administration & Society, 37*(3), 281–320.

Joshi, P. K., Jha, A. K., Wani, S. P., Sreedevi, T. K., & Shaheen, F. A. (2008). Impact of watershed program and conditions for success: A meta-analysis approach. *Global Theme on Agroecosystems, Report, 46*.

Kannan, N., Loveciya, S., Thusalini, A., & Thiruvaran, T. (2024). *Climate-smart water management strategies with special emphasize to micro-irrigation practices for sustainable use of water for irrigation in Sri Lanka: A conceptual framework* (176 p.). YSF Thematic Publication.

Kanyi, M., Lawver, D. E., Ulmer, J., & Murimi, M. (2017). Attitudes and adoption of rainwater harvesting: Influence of gender, awareness, and social status. *Journal of International Agricultural and Extension Education, 24*(1), 63–73.

Khandelwal, N., Patel, Y., & Choudhary, M. (2024). Women's engagement in soil and water conservation: A research perspective. *Gujarat Journal of Extension Education, 37*(1), 70–75.

Khurana, L., Kumar, K., & Sing, D. P. (2024). Exploring watersheds: Assessing drainage basin characteristics and management approaches for channel and overland flows. *International Journal of Landscape Planning and Architecture, 10*(1), 24–34.

Kunihira, R. (2014). *Socio-cultural factors affecting women's participation in watershed resources management in Chahi catchment, South-Western Uganda*. MSc thesis in Integrated Watershed Management.

Macherera, M., & Chimbari, M. J. (2016). A review of studies on community based early warning systems. *Jàmbá: Journal of Disaster Risk Studies, 8*(1), 206.

Manjur, K., Amare, H., HaileMariam, G., & Tekle, L. (2014). Livelihood diversification strategies among men and women rural households: Evidence from two watersheds of Northern Ethiopia. *Journal of Agricultural Economics and Development, 3*(2), 17–25.

Nimavat, S. (2018). Gendered landscapes: The imperative of women's inclusion in India's watershed management strategies. https://ijrar.com/upload_issue/ijrar_issue_20544425.pdf

Paolisso, M., Gammage, S., & Casey, L. (1999). Gender and household-level responses to soil degradation in Honduras. *Human Organization, 58*(3), 261–273.

Qianxiang, L., Igbokwe, K. N., & Jiayong, L. (2005). Community-based integrated watershed management. *Chinese Journal of Population Resources and Environment, 3*(1), 60–64.

Renganayaki, S. P., & Elango, L. (2013). A review on managed aquifer recharge by check dams: A case study near Chennai, India. *International Journal of Research in Engineering and Technology, 2*(4), 416–423.

Richardson, C. J., Flanagan, N. E., Ho, M., & Pahl, J. W. (2011). Integrated stream and wetland restoration: A watershed approach to improved water quality on the landscape. *Ecological Engineering, 37*(1), 25–39.

Rigonato, M. B., De Mello, K., Valente, R. A., & Faria, L. C. (2023). Payment for water-related ecosystem services as a strategic watershed management approach. *Journal of Environmental Protection, 14*(8), 660–684.

Ronderos, A. M. G. (1989). Alternative strategies for the increased participation of women in reforestation efforts in Hojancha, Costa Rica. *NWSA Journal*, 762–764.

Samaraweera, W., Dharmadasa, R., Kumara, P., & Bandara, A. (2024). Evidence of climate change impacts in Sri Lanka—A review of literature. *Sri Lanka Journal of Economic Research, 11*(2).

Sharda, V. N., & Venkateswarlu, B. (2007). Crop diversification and alternate land use systems in watershed management. In: *Best-bet options for integrated watershed management* (pp. 111–128). Proccedings of the Comprehensive Assessment of Watershed Programs in India. https://oar.icrisat.org/186/1/35_2010_CPE167_Bestbet_options.pdf

Singh, S. (2018). *Gender mainstreaming in integrated watershed management project: A study in hill region of Uttarakhand*.

Surasinghe, T., Kariyawasam, R., Sudasinghe, H., & Karunarathna, S. (2019). Challenges in biodiversity conservation in a highly modified tropical river basin in Sri Lanka. *Water, 12*(1), 26.

Ullah, A., Sam, A. S., Sathyan, A. R., Mahmood, N., Zeb, A., & Kächele, H. (2021). Role of local communities in forest landscape restoration: Key lessons from the Billion Trees Afforestation Project, Pakistan. *Science of the Total Environment, 772*, 145613.

Van Der Zaag, P., & Savenije, H. G. (2004). *Principles of integrated water resources management*. UNESCO-IHE Delft, The Netherlands.

Wang, Y., & Wu, J. (2018). An empirical examination on the role of water user associations for irrigation management in rural China. *Water Resources Research, 54*(12), 9791–9811.

Wang, G., Mang, S., Cai, H., Liu, S., Zhang, Z., Wang, L., & Innes, J. L. (2016). Integrated watershed management: Evolution, development and emerging trends. *Journal of Forestry Research, 27*, 967–994.

Wani, S. P., & Garg, K. K. (2009). *Watershed management concept and principles.*

Part III
Climate Crisis and Gender in East-Asia

Chapter 9
Women Empowering in Biodiversity Conservation and Climate Change Adaptation: A Case Study at the Ramsar Site of Cambodia

Leakhena Thoeun, Chealy Chet, Chansopheak Keng, and Serey Sok

Abstract Wetland management and biodiversity conservation are important for environmental sustainability and climate change adaptation. This chapter aims to adopt a concept of empowerment to explore roles and responsibilities of women attributed to wetland management and climatic impact adaptation, from the perspectives of both men and women, considering the personal, collective, and rational dimensions. Quantitative and qualitative data were collected in the Siem Bouk district, Stung Treng Province, as part of the Ramsar Site. The study site is adjacent to the Mekong River and borders Laos. A survey was conducted using a structured questionnaire to collect qualitative data from 233 households. We also collect qualitative data among key stakeholders, including focus group discussions, in-depth interviews, and key informants. The research concludes that women remain relatively underrepresented in biodiversity conservation and adaptation to climate change. The analysis of the dimensions of empowerment generates three findings. First, women and men share moderate personal empowerment. While men are more popular, women have more self-confidence in biodiversity conservation and climate change adaptation. Second, women and men assessed the neutral degree of their rational empowerment as one in which they could make decisions without considering all the implications. Third, women and men assessed a neutral degree of collective empowerment regarding teamwork and social activities, but men are more active. This empirical study helps generate knowledge about women's empowerment, anchored with three principles. Key stakeholders can use the results to empower women in the Ramsar Site of Cambodia and beyond.

Keywords Biodiversity conservation · Climate change · Women empowerment · Ramsar Site · The Mekong River · Cambodia

L. Thoeun · C. Chet · C. Keng · S. Sok (✉)
Royal University of Phnom Penh (RUPP), Toul Kork, Phnom Penh, Cambodia
e-mail: sok.serey@rupp.edu.kh

Md. A. Abedin et al. (eds.), *Climate Crisis in South and East Asia*, Sustainability Solutions, https://doi.org/10.1007/978-3-032-01012-4_9

9.1 Background

Wetlands are productive and diverse ecosystems and provide various ecological services, including water purification, flood control, shoreline stabilization, and habitat provision for numerous plant and animal species (UNEP-WCMC et al., 2018). According to the Ramsar Convention on Wetlands (2018), wetlands are the most biologically diverse ecosystems, harboring a substantial portion of the world's species. In particular, wetland management and biodiversity conservation are important for environmental sustainability and the overall well-being of ecosystems (Ramsar Convention Secretariat, 2013). As a result, efficient wetland management practices are crucial to maintaining these essential functions and preserving biodiversity. By implementing conservation strategies and adopting sustainable use approaches, we can ensure the continued availability of wetland resources and safeguard their ecological integrity (UNEP-WCMC et al., 2018).

Biodiversity conservation, for example, the protection of wetland ecosystems, plays a critical role in environmental sustainability. Biodiversity forms the foundation of ecosystem stability and resilience, contributing to ecosystems' overall health and functioning (CBD, 2020b). It supports crucial ecosystem services such as pollination, nutrient cycling, and pest control, which are vital for human well-being and agricultural productivity (IPBES, 2019). Moreover, wetland conservation and biodiversity preservation have direct benefits for human societies. Wetlands act as natural buffers against extreme weather events, absorbing and storing excess rainfall, thereby reducing the risk of floods. They also act as filters, purifying water and improving its quality (UNEP-WCMC et al., 2018). Moreover, wetlands provide valuable recreational and cultural opportunities, supporting tourism, fishing, and traditional practices. Additionally, they serve as carbon sinks, aiding in climate change mitigation by sequestering and storing carbon dioxide (IPBES, 2019). Promoting wetland management and biodiversity conservation aligns with global sustainability frameworks such as the United Nations Sustainable Development Goals (SDGs). SDG 15 specifically aims to protect, restore, and promote the sustainable use of terrestrial ecosystems, including wetlands, to halt biodiversity loss and ensure the planet's long-term health (UN, 2020).

Freshwater wetlands have promoted rural livelihoods in Cambodia through daily food consumption and income-generation activities. Biodiversity conservation and wetland management are vital for preserving wetlands' ecological, economic, and social well-being (Torrell et al., 2004). The wetlands cover over 30% of the national territory; the Mekong, Tonle Sap, and Great Lake systems dominate freshwater hydrology. According to Torrell et al. (2004), wetlands are ecologically and highly productive ecosystems that support critical services, including carbon sequestration, flood control, and water purification. With their great contribution, between 1999 and 2018, Cambodia registered five wetlands or Ramsar Sites covering 85,235 ha. In 1999, Cambodia registered three sites of 54,600 ha, one site in 2015, and one in 2018. The wetlands that have been designated as Ramsar Sites include Prek Toal Ramsar Site, Koh Kapik and Associated Islets, Boeung Chhmar, and Associated River System and Floodplain, and Middle Stretches of Mekong River North of Stung Treng, and Stung Sen Ramsar Site.[1]

However, wetlands or the Ramsar Sites have been socially and economically beneficial to rural livelihoods; they are endangered due to agriculture-related activities, land reclamation, pollution, illegal harvest of forest products, illegal fishing, overfishing, and population growth (Sophal, 2004). The local capacity to access and use wetland-related resources in the communities remained influenced by laws, norms, structures, processes, and institutional arrangements (Ritesh et al., 2011). Zheng et al. (2014) believe that community engagement and key stakeholders are significant for wetland management, but the opportunities remain limited. Moreover, the roles and responsibilities of the key stakeholders are not shared among them (Ratner et al., 2013). In addition, key stakeholders (MEA, 2005) did not sufficiently contribute to the decision-making process in accessing water-related resources and biodiversity. Simultaneously, women did not play sufficient roles in wetland management, especially at the commune, district, and national levels. Women and men could equally participate in decision-making at the household level in managing water-related resources collected from the wetlands (Thuon et al., 2023).

The chapter aims to promote gender equality and empower women in wetland management and biodiversity conservation in Siem Bouk District, Stung Treng Province. We adopted the three dimensions by Rowlands (1997) to understand women's and men's engagement in wetland management, focusing on personal empowerment, rational empowerment, and collective empowerment. The analysis of this chapter includes personal self-confidence and capacity to empower the community as personal empowerment, authority in decision-making, rational empowerment, and capacity to be involved in teamwork, and involvement in social activities as collective empowerment. The chapter is structured into (1) Background, (2) Women's Empowerment in Wetland Management, (3) Study Area and Methods, (4) Results and Findings, and (5) Conclusion and Implications.

9.2 Women's Empowerment in Wetland Management

In developing countries, women often face discrimination and inequality in accessing resources, information, and decision-making opportunities related to wetland management (UN Women, 2014). However, Duflo (2012) identified women's empowerment and economic development as complex. Wilkinson (1998) identified empowerment as a popular management concept that aims to address workplace issues but lacks a detailed discussion of implementation challenges and context specificity. In developing countries, women are crucial in wetland management, as they are often responsible for water and food-related activities, such as farming, fishing,

[1] See detail https://rsis.ramsar.org/ris-search/?language=en&f[0]=regionCountry_en_ss%3ACambodia&pagetab=2.

and collecting water (Setti et al., 2018). Empowering women in wetland management can contribute to economic development and poverty reduction, as women's access to and control over wetland resources can increase their income and improve their livelihood. Women also lead to better environmental conservation outcomes, as women often have a deeper understanding of the ecosystem and its resources and may be more likely to prioritize conservation over short-term economic gains (Chowdhury et al., 2022).

The participation of various stakeholders, including women, is crucial for effective wetland management and conservation efforts. In many societies, women play key roles in natural resource management, including wetlands. Their involvement enhances diverse perspectives, local knowledge, and community-based approaches, contributing to more inclusive and sustainable management practices (IPBES, 2019; UNEP-WCMC et al., 2018). Despite the recognized significance of women's roles in environmental conservation, their participation in wetland management and biodiversity conservation has received limited attention in the literature (UN Women, 2017). Women possess unique knowledge, skills, and perspectives essential for sustainable resource management (Agarwal, 2010; Rocheleau et al., 2013). Their traditional roles as caregivers and primary resource users often give them direct experience and a deep understanding of local ecosystems (Dankelman, 2012).

The involvement of women in wetland management and biodiversity conservation can contribute to more holistic and inclusive decision-making processes, leading to better conservation outcomes Ali & Grobicki (2016). Their perspectives can enhance the identification of community needs, improve the design of conservation strategies, and promote the sustainable use of resources (Rocheleau et al., 2013). However, numerous challenges and impediments hinder women's full and effective participation in wetland management and biodiversity conservation. These challenges include gender inequalities, limited access to resources and decision-making platforms, cultural norms and biases, and inadequate recognition of women's contributions (Agarwal, 2010). To overcome these challenges, it is crucial to address gender biases and promote gender equality in conservation initiatives. This involves creating supportive and enabling environments that empower women and provide them with equal opportunities for participation, leadership, and decision-making (UN Women, 2016).

Capacity-building programs and training opportunities tailored to women's needs can enhance their skills and confidence in engaging in conservation activities (Agarwal, 2010). Establish gender-responsive policies and frameworks ensure recognition and value of women's contributions and representation in relevant institutions and decision-making bodies (Bryan et al., 2024). The conceptual framework of this research is based on the concept of empowerment developed by Rowlands (1997). The framework incorporates three main dimensions of empowerment: personal, collective, and rational. These dimensions provide a comprehensive understanding of empowerment and highlight the strategies and mechanisms through which individuals and groups can gain control over their lives and work towards achieving their goals.

We adopted the three dimensions introduced by Rowlands (1997) for this paper to explore and understand the different dimensions of empowerment in the context of wetland management and biodiversity conservation. It recognizes the importance of addressing power imbalances, promoting participation, collaboration, and collective action to enhance women's empowerment and promote gender equality. The framework provides a theoretical foundation for analyzing the research findings and developing recommendations for promoting empowerment and sustainable practices in wetland management. According to Rowlands (1997), empowerment is a process of change that involves increasing the capacity of individuals and groups to make choices and to transform those choices into desired actions and outcomes. This process gives people more control over their lives and enables them to participate fully in the decisions that affect them. Empowerment can be seen as a bottom-up approach that emphasizes the importance of grassroots participation and community mobilization in social change. It involves building the skills, knowledge, and resources necessary for individuals and groups to take control of their lives and to work towards their goals. Empowerment also involves challenging power relations and the distribution of resources and opportunities in society, and it is closely linked to issues of social justice and human rights (Table 9.1).

Empowerment is a complex and multifaceted concept defined and operationalized in various ways in the literature. According to Kabeer, empowerment refers to "people's ability to make strategic life choices in situations where this ability was previously denied to them." This definition emphasizes giving people more control over their lives and the ability to make their own decisions (Kabeer, 2001).

Table 9.1 Conceptualizing women empowerment in wetland management

Dimensions	Description
Personal empowerment	This dimension focuses on enabling individuals to have greater control over their lives. It involves building self-esteem and confidence and developing the necessary skills to take action and make decisions. Personal empowerment emphasizes self-awareness, recognizing strengths and weaknesses, and identifying the available resources and opportunities
Collective empowerment	Collective empowerment emphasizes the importance of groups working together to achieve common goals. It involves fostering solidarity and cooperation and creating social networks and institutions that support collective action. This dimension aims to build a sense of shared identity, common purpose, and mutual support among individuals, enabling them to advocate for their rights and interests collectively
Rational empowerment	Rational empowerment focuses on the capacity of individuals and groups to make informed decisions and take action based on their own needs and interests. It emphasizes developing critical thinking skills, access to information and resources, and engagement in dialogue and negotiation with others. Rational empowerment challenges power imbalances and creates opportunities for meaningful participation and influence in decision-making processes

Adopted from Rowlands (1997)

Empowerment is “the process of increasing individuals’ or groups’ capacity to make choices and transform those choices into desired actions and outcomes. This definition highlights the importance of developing skills and capabilities that enable individuals to take action and achieve their goals” (Zimmerman, 1995). Similarly, Wallerstein defines empowerment as “a social action process that promotes participation of people, organizations, and communities in gaining control over their lives and health.” This definition emphasizes the importance of collective action and community participation in empowerment (Wallerstein & Duran, 2006).

These definitions illustrate empowerment’s multidimensional nature, including individual agency, skills and capabilities, and collective action. Empowerment is a procedure that enables individuals and communities to develop the skills, knowledge, and resources necessary to take control of their lives and work towards their goals (Labonte & Feather, 1996). Another definition of empowerment comes from Freire and Macedo (1996); within, empowerment is a critical consciousness-raising process that transforms individuals and society. This definition highlights the importance of education and awareness-raising in empowering individuals to challenge oppressive structures and norms (Continuum, 1996, n.d.). In the context of women’s empowerment, the United Nations Development Program (UNDP) defines empowerment as “the process of growing the capacity of individuals or groups to make choices and to change those choices into desired actions and outcomes” (UNDP, 2010). This definition highlights the importance of agency and the ability to exercise one’s choices and actions.

9.3 Study Area and Methods

Stung Treng Province is located in Northeast Cambodia, with its northern boundary forming the international border with Laos. The province shares borders with Mondulkiri and Kratie to the south, Ratanakiri to the east, and Preah Vihear and Kampong Thom to the west. According to the 2019 national census, the province has a population of 159,565 inhabitants or 76,472 families, with approximately 9.5% of the population being indigenous people. The research was conducted in the Siem Bouk District of Stung Treng Province. The district is enriched with biodiversity and wetland ecosystems, which include the Ramsar Site wetland management area. The Ramsar Site wetland management area is of significant ecological and social importance, as it is home to a diversity of plant and animal species and is a crucial source of food, water, and other resources for the local community. The study district was selected based on its proximity to the Ramsar Site wetland management area and its demographic characteristics. Two communes in the Siem Bouk District were chosen for the study: Koh Sampeay and Sre Krasang. These communes are located near the Ramsar Site wetland management area and the significant contribution of women in community-based organizations (CBOs) (Maps 9.1 and 9.2).

The research was conducted in two communes of Siem Bouk District, Stung Treng Province, interviewing 233 households regarding demographic information

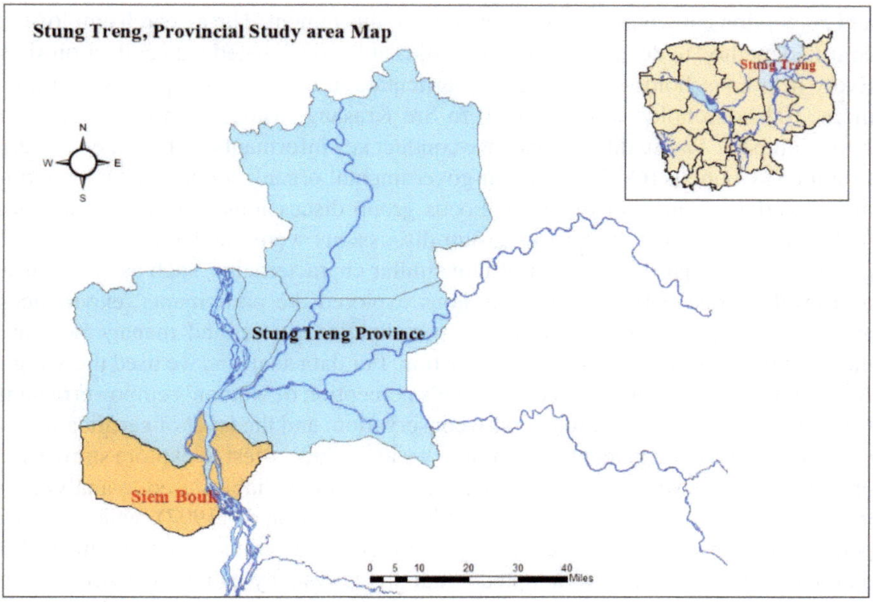

Map 9.1 Map of Stueng Treng and Siem Bouk Districts

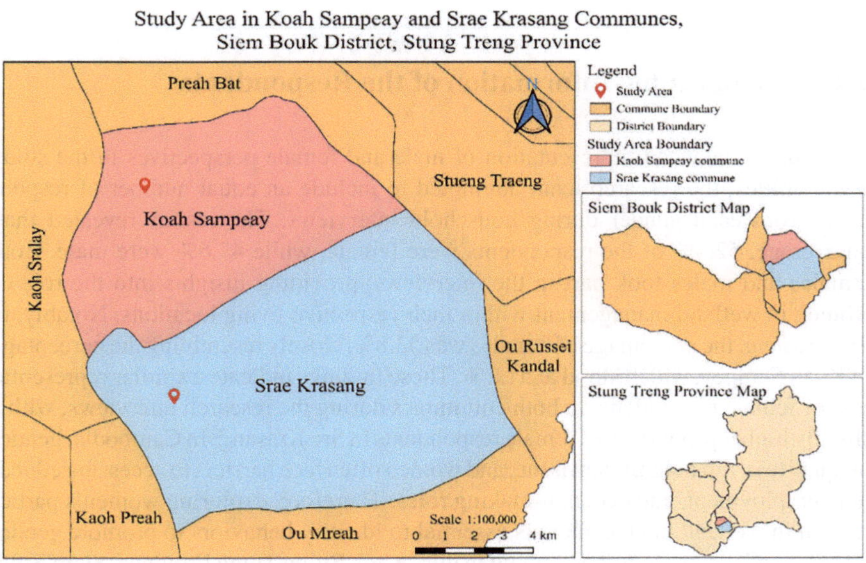

Map 9.2 Maps of Koh Sampeay and Sre Krasang communes

and perceptions of women's roles in wetland management. The research employed a stratified method to determine the households at the designated data collection sites. Every fifth household was selected to participate in the interview process. Furthermore, 104 households were sampled in Sre Krasang, while 129 were sampled in Koah Sampeay. For qualitative data, we conduct key informants with women leaders, Commune Councils (CoCs), and non-governmental organizations (NGOs) working in the wetland management sector. Focus group discussions were also conducted with women and men. The focus group discussions were conducted with a small group of women participants who share similar characteristics, such as age, education level, or occupation. The discussions explored the participants' experiences, attitudes, and perceptions of women's participation in wetland management and their experiences of obstacles to participation. For data analysis, we used the Weight Average Index (WAI) to assess the village's perception of women's empowerment in wetland management and biodiversity conservation, and the level of certification on current wetland management. In addition, we also used a *t*-test to explore similarities or differences between women and men regarding the variables used to analyze the three dimensions of empowerment introduced by Rowlands (1997), which include personal empowerment, collective empowerment, and rational empowerment. For qualitative data analysis, we adopted situation analysis by using the results of key informants and focus group discussion to describe and explain the quantitative data collected through the survey.

9.4 Demographic Information of the Respondents

To ensure a balanced representation of male and female perspectives in the study communities, the research team attempted to include an equal number of respondents from each gender during household interviews. The results revealed that, on average, 52.4% of the respondents were female, while 47.6% were male. Both females and males took part in the interviews, providing insights into the role of women in wetland management within their respective living locations. Notably, in Sre Krasang, the percentage of females was 53.8%, closely resembling the percentage in Koah Sampey, which stood at 51.2%. These findings indicate a similar representation of female respondents in both communes during the research interviews, with a slightly higher proportion of female respondents in Sre Krasang. In Cambodia, gender inequality is still a significant issue, and women often face barriers to accessing education, employment, and decision-making roles. Therefore, exploring women's participation in wetland management is essential to identify behaviors to promote gender equality and women's empowerment in this sector. Stung Treng Province, Siem Bouk District, is in northeastern Cambodia. Many essential wetlands are important ecosystems that provide many ecosystem services, such as water purification, climate regulation, and support for biodiversity. However, these wetlands face significant threats from human activities, such as land-use change, pollution, and overexploitation.

Therefore, it is important to study the impediments to women's participation in wetland management in this region to ensure that wetlands are conserved and sustainably managed (Maltby, 1991; Ramsar Convention Secretariat, 2018). Despite the importance of wetlands and gender equality in Cambodia, there is currently limited research on the impediments to women participating in wetland management in this region. Therefore, a study in Siem Bouk District can help fill this research gap and provide essential insights into the challenges and opportunities for promoting women's participation in wetland management in Cambodia. Studying the impediments to women's participation in wetland management in Siem Bouk District, Stung Treng Province, can help promote sustainable wetland management, gender equality, and women's empowerment in Cambodia.

The respondents were asked to indicate their marital status, and an independent *t*-test was conducted to examine the differences between male and female respondents. The results revealed that most respondents, 72.1%, were already married. It is interesting to note that there were differences in the marital status distribution between males and females. Among the married respondents, 75.7% were male, while 68.9% were female. This indicates that more males were married than females in the sample. In addition to married individuals, the researchers also examined other marital statuses. They found that 21.9% of the respondents were single and relied on support from their families. This suggests that a significant portion of the participants had not yet entered into a marital union and depended on their families for various forms of support. Furthermore, the study found that 3.9% of the respondents were widows or widowers, indicating that a small percentage of the participants had experienced the loss of their spouse. Additionally, 1.7% of the respondents had gone through a divorce, signifying that a minority had previously terminated their marriage. It is also noteworthy that the proportions of single, divorced, and widowed respondents were similar for both males and females.

In addition, the respondents were asked to provide information about their age, education, household members, resettlement land size, and agricultural land size. The research aimed to understand the different situations of males and females, so an independent *t*-test was applied to determine the significance of the differences between the two groups of respondents. Additionally, a one-sample *t*-test was used to compare the overall mean education of the respondents with the national data on education policy. The analysis of detailed demographic information among male and female respondents revealed that the average age of the respondents, 36 years old, did not show any significant difference between the two groups. Regarding household respondents, the average agriculture land size was 19,250 m^2 (1.9 ha), and there was no significant difference between male and female respondents. Similarly, the average resettlement land size was 1624 m^2 (0.16 ha), and there was no significant difference between the two groups of respondents. The average household size in the study area was 4.7 persons per household, and there was no significant difference between males and females.

Education is a crucial attribute of respondents' demographic information that supports social development and economic growth. The analysis showed that the overall group of respondents had an average education level of grade 7.5, and there was no significant difference between male and female respondents in the study area. This means that, on average, the respondents had completed primary education. However, a one-sample t-test indicated that the overall mean education of the respondents was significantly lower than the national education policy, which states that both females and males should complete basic education at grade 9 (p-value $= 0.000$). This result suggested that the education situation of the respondents in the study area has not yet met the national requirements. An independent t-test was used to measure the difference between the primary occupations of female and male respondents. All participants were asked about their primary job in their household, with seven suggested types of jobs. The analysis revealed that gardening was the main primary occupation in the study area, accounting for 42.9% overall, followed by rice farming at 39.1% among household respondents. The other occupations, such as fishing, livestock raising, employment as workers, self-employment, and regular employment, had smaller percentages overall.

Regarding gender differences, 45% of males identified as gardeners for crops and vegetables, which was higher than the 41% of females in the same occupation. On the other hand, 41.8% of females were engaged in rice farming as their primary job, surpassing the 36% of males in this category. In the category of employed workers, 7.4% of female respondents identified themselves as such, compared to 4.5% of male respondents. None of the female respondents identified as fishermen, whereas 0.9% of male respondents did. Regarding livestock raising, 2.5% of female respondents considered it their primary occupation, while 1.8% of male respondents shared the same job. Among employed workers, which includes individuals working as laborers or daily wage workers, 7.4% of female respondents identified themselves in this category, while 4.5% of males fell under it. Employees in jobs such as NGO staff, government officers, or other positions with monthly salaries were identified by 9.9% of males, higher than the 4.9% of females in the same category. Lastly, 2.5% of female respondents identified themselves as self-employed, while 1.8% of males did the same.

The results suggest that agriculture still holds an important role as the primary job in households in the study area. Despite the possibility of earning more income from other types of jobs, such as fishing or employment, the respondents still consider agriculture the main occupation for their households. Female and male household respondents reported five main categories of secondary jobs. Among the nine categories of secondary jobs surveyed, rice farming emerged as the most prevalent, comprising 57.5% of the secondary jobs for household respondents, followed by livestock raising at an overall rate of 55.8%, employed workers at 42.9%, gardeners at 41.6%, and fishermen at 29.6%. Furthermore, the respondents had an average of 2.54 secondary jobs in the study area. Notably, the group of males exhibited a significantly higher number of secondary jobs than females, with statistical significance at a p-value of 0.05. This finding suggests that males have more opportunities to engage in multiple secondary jobs compared to females.

9.5 Results and Findings

9.5.1 Personal Empowerment

Two questions with five-scale indicators to assess the personal empowerment of women and men in engaging biodiversity conservation and climate change adaptation: "To what extent do you agree with the following statements regarding personal self-confidence?" and "To what degree do you agree with the following statements regarding your capacity for community empowerment?" Overall, the respondents illustrated in Fig. 9.1 agreed that women and men had similar personal self-confidence, such as being popular, hard-working, honest, and dynamic, throughout the creative statements (p-value = 0.475). However, women and men rated moderate agreement regarding working with emotions. t-test shows that while men were more popular (p-value = 0.032), women shared a higher degree of being exuded with self-confidence (p-value = 0.039). Exuding confidence is all about being comfortable and strong in who you are without needing any deviation from it or distraction. Concerning the capacity to empower the community, women and men similarly assessed neutral degrees when empowering their community for wetland management (p-value = 0.982) (Fig. 9.2).

Both women and men could work together confidently to support their community development in wetland management efforts. The analysis from the interviews during the fieldwork suggests that gender did not affect their significant role in determining the capacity for community empowerment in this context. It highlights the potential for collaboration and cooperation between genders to address wetland management challenges and achieve community empowerment goals effectively. Nevertheless, men still held personal empowerment over women since they were mainly the leaders in biodiversity conservation and climate change adaptation at local government and community-based organizations (CBOs). According to a member of the women's group in Sre Krosang commune, even though women had some roles in the communities, for example, Commune Councils (CoCs) and CBOs, men still led most of the work in the communes. Men were chairs and women were deputies, mainly responsible for gender and children-related affairs. At the same time, an NGO officer revealed that some women have more capacity and talent than men in the community. However, women did not dare to challenge and claim their rights and capacity to promote their roles in community development.

In the community, women played a major role in gathering other villagers for meetings and events organized by CoCs, CBOs, and NGOs. Women were good at administrative and financial work supporting the CoCs and CBOs, such as community fishery and forestry. In contrast, men were more powerful in helping the community and their neighbors participate in and organize social events, such as the Khmer New Year and Pchum Ben, Kathin, weddings, parties, and funerals. Men could be fully involved, support the events, and even stay overnight, while women could hardly do so. A study by Sok et al. (2012) found that men could be engaged in patrols for fishery management in the riverine communities of the Tonle Sap Lake. Women were

Fig. 9.1 Personal self-confidence. *Note* Overall assessment (OA); Weight Average Index (WAI) measured on a five-point scale: 0–0.20 = strong disagree (SD), 0.21–0.40 = disagree (D), 0.41–0.60 = moderate (M), 0.61–0.80 = agreed (A), 0.81–1 = strongly agree (SA). Significant at level 0.05

also included on committees in the community fishery, but they were not assigned to be involved in patrol, especially at night, because they were not fully safe. In contrast, women exuded confidence; they stayed positive and communicated well with people in the communities. Women were not good at interacting during the meetings organized by local authorities and NGOs.

9.5.2 *Rational Empowerment*

Overall, women and men agreed they were authorized to decide on wetlands biodiversity conservation and climate change adaptation (p-value = 0.381). However, women and men similarly rated the neutral degree of their rational empowerment as one in which they could make decisions without considering all the implications. They also claimed they preferred to avoid making decisions if they could (Fig. 9.3). According to an interview with an NGO officer, effective wetland management requires inclusive decision-making processes incorporating diverse perspectives. Engaging in women's decision-making was crucial to integrating women and children-related issues into planning and policy implications related to biodiversity conservation

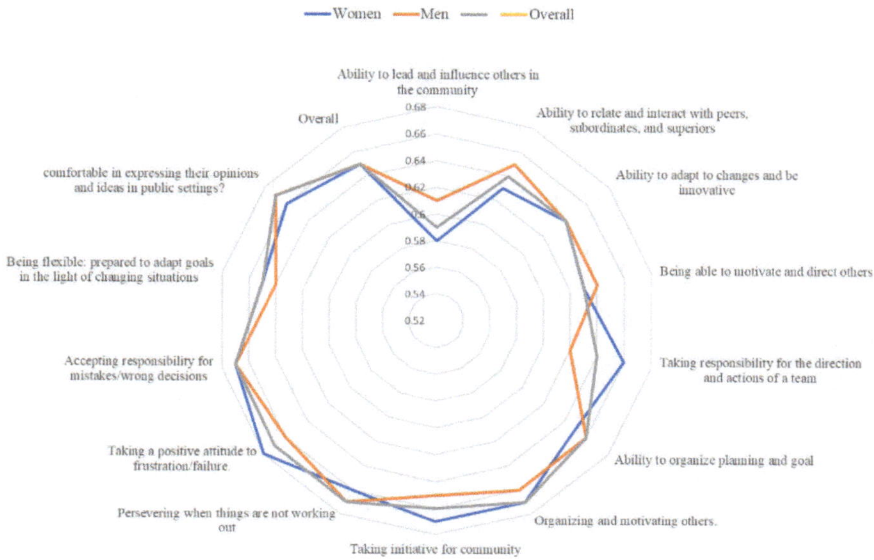

Fig. 9.2 Capacity to empower the community. *Note* Overall assessment (OA); Weight Average Index (WAI) measured on a five-point scale: 0–0.20 = strong disagree (SD), 0.21–0.40 = disagree (D), 0.41–0.60 = moderate (M), 0.61–0.80 = agreed (A), 0.81–1 = strongly agree (SA). Significant at level 0.05

and climate change adaptation. However, women's participation in wetland-related decision-making remained limited in many contexts. Barriers such as socio-cultural norms, restricted access to education and resources, and gender inequalities hinder women's engagement. Promoting gender-inclusive participation and empowering women to contribute to decision-making processes actively is essential for sustainable wetland management. Community-based management initiatives have emerged as successful models for sustainable forest practices. Women's active participation in these initiatives has led to positive social, economic, and environmental outcomes.

Through collective decision-making, women contributed to improved forest governance, sustainable livelihoods, and the preservation of local ecosystems. Recognizing and supporting women's leadership roles in community wetlands can enhance sustainable development and biodiversity conservation in the wetland community (personal community, CoC) Adopting gender-responsive approaches is crucial for addressing gender disparities in forest management. This involves recognizing and addressing women's needs, interests, and roles within forest management policies and practices. Gender-responsive approaches can facilitate women's empowerment, increase access to resources and training, and promote equitable distribution of benefits derived from forest resources (personal communication, focus group discussion). Traditional gender roles often assign men as leaders and decision-makers, while women are expected to be more nurturing and supportive [per-com]. Men might have

Fig. 9.3 Authority in decision-making by gender

been socialized to believe they are naturally better suited for leadership and collaboration, leading to a more positive self-assessment in these areas. The gender biases and stereotypes contributed to women being less involved in teamwork, especially wetland management activities (personal communication, NGOs official).

The focus group discussion reveals that men had stronger leadership skills and influence, and women hesitated to share their concerns and decision-making. However, men were always dominant in their authority for policy implementation, planning, and program interventions; women seldom used their opportunity and available space to negotiate with men. In most cases, women preferred to work based on what men or a group discussion rather than their brothers to challenge during decision-making. In the meantime, young and dynamic women preferred not to be involved in biodiversity conservation and climate change adaptation. In the Ramsar Site, NGOs trained women to work for the community development; they were mainly older in age group and had a lower level of education. The interviews with key stakeholders described decision-making involving personal capacity, decentralized authorities, and resources available to ensure interventions happen. Women and men had opportunities to rank the prioritized activities into the annual investment plan to respond to the local needs for program intervention. However, the villagers were too exhausted to get involved in the decision process for the development of their communities because the CoCs annually allocated the available budget for infrastructure development, especially road construction.

9.5.3 Collective Empowerment

The respondents assessed, as described in Fig. 9.4, a moderate degree of their capacity to be involved in teamwork; men were more capable than women (p-value = 0.000). Simultaneously, the respondents evaluated a high degree of teamwork engagement in achieving outcomes, learning and improving opportunities within the teams engaged, enhancing teamwork skills to accomplish these goals, and closely working with relevant key stakeholders. t-test analysis reveals men were more capable than women about teamwork skills (p-value = 0.000), involvement in teamwork to achieve outcomes (p-value = 0.022), challenges and benefits of collaborating with different stakeholders (p-value = 0.000), open and effective communication within the team (p-value = 0.000), positively perceive your roles and contributions within the teams (p-value = 0.000), approach conflict resolution and problem-solving within the team (p-value = 0.036), contribution to the team's learning and development (p-value = 0.000), crucial part to create the good cooperation (p-value = 0.000), and the leaders bring concerns of community for policy and planning (p-value = 0.000). In the meantime, women and men shared a similar capacity regarding fostering a collaborative and supportive environment within the team (p-value = 0.688), learning and improvement opportunities within the teams engaged (p-value = 0.599), enhanced teamwork skills to achieve these goals (p-value = 0.225), and working closely with relevant stakeholders (p-value = 0.514).

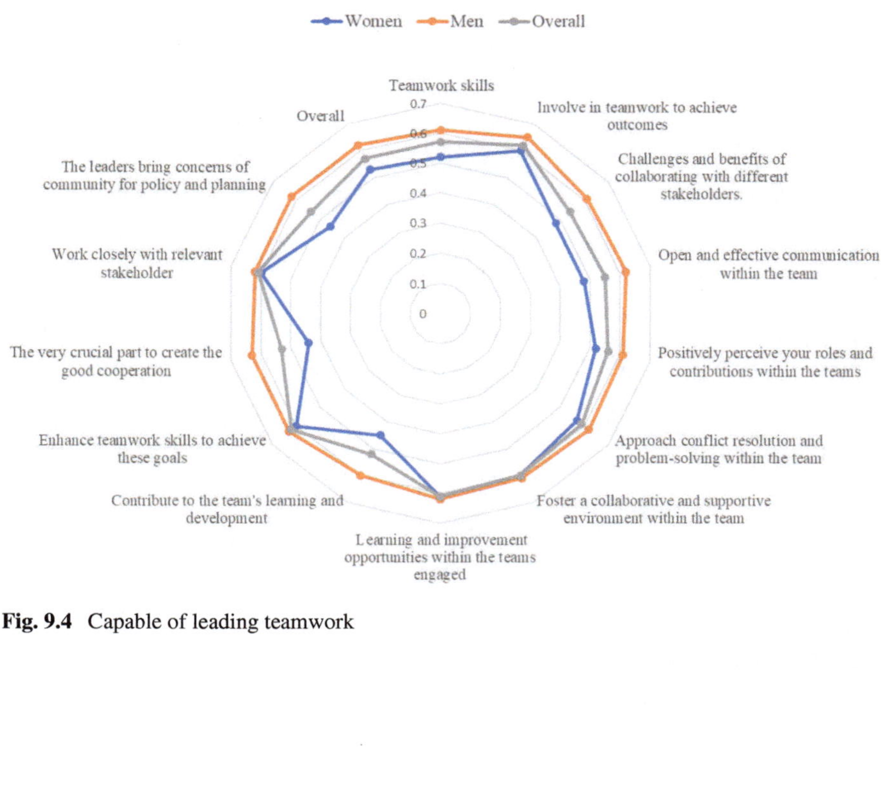

Fig. 9.4 Capable of leading teamwork

Figure 9.5 describes the respondents' involvement in social activities; they moderately assessed a higher percentage of men than women (p-value = 0.000). The respondents also assessed a high degree of promoting gender equality, embracing transparency and accountability in financial matters, motivating and inspiring other community leaders, and helping the welfare of girls/boys/and the elderly. Men shared a higher degree of involvement in social activities regarding participating in all activities in the community in wetland management (p-value = 0.013), protect the interests of the poor (p-value = 0.001), strengthening the quality of public services in the community (p-value = 0.000), embracing transparency and accountability in financial matters (p-value = 0.006), motivation and inspiration to other leaders in community (p-value = 0.0350), ability to join with the community even and solidary for society engagement (p-value = 0.000), having ability to support the community to wetland management (p-value = 0.000). t-test analysis shows that Promoting gender Equality (p-value = 0.126), helping with the welfare of girls/boys/the elderly in the community (p-value = 0.806), and delivering social service to improve quality of life (p-value = 0.058).

An officer at an NGO reveals that it was hard to call for voluntary work to promote biodiversity conservation and climate change adaptation, making a living was the main barrier for them to get involved. In promoting community development, NGOs have worked to build the capacity of the villagers. The community meetings organized

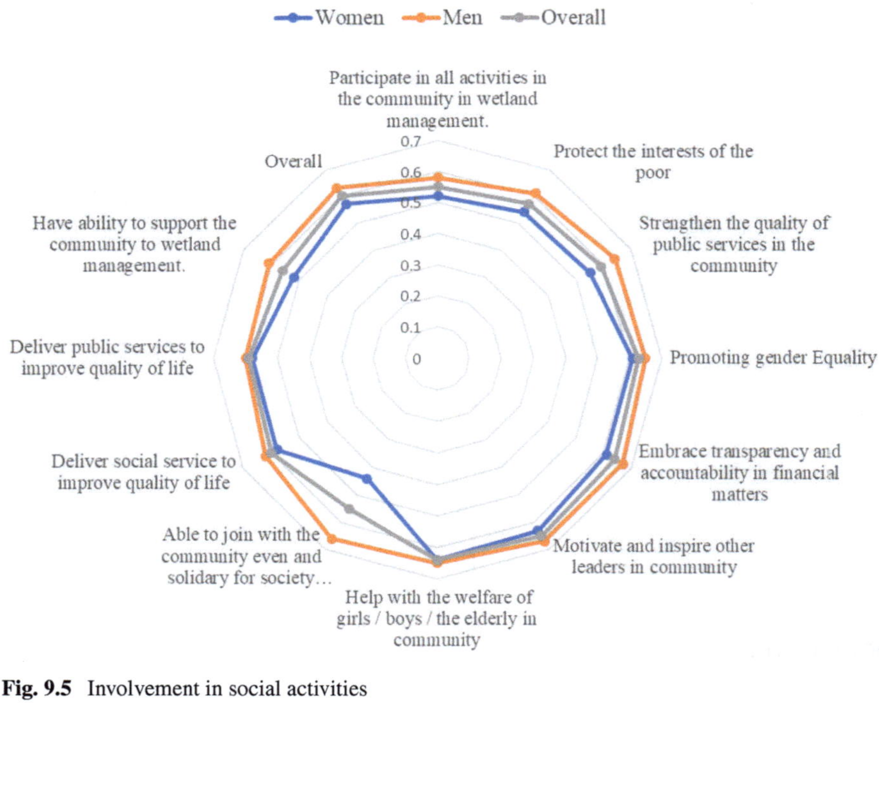

Fig. 9.5 Involvement in social activities

by NGOs created a platform for the local people to discuss solutions collectively. In general, women and men were the best partners for teamwork and carrying out activities to support social work. In promoting biodiversity conservation and climate change adaptation, women and men jointly worked to conduct outreach programs, raise awareness, and share information about the villagers in the communities.

9.6 Conclusion

Based on the findings and general understanding of Cambodia's context, we conclude that women have some opportunities to engage in biodiversity conservation and adaptation to climate change. However, the study concludes that women remain relatively underrepresented in these activities. The research has three findings based on the principles of empowerment by Rowlands (1997): First, women and men share moderate personal empowerment in biodiversity conservation and climate change adaptation. The respondents assessed a high degree of personal self-confidence, but the neutral degree of capacity to empower the community. However, women and men similarly agree they have high personal confidence, but they rated a neutral degree of having capacity and agreed that their capacity to empower the community. While men are more popular, women have a higher degree of self-confidence. Second, the study reveals that women and men agree they were authorized to agree on biodiversity conservation and climate change adaptation in wetlands. Women and men similarly assessed the neutral degree of their rational empowerment as one in which they could make decisions without considering all the implications. They also claim they preferred to avoid making decisions if they could. Third, women and men assessed the neutral degree of collective empowerment regarding teamwork and social activities. Comparatively, men are more active than women in teamwork and social activities.

9.7 Planning and Policy Implications

The engagement, especially women in biodiversity conservation and adaptation to climate change, should be promoted because Cambodia's Ramsar Sites are crucial for rural livelihoods. Therefore, the three dimensions of empowerment introduced by Rowlands (1997) are important and improved with a focus on improved women's personal, collective, and rational empowerment development. The empowerment should address improving their capacity and authority in decision-making. Moreover, the capacity for building teamwork and the opportunities to engage in social activities should be promoted at all levels of villagers, especially women. The increased internal resources for implementing social activities, including capacity building, awareness raising, campaigns, community meetings, and patrols to manage the biodiversity and natural resources. Since the 1990s, NGOs and CBOs have empowered communities, especially women, to engage in biodiversity conservation and climate adaptation. In recent years, the financial and human resources of NGOs and CBOs have gradually shrunk. Therefore, the annual budget available from the commune investment plan (CIP) will increase the number of social activities and support for the operation of the CBOs. At the same time, the contribution of NGOs remains useful in providing capacity building and technical support for the CBOs to work with the communities to promote empowerment in participating in biodiversity conservation and climate change adaptation.

References

Agarwal, B. (2010). *Gender and green governance: The political economy of women's presence within and beyond community forestry*. OUP.

Ali, M., & Grobicki, A. (2016). Women's roles in managing wetlands. *Solutions, 7*(6), 58–63.

Bryan, E., Alvi, M., Huyer, S., & Ringler, C. (2024). Addressing gender inequalities and strengthening women's agency to create more climate-resilient and sustainable food systems. *Global Food Security, 40*, 100731.

Chowdhury, E. K., Dhar, B. K., & Stasi, A. (2022). Volatility of the US stock market and business strategy during COVID-19. *Business Strategy & Development, 5*(4), 350–360.

Continuum, P. F.-N. Y. (1996, undefined, n.d.). *Pedagogy of the oppressed* (revised). Academia.Edu. Retrieved February 25, 2023, from https://www.academia.edu/download/34621170/EDST_565_Freire__Illich__and_Liberation_Theology.pdf

Dankelman, I. (2012). Women advocating for sustainable livelihoods and gender equality on the global stage. *Women reclaiming sustainable livelihoods: Spaces lost, spaces gained* (pp. 21–41). Palgrave Macmillan UK.

Duflo, E. (2012). Women empowerment and economic development. *Journal of Economic Literature, 50*(4), 1051–1079.

CBD. (2020b). Estimation of resources needed for implementing the Post- 2020 global biodiversity framework. Preliminary second report of the panel of experts on resource mobilization (CBD/SBI/3/5/Add.28 June 2020). *Convention on Biological Diversity*. https://www.cbd.int/doc/c/2c34/9558/f1487764d65e89bafb74d8fa/sbi-03-05-en.pdf

Freire, P., & Macedo, D. (1996). A dialogue: Culture, language, and race. In *Breaking free: The transformative power of critical pedagogy* (pp. 199–228).

IPBES. (2019). *Global assessment report of the Intergovernmental Science-Policy Platform on Biodiversity and Ecosystem Services*. In E. S. Brondízio, J. Settele, S. Díaz, & H. T. Ngo (Eds.). IPBES Secretariat. ISBN: 978-3-947851-20-1.

Kabeer, N. (2001). *Reflections on the measurement of women's empowerment*. https://eprints.lse.ac.uk/53116/

Labonte, R. N., & Feather, J. (1996). *Handbook on using stories in health promotion practice*. Ottawa: Health Canada.

Maltby, E. (1991). Wetland management goals: Wise use and conservation. *Landscape and Urban Planning, 20*(1–3), 9–18. https://doi.org/10.1016/0169-2046(91)90085-Z

Millennium Ecosystem Assessment (MEA). (2005). *Ecosystems and human well-being: Wetlands and water synthesis*. World Resources Institute.

Ramsar Convention on Wetlands. (2018). State of the world's wetlands and their services to the people. In *Global wetland outlook* (pp. 1–84). Ramsar Convention Secretariat. https://www.global-wetland-outlook.ramsar.org/outlook

Ramsar Convention Secretariat. (2013). *The Ramsar Convention manual: A guide to the convention on wetlands (Ramsar, Iran, 1971)* (6th ed., pp. 1–109).

Ratner, B. D., Cohen, P., Barman, B., Mam, K., Nagoli, J., & Allison, E. H. (2013). Governance of aquatic agricultural systems: Analyzing representation, power and accountability. *Ecology and Society, 18*(4), 59.

Ritesh, K., Pierre, H., Randy, M. G., Sonali, S., Sellamuttu, S. T., & Nick, C. D. (2011). Assessing wetland ecosystem services and poverty inter-linkages: A general framework and case study. *Hydrological Sciences Journal, 56*(8), 1602–1621.

Rocheleau, D., Thomas-Slayter, B., & Wangari, E. (Eds.). (2013). *Feminist political ecology: Global issues and local experience*. Routledge.

Rowlands, J. (1997). *Questioning empowerment: Working with women in Honduras*. Oxfam.

Setti, M., Banchelli, F., Falasconi, L., Segrè, A., & Vittuari, M. (2018). Consumers' food cycle and household waste. When behaviors matter. *Journal of Cleaner Production, 185*, 694–706.

Sok, S., Yu, X., & Wong, K. K. (2012). Impediments to community fisheries management: Some findings in Kompong Pou commune, Krakor District in Cambodia's Tonle Sap. *Singapore Journal of Tropical Geography, 33*(3), 398–413.

Sophal, M. (2004). Sustainable use and management of Cambodian Wetlands: Social and institutional challenges. In *Wetlands management in Cambodia: Socioeconomic, ecological, and policy perspectives* (pp. 22–24). World Fish Center Press.

Thuon, T., Rath, S., Duong, C., & Pok, K. (2023). Toward more inclusive wetland governance and the role of women and other marginalized groups in the Cambodian Mekong Delta: A case study of Boeung Preklapov Wetland in Takeo Province. *Cambodia Journal of Basic and Applied Research, 5*(1), 48–58.

Torrell, M., Salamanca, A. M., & Ratner, B. D. (Eds.). (2004). *Wetlands management in Cambodia: Socioeconomic, ecological and policy perspectives* (55 pp.). Sida.

UNDP (United Nations Development Programme). (2010). *Human Development Report 2010: The Real Wealth of Nations: Pathways to Human Development*. New York.

UN. (2020). *Sustainable Development Goal 2015 – 2030*. Goal 15 Life on Land. https://sdgs.un.org/goals/goal15

UN Women. (2014). *World survey on the role of women in development 2014: Gender equality and sustainable development*. UN Women.

UN Women. (2016, 2017). *UN Women Annual Report 2016-2017*. UN-Women-Annual-Report-2016-2017-en.pdf

UNEP-WCMC, IUCN, & NGS. (2018). *Protected planet report 2018*. United Nations Environment Programme-World Conservation Monitoring Centre (UNEP-WCMC), International Union for the Conservation of Nature (IUCN), & NGS. https://livereport.protectedplanet.net/pdf/Protected_Planet_Report_2018.pdf

Wallerstein, N. B., & Duran, B. (2006). Using community-based participatory research to address health disparities. *Health Promotion Practice, 7*(3), 312–323.

Wilkinson, A. (1998). Empowerment: Theory and practice. *Personnel Review, 27*(1), 40–56.

Zheng, D., Mi, J., Ravesteijn, W., & Qiu, F. (2014). Responsible resource management: The predicament and reform path for Chinese wetland conservation. *Wetlands Ecology and Management, 22*, 509–521. https://doi.org/10.1007/s11273-014-9348-4

Zimmerman, M. A. (1995). Psychological empowerment: Issues and illustrations. *American Journal of Community Psychology, 23*(5), 581–599. https://doi.org/10.1007/BF02506983

Chapter 10
Women's Household Roles, Perceived Dangerousness, and Household Resilience: Empirical Evidence from Geological Disasters in Sichuan, China

Bingwei Tian, Sainan Li, and Li Peng

Abstract Climate change has led to an increase in extreme precipitation and a marked rise in the global incidence of geological disasters, making it crucial to improve risk resilience. This chapter discusses the family's perceived dangerousness, perception bias and risk resilience in geological disasters from the perspective of gender. First, a competency-based household resilience assessment framework was constructed, and perceived dangerousness and perceived bias were quantified. The impact of female ownership on the formation of low perceived dangerousness and high resilience families was analyzed by combining propensity score matching (PSM). The findings are as follows: (1) There are significant differences in the overall perceived dangerousness scores of different villages, and there are gender differences, and the proportion of females with small perception bias is the highest. (2) The household resilience score mainly showed a medium high level, but the distribution of the three core abilities was uneven. (3) The coupling type distribution of perceived dangerousness and household resilience showed regional diversity. Most villages were dominated by families with low perceived dangerousness and a low proportion of families with high resilience. (4) The PSM results showed that female heads of household significantly positively affected the formation of families with low perceived dangerousness and high resilience to some extent, and the results passed the robustness test. However, women's role in family decision-making is still limited by gender roles and social culture. The paper highlights the importance of gender factors in building household resilience and suggests that policies should empower women with more decision-making power and resource support to make disaster management more gender-sensitive and communities more resilient to disasters.

B. Tian · S. Li
Institute for Disaster Management and Reconstruction, Sichuan University—The Hong Kong Polytechnic University, Chengdu, China

L. Peng (✉)
College of Geography and Resource Science, Sichuan Normal University, Chengdu, China
e-mail: pengli@imde.ac.cn

Md. A. Abedin et al. (eds.), *Climate Crisis in South and East Asia*, Sustainability Solutions, https://doi.org/10.1007/978-3-032-01012-4_10

Keywords Climate change · Geological disasters · Household resilience · Perceptual bias · Gender differences

10.1 Introduction

Global climate change is increasing the frequency and destructive power of geological disasters. Climate change is not only triggering an increase in extreme weather events, but also directly affecting the occurrence and scale of geological disasters by altering rainfall patterns, snowmelt rates and soil stability (Dhakal, 2013; Huang & Cheng, 2012). In Sichuan Province of China, due to its special geological structure and climate conditions, geological disasters such as landslides and debris flows occur frequently, posing a great challenge to the safety of life and property of local residents (Peng et al., 2024), and seriously affecting the social stability and development ability of the community (Li et al., 2023). Under these threats, women are particularly at risk from disasters (Bradshaw & Fordham, 2015). Women are often disadvantaged in social structures, with limited access to economic resources, educational opportunities and social support (Ronan et al., 2015). In addition, cultural and customs barriers often prevent women from equally participating in resource allocation and reconstruction decisions during post-disaster recovery, which makes their resilience significantly lower than that of men (Cutter, 2017).

In response to threats such as climate change and geohazards, improving risk resilience is considered an effective mitigation tool that can help cities and communities better prevent, respond, and recover from disaster shocks (Meerow & Newell, 2019). However, current research has focused on urban resilience and community resilience, exploring how to improve overall disaster resilience through urban planning (Masnavi et al., 2018), infrastructure development (Zang et al., 2025) and social organization optimization (Carmen et al., 2022). However, there are relatively few studies on household resilience, especially on family-level risk resilience in the context of geological disasters. In fact, the improvement of resilience in large-scale regions usually requires a long time and a large amount of resources (Zhang et al., 2024), and it is difficult for individuals to rapidly enhance resilience. In contrast, as the basic unit of the society, the family can more easily improve its resilience by using effective means, so as to further enhance the resilience of the community and even the whole society.

Perceived dangerousness is an individual's subjective judgment of potential disasters. To a certain extent, the stronger the perception, the greater the individual's negative emotional reactions such as anxiety and fear. However, due to the influence of some individual characteristics, disaster experience, or the same group effect, residents may have some irrational judgments, resulting in perception bias (Caparos et al., 2020), that is, the difference between the subjective perception of risk level and the objective risk level of residents. This can affect the normal life of residents, for example, excessive perception may lead to residents in a state of excessive prevention or complete abandonment at times. In addition, there may be significant gender differences in perception bias. For example, it has been shown that women generally overestimate the actual level of social crime rate and AI perception (Ragot et al., 2020; Zhang et al., 2021). However, the research on the perception bias of geological disaster risk from the perspective of gender difference is still very limited. In addition, in the field of household resilience, although scholars have pointed out that gender factors may play a key role in the process of family response to disasters and recovery, there is a lack of systematic research on the specific impact of women in the formation of a "double advantage" family with low perceived dangerousness and high resilience.

The remainder of this chapter is organized as follows (Fig. 10.1): Sect. 10.2 introduces the research area and data sources, focusing on the characteristics of geological disasters and female-dominated family culture in Sichuan Province. Section 10.3 describes the research methods, including the household resilience assessment framework and propensity score matching (PSM) analysis. Section 10.4 shows the research results, analyzes the perceived dangerousness, perception bias, household resilience and the impact of female ownership on low perceived dangerousness and high resilience family; and Sect. 10.5 is the discussion, and Sect. 10.6 is the conclusion.

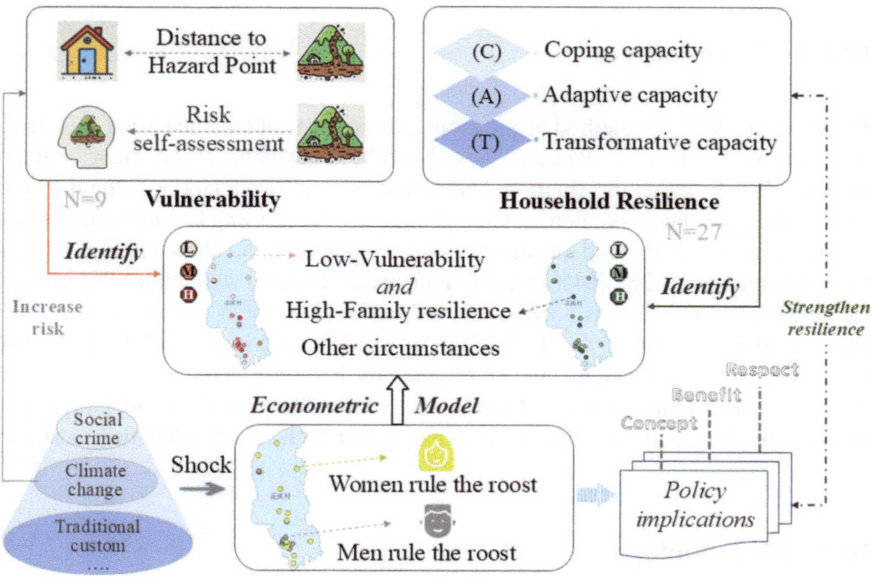

Fig. 10.1 Research framework

10.2 Study Area and Data Sources

10.2.1 Study Area

Sichuan Province is located in Southwest China, with a total area of 486,000 km^2, mainly mountainous areas, accounting for 74.2% of the total area of the province. By the end of 2021, 30,861 potential geological disaster sites such as landslides and debris flows had been identified, accounting for about 11% of the country's total. These disasters directly threaten the lives and property of more than 1.23 million people.

In addition to the unique natural environment, Sichuan also has a unique culture of "*Raking ears*." In traditional Chinese families, most family decisions are dominated by men, while Sichuan has formed a different family structure, with a large number of female-dominated families (Yang et al., 2019). This unique phenomenon provided an adequate sample of female-headed households for this study, which helped ensure the authenticity and representativeness of the findings.

10.2.2 Data Sources

This chapter adopts the method of questionnaire survey to obtain the nested data structure of family-perceived dangerousness-household resilience. The main contents included family basic information, individual perception, adaptive behavior, and village ability of the respondents. After eliminating the missing key information and irrelevant content, 536 valid questionnaires were obtained. As shown in Fig. 10.2, of the 536 respondents, 287 (64%) were female and 251 (36%) were male. The average age of the respondents was 55.73 years old, and 355 of them were under 60 years old, accounting for 66.23%. Most of the respondents were married (95.71%) and had stable families. Respondents were generally less educated. Among them, 173 (32.28%) had primary school education or below, 283 (52.80%) had junior high school education, and only 80 (14.92%) had senior high school education or above.

10.3 Methodology

10.3.1 Building a Competency-Based Household Resilience Assessment Framework

Based on the capacity framework, this paper divides household resilience into three core competencies: coping capacity (C), adaptation capacity (A) and transformative capacity (T), and constructs a composite disaster resilience index to visually present

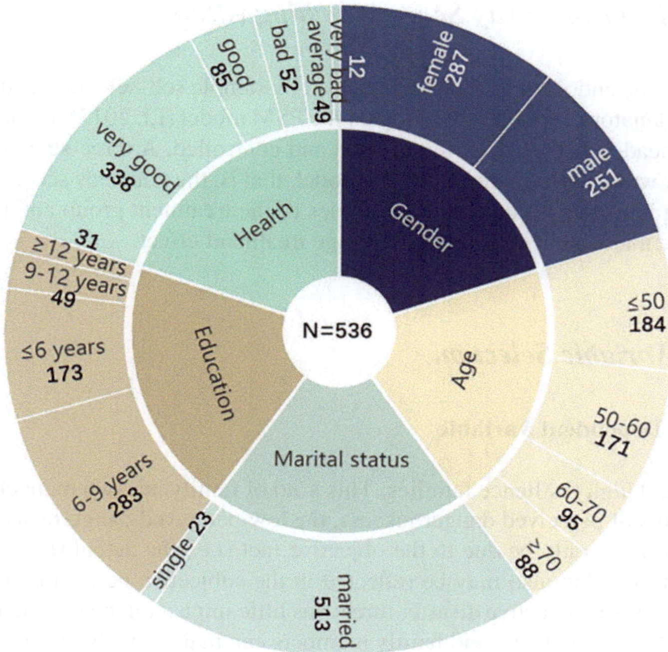

Fig. 10.2 Sample statistics

the overall family capacity (Tan et al., 2020). Since disaster recovery is a dynamic process, C, A, and T are involved in each stage, and all three are indispensable and equally important. Therefore, the weight of each dimension in the index was 1/3. The calculation formula is as follows:

$$\text{Household Resilience} = 1/3 * (\text{C} + \text{A} + \text{T})$$

10.3.2 Low Risk and High Resilience Family Identification

In this study, the entropy weight method was used to eliminate the data dimension difference by standardized processing, and the information entropy of each index was calculated according to the entropy value formula. After determining the weight, the perceived dangerousness index and resilience index were calculated respectively. Families with low perceived dangerousness and high resilience were further identified by dividing the results into three three-level categories of low, medium, and high using the natural breakpoint method.

10.3.3 The Propensity Score Matching (PSM)

To mitigate the endogeneity problems caused by sample self-selection and the omission of explanatory variables, we adopted the PSM model (Li, 2012). Female-headed and male-headed households were treated and controlled. A three-stage processing framework was used to model the PSM model, that is, the tendency score was calculated using logit regression, and the samples of the treatment group and the control group were matched to estimate the average treatment effect.

10.3.4 Variable Selection

10.3.4.1 Dependent Variable

Low risk and high resilience families: This kind of family has two main characteristics: In terms of perceived dangerousness, the low perceived dangerousness of such families may not only be due to the objective fact (i.e., the actual risk of the environment is low), but also may be reflected in the subjective perception of low risk degree, that is, the objective disaster threat has little impact on the psychological and emotional life of residents, and family members can maintain a high sense of happiness. In terms of resilience, such households have a strong ability to withstand and mitigate the impact of geological disasters, reflecting a high level of resource reserves, resilience and social support. Therefore, families with low perceived dangerousness and high resilience are seen as "double advantaged" families, with both psychological strength and coping resilience.

10.3.4.2 Independent Variable

A family in which the woman heads the household. In this article, the female master means that in the family, women (such as wives, daughters-in-law, etc.) assume the decision-making responsibility of major family affairs. These major decisions include important economic aspects such as production and operation and disaster insurance, as well as decision-making and management in key areas of family life such as the allocation of family resources and the education of children. The female-headed family structure is not only limited to the case of husband and wife living together, but also covers the case of divorce or widow, in which women as the main manager or decision maker have the dominant power over family affairs.

10.3.4.3 Control Variables

In addition to the above variables, the model also sets 12 control variables from the four levels of basic personal characteristics, village resources, sense of place, and basic situation of people in the same village (Table 10.1). Basic personal characteristics included health status, ethnicity and number of disasters experienced to reflect the potential impact of individual differences on family perceived dangerousness and resilience. Village resources included the ability of village cadres, village population changes and the availability of disaster prevention and mitigation resources in the village where the respondents were located, reflecting the role of external support and environmental conditions on families. The sense of place mainly measures the degree of dependence of the respondents on the village, reflecting their psychological and emotional connection with the place of residence. The basic situation of the villagers included the degree of information mastery and the relative level of income.

10.4 Results

10.4.1 Perceived Dangerousness and Perception Bias

There were significant differences in the distribution of perceived dangerousness scores among different villages (Fig. 10.3). Some villages (such as Shexi and Renjiaping) had a lower median perceived dangerousness score, suggesting that households in these villages were less at risk as a whole and were more secure or less negatively affected. The higher median in other villages, such as Xingfu and Longtan, indicates that households in these villages face greater perceived dangerousness, which may be related to their higher geological risks or poor disaster prevention and mitigation capabilities. In addition, scores in some villages (such as Longping and Laochang) have a large distribution range and multiple outliers, reflecting significant differences in household perceived dangerousness in these villages, which may be affected by differences in individual experience, resource distribution or information access.

In this chapter, perceived bias is the contrast between actual risk and perceived risk. The actual hazard was classified into ten levels based on the distance from the house to the nearest hazard point using the natural breakpoint method, each corresponding to 1–10 points. The perceived risk was also scored on a scale of ten, each again corresponding to a score of 1–10. By comparing the scores of actual risk and perceived risk, if the perceived risk score was greater than the actual risk score, the perception bias was large. Otherwise, the perception bias is small. If the scores agree, they are considered to be free of bias.

The results are shown in Fig. 10.4, the distribution of different gender on perceived bias exist significant differences. In general, most families had perception bias, among which the perception bias was lower than the actual bias (perception bias was

Table 10.1 Indicators

Variable		Definition	Measure	Mean	SD
Disaster perceived dangerousness					
Objective	O1	The signs of disaster have become more and more obvious	1–5	2.364	1.319
	O2	Distance from your home to the nearest disaster site (km)	ANV	1.767	4.571
Subjective	S1	In the next 10 years, there may be disasters near your home	1–5	2.657	1.305
	S2	Your home is more likely to suffer a disaster	1–5	2.132	1.198
	S3	Disasters can damage houses	0, 1	0.360	0.484
	S4	Your home is more likely to be damaged than other homes	1–3	2.257	0.661
	S5	Now you live in a disaster zone	0, 1	0.399	0.490
	S6	Whether crops have been reduced due to disasters	0, 1	0.276	0.447
	S7	Whether the disaster has an impact on your family's production	0, 1	0.821	0.383
Indicators of household resilience					
Coping capacity	C11	Proportion of persons under 16 and over 65 years of age in total household population (%)	ANV	0.273	0.265
	C12	Number of children under 6 years old	ANV	1.239	1.198
	C13	Age of head of household (years)	ANV	55.989	13.312
	C14	Years of schooling of head of household (years)	ANV	6.901	4.142
	C15	Number of persons unable to work in the household	ANV	0.252	2.488
	C21	(Human capital) working-age population	ANV	2.576	1.473
	C22	(Social capital), number of relatives and friends visited during the spring festival	ANV	7.507	8.959

(continued)

Table 10.1 (continued)

Variable		Definition	Measure	Mean	SD
	C23	(Social capital) interpersonal interactions with people around you	1–5	4.500	0.748
	C24	(Economic capital) annual household income	ANV	7.965	7.278
	C25	(Economic capital) total amount of loans	ANV	6.697	2.109
	C26	(Physical capital) building area (m^2)	ANV	326.272	396.661
	C27	(Physical capital), the quality of cultivated land	1–5	2.832	0.990
	C28	(Physical capital) operating land area	ANV	2.016	3.038
	C29	(Physical capital) major livelihood strategies	0, 1	0.224	0.417
	C31	Whether the Internet is installed in the home	0, 1	0.869	0.337
	C32	Distance from your home to the nearest main road	ANV	0.425	2.198
	C33	Distance from your home to the market (km)	ANV	5.820	14.227
Adaptive capacity	A11	Whether to prepare disaster emergency supplies	0, 1	0.653	0.476
	A12	Family members learn about disaster prevention and mitigation	0, 1	0.713	0.453
	A13	The family attended disaster training	0, 1	0.692	0.462
	A14	Family members have participated in disaster evacuation drills	0, 1	0.700	0.458
	A15	Whether the home is reinforced or maintained	0, 1	0.569	0.495
Transformative capacity	T11	The villagers are hopeful about the future development of the village	1–5	3.994	1.066
	T21	Villagers exchange experience in village development	1–5	3.647	1.202
	T22	Pay attention to changes in disaster hazards	0, 1	0.666	0.472

(continued)

Table 10.1 (continued)

Variable		Definition	Measure	Mean	SD
	T23	Pay attention to the surrounding geology	0, 1	0.879	0.326
	T24	Whether they will voluntarily move out	1–5	3.425	1.473
Explaining variable					
Gender	G1	Gender makes major decision makers in the home	0, 1	0.563	0.496
Control variable					
Personal information	K1	Physical fitness level	1–5	1.722	1.111
	K2	Minority nationality or not	0, 1	1.198	0.398
	K3	Number of disasters experienced	ANV	3.237	3.821
Village condition	V1	Village leadership	1–5	3.800	1.139
	V2	Village resources	1–5	3.612	1.114
	V3	Village disaster prevention and mitigation	1–5	4.166	0.960
	V4	Whether the number of village population will increase	ANV	3.140	1.461
Sense of place	D1	When you hear people from other villages say "bad things" about your village, you will defend it	1–5	3.627	1.198
	D2	You involved in the decision-making of the village's public affairs	1–5	3.931	1.163
	D3	The village is the main source of family income	1–5	3.485	1.502
Villager information	E1	Richer people in the village have more resources, more information	1–5	3.597	1.305
	E2	The income level of the family is similar to that of other	1–3	1.888	0.709

ANV actual numerical value

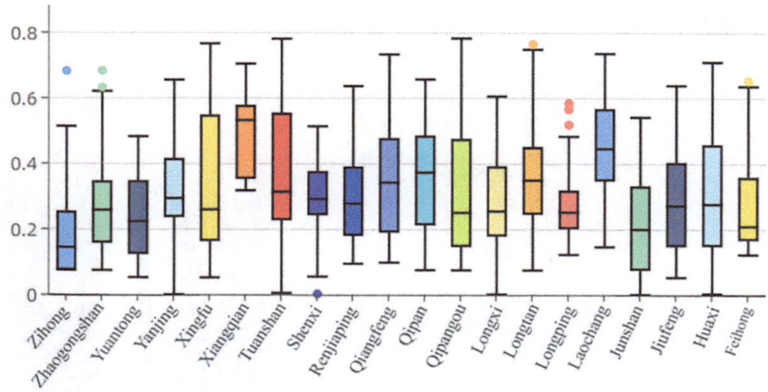

Fig. 10.3 Distribution of perceived dangerousness scores in different villages

small) accounted for the highest proportion (50.75%). In this category, the proportion of females was 38.62%, which was significantly higher than 12.13% of males. The higher proportion of women in the small perceived bias category may be due to their higher perceived dangerousness ability, more positive psychological regulation ability, and greater emotional resilience. This tendency enables women to pay more attention to the daily quality of life of the family, reduce the impact of negative emotions, and in turn create a more stable and happy living environment for family members. In the category of perceived deviation higher than actual deviation (large perceived deviation), the total proportion was 36.19%, and the proportion of women was 18.28%, which was not significantly different from 17.91% of men. The proportion of families without bias was the smallest, only 13.06%, and the difference between the proportion of women and men was also small.

10.4.2 Household Resilience

As shown in Fig. 10.5, household resilience scores are mainly concentrated in the 0.35–0.45 range, among which the number of families with scores between 0.4 and 0.45 is the largest, with 89 households, followed by 0.35–0.4 and 0.45–0.5 ranges, with 78 and 74 households, respectively. This indicates that the resilience scores of most families are at a medium to high level, reflecting that the sample families have gradually adapted and have a certain response and recovery ability under the long-term disaster threat. In contrast, the number of households with scores below 0.25 and above 0.6 was small, ranging from 3 to 22, respectively, indicating the rarity of households at very low or very high levels of resilience. Overall, resilience scores showed a trend toward a close to normal distribution, indicating that the distribution of household resilience scores in the study sample was relatively balanced.

Fig. 10.4 Perception bias

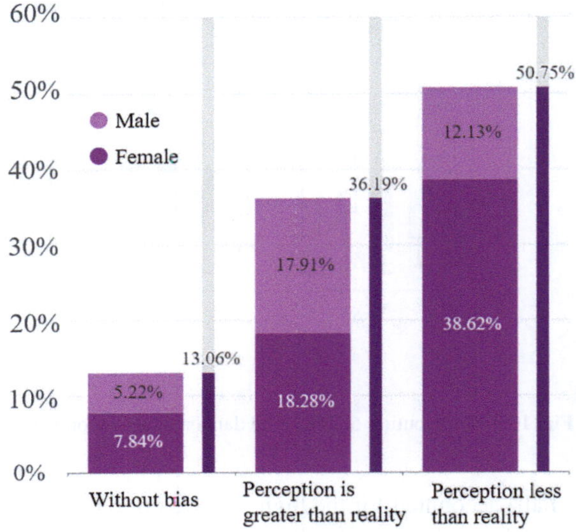

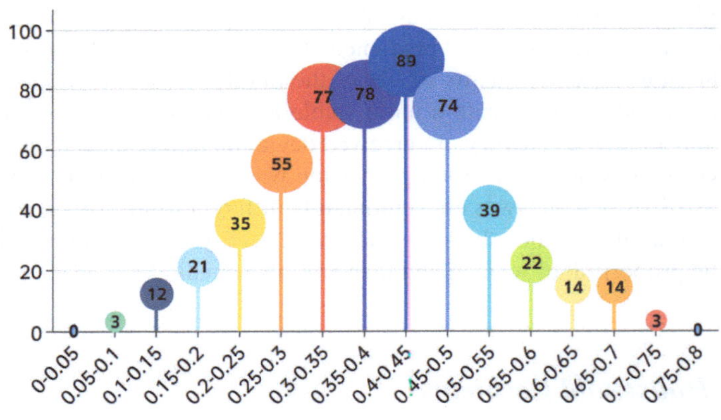

Fig. 10.5 Household resilience score interval distribution

10.4.3 Coupling Condition

The statistical results of coupling type distribution of perceived dangerousness and family toughness in various local villages are shown in Fig. 10.6. The coupling type distribution of perceived dangerousness and family toughness in sample villages presents significant diversity and regional differences, and families are divided into nine coupling types. From the overall distribution, households with low perceived dangerousness (L) type dominated, with L-L and L-M type accounting for a high proportion in most villages, such as Yanjing and Renjiaping villages, where L-L type accounted for nearly half. This suggests that households in many villages, while

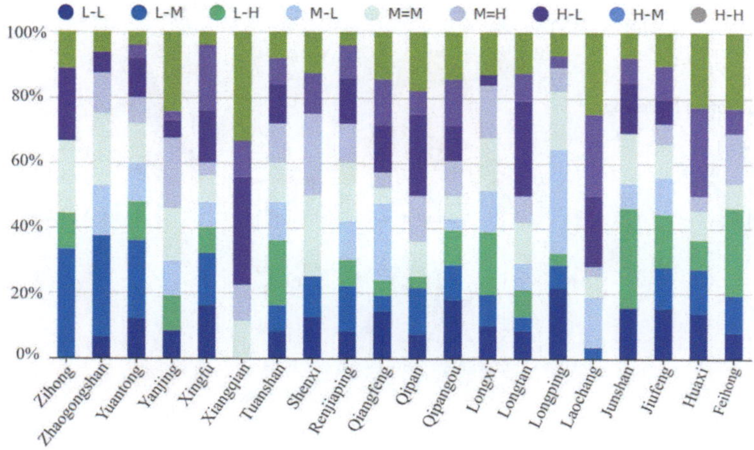

Fig. 10.6 Household types in the sample villages. *Note* L: low; M: middle; H: high

having low perceived danger, have relatively limited levels of resilience and may lack sufficient resources or capacity to deal with potential risks. On the other hand, the distribution of high hazard perception (H) types showed more concentrated and differentiated characteristics. For example, in Zhong, Qipan and other villages, the proportion of families with high perceived dangerousness is significantly higher, and there are more H–L (high perceived dangerousness–low resilience) families, which reflects that these villages may face high risk of geological disasters, but the resilience level of families is not enough to effectively cope with these threats. In addition, the proportion of households of the L–H (low perceived dangerousness–high resilience) type was low in all villages, showing scarcity, indicating that households with both high resilience and low perceived dangerousness were rare.

10.4.4 Regression Result

10.4.4.1 Results of PSM

In this study, proximity matching, nearest neighbor matching with calipers, radius matching, and kernel matching were used for propensity score matching analysis. Before calculating ATT, co-support was tested to assess the matching effect. Figure 10.7 illustrates the density distribution of *P*-scores for the experimental and control groups before and after matching. The results showed that after matching, the propensity score distributions of the experimental and control groups were much closer, indicating that matching significantly reduced the differences between the two groups. In particular, after matching, the propensity score intervals of the two groups showed a greater degree of overlap, forming a common support domain. In

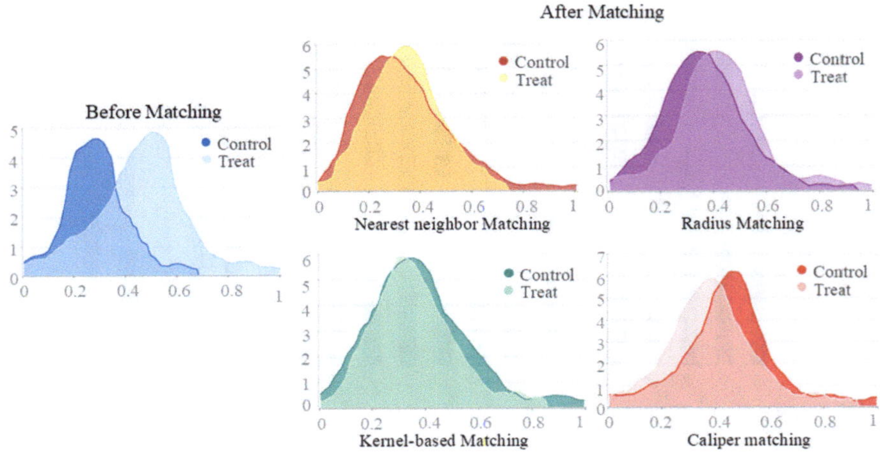

Fig. 10.7 Kernel density distribution before and after matching

addition, most of the observations were within the common support range, indicating a high matching quality.

In terms of the proportion of low risk and high resilience families, each matching method showed a significantly higher proportion in the treatment group than in the control group. Although the ATT values generated by the different matching methods were slightly different, all values were positive and the first three methods reached significance levels. This result suggests that, even when controlling for systematic differences across samples, female dominance still significantly and positively affected the conversion of families to low perceived dangerousness and high resilience types.

Further refining the empirical analysis and examining the specific effects of female homemaking on family disaster perceived dangerousness and household resilience separately, the results show some interesting phenomena. In terms of disaster perceived dangerousness, each matching method showed that the scores of the treatment group were significantly lower than those of the control group, that is, the ATT values were negative, and the results of the first three methods were significant. This suggests that female homemakers can significantly reduce the family's perception of disaster risk. However, in terms of household resilience, none of the ATT values of the four methods reached the level of statistical significance, which means that female homemaking does not directly improve household resilience. This is not inconsistent with the previous result, which focuses on the comprehensive family type transformation, that is, the composite characteristics of low perceived dangerousness and high resilience family, while the single dimension of household resilience results reveal that female ownership has different effects on different family traits. In summary, the main effect of female household ownership was to reduce the family's perception of disaster risk, thereby promoting the transformation of family types, rather than directly improving household resilience (Table 10.2).

Table 10.2 Mean treatment effects obtained from different matching approaches

Matching algorithms	Influencing factors	Treated	Controls	SE	ATT
Nearest neighbor matching	Disaster perceived dangerousness	0.3010	0.3366	0.0208	– 0.0356*
	Household resilience	0.3782	0.3758	0.0139	0.0024
	Low risk and high resilience	0.2333	0.1567	0.0454	0.0767*
Radius matching	Disaster perceived dangerousness	0.3010	0.3262	0.0161	– 0.0252*
	Household resilience	0.3782	0.3768	0.0113	0.0013
	Low risk and high resilience	0.2333	0.1717	0.0360	0.0617*
Kernel-based matching	Disaster perceived dangerousness	0.3000	0.3278	0.0157	– 0.0278*
	Household resilience	0.3789	0.3773	0.0111	0.0016
	Low risk and high resilience	0.2351	0.1713	0.0351	0.0637*
Caliper matching	Disaster perceived dangerousness	0.3000	0.3289	0.0207	– 0.0289
	Household resilience	0.3789	0.3766	0.0142	0.0022
	Low risk and high resilience	0.2351	0.1821	0.0459	0.0530

* $p < 0.1$; ** $p < 0.05$; *** $p < 0.01$

10.4.4.2 Robustness Test

To verify the reliability of the conclusion, this chapter adopts the Logit and Probit models for robustness test, and keeps the consistency of dependent variables, explanatory variables and control variables. The fitting degrees of the model were 80.60% (Logit) and 80.78% (Probit), respectively, indicating that the model had good explanatory power. Specifically, G1 (Logit: 0.0613*, Probit: 0.0670**) shows in both models that female ownership has a significant positive impact on low-risk and high-resilience families, which fully confirms the robustness and reliability of the conclusions in this chapter (Table 10.3).

Table 10.3 Results of logit and probit models for robustness test

Variables	Logit	SE	Probit	SE
G1	0.0613*	0.0336	0.0670**	0.0334
K1	– 0.0095	0.0160	– 0.0080	0.0158
K2	– 0.0867*	0.0867	– 0.0921*	0.0472
K3	0.0093**	0.0040	0.0093**	0.0040
C1	– 0.0083	0.0201	– 0.0098	0.0197
C2	0.0097	0.0179	0.0115	0.0178
C3	0.0549**	0.0239	0.0534**	0.0227
C4	– 0.0169	0.0109	– 0.0170	0.0111
D1	0.0244*	0.0148	0.0240	0.0146
D2	0.0536***	0.0189	0.0503***	0.0174
D3	0.0297**	0.0115	0.0303	0.0114
E1	– 0.0326***	0.0119	– 0.0319***	0.0121
E2	– 0.0313	0.0236	– 0.0318	0.0237
Degree of fitting (%)	80.60		80.78	

* $p < 0.1$; ** $p < 0.05$; *** $p < 0.01$

10.5 Discussion

10.5.1 Gendered Risk Perception and Role Responsiveness in the Household

This study finds that female-headed households are more likely to exhibit lower perceived disaster risk and more accurate risk judgment, thereby increasing the likelihood of being categorized as low-risk, high-resilience families. Rather than viewing women solely as disaster victims, our findings suggest that they often act as micro-level coordinators of safety and preparedness, especially within the domestic domain—a role shaped by and reinforcing gender norms (Nurdin & Handayani, 2021; Scabin et al., 2024).

The social positioning of women as caregivers makes them more attuned to environmental signals and everyday vulnerabilities. This explains the behavioral pattern observed in our data: women engaging more actively in risk prevention tasks such as household reinforcement or emergency stockpiling (Alyssa Mari et al., 2021). However, their preparedness is frequently framed as "emotional responsibility" rather than as a form of operational leadership, which limits how it is institutionally recognized or supported (Danielsson & Eriksson, 2022).

Such framing also affects how resilience is conceptualized in disaster literature. Much of the discourse centers on infrastructural or policy-based resilience, while overlooking the cognitive and emotional labor women contribute to sustaining household stability. This study's evidence underscores the need to expand our understanding of resilience beyond its material dimensions, to include the less visible—but equally vital—psychosocial processes driven by gendered roles and perceptions.

10.5.2 Structural Constraints and the Undermining of Female Adaptive Capacity

While female-headed households exhibit stronger disaster awareness and more consistent preparedness behaviors, their efforts to enhance actual household resilience remain constrained by persistent structural and institutional limitations. This points to a broader disconnect between individual capability and systemic opportunity.

Social and cultural norms continue to shape the boundaries of female authority. Women may act as household decision-makers informally, yet lack formal influence in public risk governance spaces—whether in village committees, disaster response planning, or resource allocation (Huang & Cheng, 2012; Zaidi & Fordham, 2021). This "informal agency/formal exclusion" dichotomy creates an environment where women's capacity is routinely exercised but rarely scaled or legitimized.

Moreover, post-disaster periods tend to reinforce traditional hierarchies. Relief channels often privilege male heads of household by default, regardless of actual caregiving responsibility. Simultaneously, the emotional and physical burden on women increases, especially in contexts of displacement or financial hardship. Research has documented rises in domestic violence and reduced access to support services in such contexts, further limiting women's long-term coping and recovery potential (Arora, 2022; Duvvury et al., 2024).

Rather than viewing this as a failure of individual agency, it should be understood as the outcome of institutional arrangements that prioritize certain forms of authority and overlook others. What this study reveals is not simply the presence of gendered vulnerability, but a misalignment between where resilience is generated (within the household) and where it is formally enabled or resourced (within public systems). This insight opens space for further investigation into how informal resilience practices might be better recognized and integrated into formal disaster governance frameworks—without assuming that inclusion automatically leads to empowerment.

10.5.3 Policy Implications

The empirical results reveal that female-headed households are more likely to form low-risk, high-resilience families due to their stronger disaster risk perception and emotional regulation capacities. However, as discussed in Sects. 10.5.1, 10.5.2 and

10.5.3, these advantages are often undermined by structural constraints such as limited participation in decision-making, unequal access to financial and institutional resources, and traditional gender norms that increase caregiving burdens. These factors hinder the translation of women's perceptual strengths into tangible improvements in household resilience. Effective policy responses should therefore focus on building institutional pathways that enable this transformation. Efforts could include improving women's access to disaster-related knowledge and preparedness resources, embedding gender-sensitive practices into local governance structures, and ensuring that recovery programs explicitly support the roles and needs of female-headed households. It is also important to provide financial incentives for women's participation in formal risk planning processes, such as subsidies for training or leadership roles (Oktari et al., 2021). Moreover, promoting public narratives that legitimize and normalize women's leadership in disaster response can help shift underlying social attitudes. By integrating women's capacities into the formal risk governance framework—rather than treating them as passive beneficiaries—policy-makers can foster more equitable, community-based resilience that reflects the social realities uncovered in this study.

10.6 Conclusion

With the intensification of global climate change, the frequency and destructive power of geological disasters are increasing, posing a serious challenge to human society, especially vulnerable groups such as women and poor families. Women are often at higher risk of disasters due to their weaker ability to access resources, respond to disasters, and recover from disasters. However, their unique strengths and potential in household resilience building have not received sufficient attention. This study explored the spatial differences and gender effects of perceived dangerousness, household resilience, and their coupling characteristics using women in charge as the entry point. The results showed that perceived dangerousness scores varied significantly across villages, with some villages (such as Shexi) having lower overall perceived risk of households, while others (such as Xingfu) were at higher risk. There is a significant gender difference. Women account for the highest proportion of families with small perception bias, reflecting women's stronger psychological adjustment ability and emotional resilience. Household resilience scores were concentrated between 0.35 and 0.45, showing a medium to high level. The distribution of coupling types between perceived dangerousness and household resilience showed regional diversity. Most villages were dominated by families with low perceived dangerousness, but the proportion of families with high resilience was low. The results of PSM showed that female homemaking significantly reduced the family's disaster perceived dangerousness, and promoted the formation of low perceived dangerousness and high resilience families to a certain extent, but did not significantly improve the overall household resilience. The robustness test further confirmed the reliability of the findings, revealed the key role of gender in household resilience, and provided an important basis for the development of gender-sensitive disaster reduction policies. Female homemakers can significantly reduce the family's disaster perceived dangerousness, and promote the formation of low perceived dangerousness and high resilience family to a certain extent.

References

Alyssa Mari, T., Heidi, S., & Meghna, R. (2021). Natural hazards, disasters and violence against women and girls: A global mixed-methods systematic review. *BMJ Global Health, 6*(4), e004377. https://doi.org/10.1136/bmjgh-2020-004377

Arora, S. (2022). Intersectional vulnerability in post-disaster contexts: Lived experiences of Dalit women after the Nepal earthquake, 2015. *Disasters, 46*(2), 329–347. https://doi.org/10.1111/disa.12471

Bradshaw, S., & Fordham, M. (2015). Double disaster. In *Hazards, risks, and disasters in society* (pp. 233–251).

Caparos, S., Linnell, K. J., & Blanchette, I. (2020). The local perceptual bias of a non-remote and educated population. *Psychological Research, 84*(5), 1211–1222. https://doi.org/10.1007/s00426-019-01158-6

Carmen, E., Fazey, I., Ross, H., Bedinger, M., Smith, F. M., Prager, K., McClymont, K., & Morrison, D. (2022). Building community resilience in a context of climate change: The role of social capital. *Ambio, 51*(6), 1371–1387. https://doi.org/10.1007/s13280-021-01678-9

Cutter, S. L. (2017). The forgotten casualties redux: Women, children, and disaster risk. *Global Environmental Change, 42*, 117–121. https://doi.org/10.1016/j.gloenvcha.2016.12.010

Danielsson, E., & Eriksson, K. (2022). Women's invisible work in disaster contexts: Gender norms in speech on women's work after a forest fire in Sweden. *Disasters, 46*(1), 141–161. https://doi.org/10.1111/disa.12464

Dhakal, S. (2013). Geological hazards in Nepal and triggering effect of climate change. *Journal of Nepal Geological Society, 30*, 75–80.

Duvvury, N., Erten, B., & Ramnarain, S. (2024). Feminist perspectives on conflict, disaster, and violence against women: Introduction to the special issue. *Feminist Economics, 30*(4), 1–15. https://doi.org/10.1080/13545701.2024.2413364

Huang, Y., & Cheng, H. (2012). The impact of climate change on coastal geological disasters in southeastern China. *Natural Hazards, 65*(1), 377–390. https://doi.org/10.1007/s11069-012-0370-7

Li, M. (2012). Using the propensity score method to estimate causal effects: A review and practical guide. *Organizational Research Methods, 16*(2), 188–226. https://doi.org/10.1177/1094428112447816

Li, S., Peng, L., Wang, X., Huang, K., & Tang, J. (2023). Do geohazards inhibit urban expansion at the regional scale? Evidence from a counterfactual analysis in Southwest China. *Cities, 142*, https://doi.org/10.1016/j.cities.2023.104558

Masnavi, M. R., Gharai, F., & Hajibandeh, M. (2018). Exploring urban resilience thinking for its application in urban planning: A review of literature. *International Journal of Environmental Science and Technology, 16*(1), 567–582. https://doi.org/10.1007/s13762-018-1860-2

Meerow, S., & Newell, J. P. (2019). Urban resilience for whom, what, when, where, and why? *Urban Geography, 40*(3), 309–329.

Nurdin, N., & Handayani, A. (2021). Understanding women's psychological well-being in post-natural disaster recovery. *Medico-Legal Update, 21*. https://doi.org/10.37506/mlu.v21i3.2977

Oktari, R. S., Kamaruzzaman, S., Fatimahsyam, F., Sofia, S., & Sari, D. K. (2021). Gender mainstreaming in a disaster-resilient village programme in Aceh Province, Indonesia: Towards disaster preparedness enhancement via an equal opportunity policy. *International Journal of Disaster Risk Reduction, 52*. https://doi.org/10.1016/j.ijdrr.2020.101974

Peng, L., Li, S., Tan, J., & Zhang, X. (2024). Peer effects affect the adaptation behaviour of rural residents to geohazards in China: A new policy pathway. *Climate Policy*, 1–17. https://doi.org/10.1080/14693062.2024.2401852

Ragot, M., Martin, N., & Cojean, S. (2020). AI-generated vs. human artworks. A perception bias towards artificial intelligence? In *Extended Abstracts of the 2020 CHI Conference on Human Factors in Computing Systems* (pp. 1–10). Association for Computing Machinery, Honolulu, HI, USA.

Ronan, K. R., Alisic, E., Towers, B., Johnson, V. A., & Johnston, D. M. (2015). Disaster preparedness for children and families: A critical review. *Current Psychiatry Reports, 17*(7), 58. https://doi.org/10.1007/s11920-015-0589-6

Scabin, F., Ferraz, J., Garcia, M., & Passoni, C. (2024). A gendered approach for meaningful stakeholder engagement: The case of women affected by a mining disaster in Brazil. In *The Routledge handbook on meaningful stakeholder engagement* (pp. 280–294). Routledge.

Tan, J., Peng, L., & Guo, S. (2020). Measuring household resilience in hazard-prone mountain areas: A capacity-based approach. *Social Indicators Research, 152*(3), 1153–1176. https://doi.org/10.1007/s11205-020-02479-5

Yang, Y., Zhou, J., Feng, P., Jiang, G., Long, Y., & Zheng, Y. (2019). Values and behavior among minorities in Southwest China: A cross-cultural validation of the refined value theory. *Frontiers in Psychology, 10*, 1750. https://doi.org/10.3389/fpsyg.2019.01750

Zaidi, R. Z., & Fordham, M. (2021). The missing half of the Sendai framework: Gender and women in the implementation of global disaster risk reduction policy. *Progress in Disaster Science, 10*, 100170. https://doi.org/10.1016/j.pdisas.2021.100170

Zang, Y., Wang, H., Huang, J., Liu, G., & Li, J. (2025). Community disaster resilience assessment and external association characteristics analysis based on accessibility of service facilities: From the perspective of 15-minute life circle. *Sustainable Cities and Society, 118*, 106051. https://doi.org/10.1016/j.scs.2024.106051

Zhang, F., Fan, Z., Kang, Y., Hu, Y., & Ratti, C. (2021). "Perception bias": Deciphering a mismatch between urban crime and perception of safety. *Landscape and Urban Planning, 207*. https://doi.org/10.1016/j.landurbplan.2020.104003

Zhang, J., Wang, T., Goh, Y. M., He, P., & Hua, L. (2024). The effects of long-term policies on urban resilience: A dynamic assessment framework. *Cities, 153*, 105294. https://doi.org/10.1016/j.cities.2024.105294

Chapter 11
Climate Change Riskscapes and Strategies for Women and Girls of Indonesia

Aprillia Findayani, Rina Suryani Oktari, and Juhadi

Abstract Climate change presents a global crisis, disproportionately impacting women and girls who often bear the brunt of its devastating consequences. In Indonesia, a nation which is acutely vulnerable to climate change's effects, these vulnerabilities are amplified, demanding tailored solutions that center gender equity. This chapter has explored the gendered impacts of climate change, highlighting the specific ways in which women and girls are uniquely affected and marginalized. This chapter has showcased the transformative power of gender-responsive programs. These programs have not only enhanced the resilience of women and girls to climate change impacts but have also demonstrated the crucial role they play in driving sustainable development. The insights gained from these case studies illuminate the importance of community participation, partnerships, and a holistic approach that integrates climate change mitigation and adaptation strategies with broader efforts to promote gender equality. The evidence presented underscores the critical need to move beyond a singular focus on climate change mitigation or adaptation to a more integrated approach that explicitly addresses gender inequality. Empowering women and girls is not merely a matter of social justice; it is essential for building climate resilience at the community, national, and global levels. Their active participation is not optional; it is a fundamental requirement for creating a sustainable and equitable future in the face of climate change. The successful programs highlighted in this chapter serve as powerful examples of what is possible when gender equity is prioritized in climate action.

A. Findayani (✉) · Juhadi
Geography Department, Universitas Negeri Semarang, Semarang, Indonesia
e-mail: aprilliafinda@mail.unnes.ac.id

A. Findayani
U-Inspire Indonesia, Jakarta, Indonesia

R. S. Oktari
Family Medicine Department, Medical Faculty Syiah Kuala University, Banda Aceh, Indonesia

Tsunami and Disaster Mitigation Research Center (TDRMC), Banda Aceh, Indonesia

Md. A. Abedin et al. (eds.), *Climate Crisis in South and East Asia*, Sustainability Solutions, https://doi.org/10.1007/978-3-032-01012-4_11

Keywords Climate change · Gender inequality · Resilience · Indonesia · Gender-responsive solutions

11.1 Introduction

Climate change is an undeniable reality, fundamentally reshaping our planet and profoundly impacting billions of people worldwide (Calvin et al., 2023). The scientific evidence is overwhelming, with rising sea levels, increasingly frequent and intense extreme weather events (heatwaves, droughts, floods, storms), and significant shifts in agricultural patterns becoming increasingly apparent and devastating (Sivakumar, 2011). These changes are not merely environmental; they represent a profound threat to global food security, water resources, and human health, triggering mass migrations and exacerbating existing social inequalities (IPCC, 2021). While the impacts of climate change are global in scope, their distribution is profoundly uneven, with vulnerable populations disproportionately affected (Doherty & Clayton, 2011). This inequitable distribution is further compounded by existing social, economic, and political disparities, creating a complex web of interconnected challenges (Benevolenza & DeRigne, 2019).

The consequences of climate change are far-reaching and deeply interconnected, affecting communities on a global scale (Jha & Dev, 2024). These consequences extend beyond immediate environmental damage to encompass economic instability, political unrest, and widespread human suffering. The disruption of established livelihoods, displacement of communities, and increased competition for dwindling resources contribute to societal instability and conflict (Caruso et al., 2016). Within this context, women and girls often bear a disproportionate burden of the effects of climate change (Abedin et al., 2013). They face unique vulnerabilities stemming from existing gender inequalities that are amplified by climate-related stressors. This includes increased responsibilities for water collection, food provision, and childcare in the face of resource scarcity and environmental disasters. Their limited access to education, land ownership, and decision-making processes further exacerbates their vulnerability.

Indonesia, an archipelagic nation of immense biodiversity and a large coastal population, exemplifies this vulnerability. Ranked third globally in climate change risk, behind only India and Nigeria, Indonesia faces a multitude of climate-related threats (ADB, 2021). Historical data reveals a consistent temperature increase, with an average annual rise of approximately 0.9 °C between 1981 and 2018, culminating in a record high in 2016 (1.2 °C above pre-industrial levels). Projected warming rates of 0.2–0.3 °C per decade suggest significant changes in weather patterns, with increased rainfall in Sumatra, prolonged dry seasons in Java, Bali, and Nusa Tenggara, and a significant rise in sea levels threatening coastal cities. These changes, often exacerbated by El Niño events, intensify droughts and floods, directly impacting food production, water security, and overall economic stability (Margono et al., 2014). The resulting food insecurity, displacement, and increased strain on resources disproportionately impact women and girls, further deepening existing inequalities. In the Indonesian context, these riskscapes intersect with gender dynamics, influencing the vulnerabilities and capacities of women and girls in adapting to environmental shifts (Tickamyer & Kusujiarti, 2020).

Understanding the specific risks faced by women and girls in Indonesia is crucial for developing effective strategies to enhance their resilience. Women often play critical roles in resource management and household well-being, making them particularly susceptible to the impacts of climate change. Factors such as limited access to resources, restricted mobility, and social norms can further compound these vulnerabilities, creating complex riskscapes that require targeted interventions.

In response to these challenges, various strategies have been proposed to empower women and girls in Indonesia to mitigate and adapt to climate change. These strategies encompass a range of approaches, including enhancing access to education and economic opportunities, promoting gender-responsive climate policies, and fostering community resilience through sustainable practices. By addressing the intersecting issues of gender and climate change, these strategies aim to build the adaptive capacity of women and girls, ultimately contributing to more sustainable and equitable outcomes.

This chapter aims to illuminate the gendered impacts of climate change in Indonesia, focusing specifically on the unique vulnerabilities faced by women and girls. We will explore the complex interplay between gender, climate change, and resilience, highlighting the specific challenges faced by women and girls and advocating for tailored solutions that address these unique vulnerabilities. Through an examination of case studies and best practices, this chapter will showcase successful program that have enhanced resilience among women and girls in Indonesia, providing valuable insights into effective gender-responsive approaches to climate change adaptation. Ultimately, this work calls for a fundamental shift towards a more equitable and sustainable future, one where women and girls are empowered to play a central role in building resilience to climate change and shaping a more just and sustainable world.

11.2 The Gendered Impacts of Climate Change in Indonesia

Climate change, a global phenomenon with profound consequences, disproportionately affects women and girls in Indonesia, a nation highly susceptible to its effects. While the impacts of climate change are felt by all, women and girls often bear the brunt of its consequences due to their unique vulnerabilities (Leiserowitz et al., 2023). These vulnerabilities stem from traditional gender roles, limited access to resources, and social and cultural norms that restrict their mobility and decision-making power.

Women and girls face a disproportionate burden from climate change, which worsens the gender inequalities that already exist and highlights their disadvantaged status in various dimensions (Khandelwal et al., 2025). Access to nutritious food, clean water, and safe migration, along with threats like gender-based violence and infectious diseases, creates a landscape where women and girls are particularly vulnerable due to complex social dynamics and biological factors (Wright et al., 2023). Besides that, the impacts on women due to climate change can lead to an increased risk of illness, malnutrition, exposure to sexual violence, and mental health issues (Stone et al., 2022), limited reproductive autonomy, adverse pregnancy outcomes, and in severe cases, death (ACOG, 2018; Rothschild & Haase, 2023a).

In Indonesia, women often bear the primary responsibility for household tasks, including water collection, food preparation, and childcare (Fahrudin et al., 2022). Climate change-induced water scarcity, food insecurity, and increased workloads disproportionately affect women and girls, leading to longer working hours and increased stress (Rothschild & Haase, 2023b). Furthermore, limited access to education, resources, and decision-making power hinders their ability to adapt to climate change impacts, access information about climate-resilient practices, and advocate for their needs. Social and cultural norms can restrict women's mobility and access to information and opportunities, further limiting their ability to participate in climate change mitigation and adaptation efforts. These factors exacerbate their vulnerability to climate change impacts, leading to increased health risks, economic hardship, and disruption to education. Addressing these vulnerabilities requires a shift towards gender-responsive solutions that empower women and girls to participate in climate change adaptation and mitigation efforts.

Women are often considered incapable of saving themselves in disaster conditions and are viewed as vulnerable communities (Ahmad, 2011). As stated in the Disaster Management Law No. 24 of 2007, as referred to in Article 55 paragraph (2) vulnerable groups consist of infants, toddlers, children, pregnant or breastfeeding mothers, sufferers and the elderly. Disasters for women are inseparable from various contexts of life dimensions (Widjanarko & Indaryani, 2015). The vulnerability factor for women in facing disasters is because of their social roles in it (Ejem & Ben-Enukora, 2025). Various general views on women in disaster conditions according to (Smyth & Hai, 2012), include: (a) Women's capacity is not seen; (b) Women are not involved in the planning and decision-making process; (c) Women's issues are not a focus in disaster mitigation; and (d) Most disaster victims are women.

The occurrence of an imbalance in the division of roles that will have an impact on activities in responding to flooding that occurs as a form of adaptation response. KPPPA (2015) stated that climate change adaptation is not gender neutral due to differences in the capacities of men and women, where men can contribute more in responding to tidal flooding as a form of adaptation. Gender injustice that occurs causes discrimination against women. Gender injustice against women includes marginalization in the context of earning a living, women are often considered to have a lower degree than men who earn a living for their families. Subordination is the assumption that women do not need to have a higher education because they only need to work at home. Furthermore, there is a stereotype of women who are physically weaker than men.

Indonesia's vast and diverse archipelago makes it exceptionally vulnerable to climate change. Rising sea levels, intensified flooding, and more frequent droughts pose significant environmental threats, worsening existing social and economic inequalities. Women and girls in Indonesia are particularly vulnerable to these impacts due to several factors:

A. *Traditional Gender Roles*

Climate change is a global phenomenon with far-reaching impacts that transcend environmental, economic, and social dimensions (Dietz et al., 2020). Within the social sphere, traditional gender roles play a significant yet often overlooked role in shaping vulnerability, adaptation, and mitigation strategies. Historically, gender roles have been deeply embedded in cultural, economic, and political structures, assigning distinct responsibilities and expectations to men and women. These roles influence how individuals and communities experience and respond to climate change, often exacerbating existing inequalities. Consequently, women and girls often bear the primary responsibility for household tasks, including water collection, food preparation, and childcare, making them particularly vulnerable to climate change-induced water scarcity, food insecurity, and increased workloads (Fig. 11.1).

In many societies, traditional gender roles dictate that men are primarily responsible for income-generating activities, such as agriculture, fishing, or resource extraction, while women are often tasked with caregiving, household management, and subsistence farming. This division of labor creates gendered vulnerabilities to climate change. For instance, men in agrarian communities may face economic stress due to crop failures or declining fish stocks, while women, who are often responsible for water and fuel collection, may experience increased burdens as these resources become scarcer due to climate-induced droughts or deforestation.

Moreover, traditional gender roles can limit women's access to resources, education, and decision-making power, further constraining their ability to adapt to climate change. In many contexts, women are excluded from land ownership, financial services, and technological innovations, which are critical for implementing adaptive

Fig. 11.1 Women are taking their task on collecting clean water. *Credit* left: Author; right: UNFPA

measures. Conversely, men may face societal pressure to fulfill the role of "provider," leading to risky behaviors such as migration or overexploitation of natural resources in the face of climate stressors.

However, traditional gender roles also present opportunities for climate change mitigation and adaptation. Women, as primary managers of household resources in many cultures, often possess valuable knowledge about sustainable practices, such as water conservation, agroecology, and biodiversity preservation. Empowering women through education, resource access, and leadership roles can enhance community resilience and foster innovative solutions to climate challenges. Similarly, redefining traditional male roles to include caregiving and community stewardship can promote more equitable and sustainable climate responses.

B. *Limited Access to Resources*

Women and girls in Indonesia are often have limited access to education, resources, and decision-making power. This can hinder their ability to adapt to climate change impacts and advocate for their needs. While Indonesia has made progress in education, disparities remain between genders, particularly in rural areas and marginalized communities. Girls may face barriers to completing their education due to cultural norms, economic constraints (e.g., needing to work to support the family), and a lack of access to schools, especially in remote islands or mountainous regions. This limits their knowledge and skills to understand and adapt to climate change impacts.

On the other hand, women are often having limited access to land ownership, credit, and financial resources. This economic vulnerability is amplified by climate change impacts like droughts, floods, and crop failures. Their limited economic independence hinders their ability to invest in adaptation measures or recover from climate-related disasters. Traditional patriarchal structures can further limit their control over household resources. Economic inequality and gender norms intensify women's vulnerability to climate change, as they frequently lack secure land tenure, access to credit, and financial resources needed for climate-resilient practices or post-disaster recovery. This economic marginalization, combined with increased unpaid care burdens, such as fetching water or securing food during droughts or floods, limits their autonomy, heightens stress, and reduces opportunities for education, paid work, and community participation. Consequently, women are often excluded from formal adaptation planning and disaster risk reduction programs, weakening overall community resilience. Ensuring equitable access to resources, education, and decision-making is therefore essential for both gender justice and effective climate adaptation.

C. *Social and Cultural Norms*

Social and cultural norms are deeply ingrained rules and expectations that guide behavior and shape perceptions within a particular society. They are often unspoken but exert a powerful influence on how individuals interact with each other, make decisions, and navigate their lives. In Indonesia, these norms play a significant role in shaping the country's social fabric and influencing how individuals, especially women and girls, experience the impacts of climate change.

Gender norms and social expectations can restrict women's mobility and access to information and opportunities. This can limit their ability to participate in climate change mitigation and adaptation efforts. Indonesia's cultural context often places women in subordinate roles within households and communities. Their voices are frequently marginalized in decision-making processes related to resource management, disaster preparedness, and climate adaptation strategies. This lack of agency prevents them from influencing policies and initiatives that directly affect their lives and well-being.

Gender equality has been placed more on procedural matters, with substantive matters usually overlooked (Oktari et al., 2021). Women's participation is statistically accounted for in terms of, say, filling a 30% quota without necessarily being ensured that they are engaged and empowered in the decision-making process. Policies and programs need to go beyond the guaranty of equal participation and focus on empowering women with necessary skills, knowledge, and resources to input into climate adaptation and mitigation policies in a relevant manner. This involves integrating gender-sensitive strategies within climate policies wherein women are granted equal opportunities as compared to decision-making frameworks and women's roles in environmental leadership are enhanced. By gender mainstreaming in climate policy, Indonesia can construct more equal and sustainable climate responses to climate issues. By so doing, not only is resilience constructed, but women and men are also ensured to contribute and benefit from climate solutions equally.

11.3 Building Resilience: Gender-Responsive Solutions

Building resilience through gender-responsive solutions requires a concerted effort from governments, civil society organizations, and international institutions. Addressing the specific vulnerabilities of women and girls in Indonesia requires a shift towards gender-responsive solutions that empower them to participate in climate change adaptation and mitigation efforts.

In Bangladesh, the training of women in early warning systems for cyclones is a testament to the effectiveness of gender-responsive solutions. This initiative recognizes that women are often disproportionately affected by disasters due to their roles as caregivers and their limited access to information and resources. By equipping them with the knowledge to interpret and act upon early warnings, their safety and that of their families is significantly enhanced.

The success of this program is evident in the declining death toll from cyclones in Bangladesh. While cyclones remain a significant threat, the country has made remarkable progress in disaster preparedness, thanks in part to the robust early warning system and the active participation of women in its implementation (Davison, 2022).

Similarly, in Nepal, the training of women as community health workers demonstrates the effectiveness of investing in women's leadership in disaster risk reduction. These women are crucial first responders, providing vital medical and support services during emergencies. The experience of Nepal's 2015 earthquake highlights the critical need for gender-sensitive disaster preparedness and response. While women played a vital role in providing immediate assistance, they also faced unique challenges, including increased risks of violence and trafficking, limited access to resources, and disproportionate impacts on their livelihoods (Fredricks et al., 2017).

Gender mainstreaming is a national development strategy in Indonesia aimed at addressing gender disparities in access, participation, control, and benefits from development (Suroso et al., 2021). Presidential Instruction No. 9 of 2000 mandates its implementation across sectors to reduce inequality and promote social inclusion for vulnerable groups. As shown in Table 11.1, Indonesia has introduced various policies and programs to mainstream gender considerations within climate resilience and disaster risk management. Gender equality and justice are crucial for mitigating climate change impacts, as responding to conflict, disasters, and vulnerabilities requires everyone's involvement to achieve the "Leave No One Behind" commitment and the Sustainable Development Goals' focus on the inherent rights of nature.

To empower women in responding to climate change impacts and enhance their role in community resilience, several key strategies can be implemented. Key strategies include:

- **Knowledge and Skills Development**

The development of knowledge and skills is crucial for women to effectively participate in climate change mitigation and adaptation efforts. This involves equipping women with the necessary understanding of climate science, environmental issues, and sustainable practices. Furthermore, training programs should focus on building practical skills, such as renewable energy technologies, disaster preparedness, and climate-resilient agriculture.

Investing in women's education and training not only enhances their individual capabilities but also contributes to a more inclusive and equitable society. By fostering women's leadership and decision-making roles in climate action, we can unlock their potential to drive sustainable development and create resilient communities.

- **Empowerment and Leadership**

Empowering women to take on leadership roles in climate action is not just about achieving gender equality: It's about unlocking a critical force for change. Women, often disproportionately affected by climate change, bring unique perspectives, experiences, and solutions to the table. Their leadership is essential for building resilience, driving sustainable development, and ensuring a just transition to a greener future.

- **Community Mobilization**: Fostering community participation and partnerships that prioritize the needs of women and girls. This includes working with community organizations, local leaders, and government agencies to develop and implement gender-sensitive climate change adaptation programs.

Table 11.1 Indonesia's commitment to gender equality in climate and disaster resilience policies

Indonesia's commitment	Gender equality strategy	Agency responsible
Paris agreement: updated NDC 2015–2030	Strengthen community involvement in development planning processes at all levels, considering gender participation, gender equality, gender balance, and vulnerable groups, as well as cross-generational needs	Ministry of environment and forestry (KLHK)
The Sendai framework	Targeting the different needs of specific groups in society such as children and youth, people with disabilities, low-income families, migrants, indigenous peoples, and the elderly. The Sendai Framework mandates gender analysis in DRR (disaster risk reduction) work to identify gender-based differences in needs and to help reduce intersectional risks by understanding vulnerability and what is needed for it—strengthening disaster resilience	National disaster management agency (BNPB)
Presidential Regulation No. 98 of 2021 on Carbon economy implementation value	Preparation of national climate resilience targets and climate resilience action plans implemented based on gender equality and vulnerable groups	Government of Indonesia, coordinating ministry for economic affairs, ministry of environment and forestry (KLHK)
COP26in Glasgow, COP27 and COP28	Gender mainstreaming in gender-responsive climate change adaptation activities and actions	Ministry of environment and forestry (KLHK)
Bali agenda for resilience	It is important to implement policies and programs that consider gender and inclusiveness in disaster risk. Greater investment is needed in women-led civil society organizations and networks to meaningfully participate and influence decision-making	National disaster management agency (BNPB)
Indonesian national disaster management plan 2020–2024	Guidelines for gender mainstreaming, disability mainstreaming and child protection in disaster management	National disaster management agency (BNPB)
National action plan on gender mainstreaming in climate change	Outlines procedures for gender considerations mainstreaming into climate policies and promoting equal access to climate adaptation opportunities	Ministry of Women's empowerment and child protection (KPPPA)

(continued)

Table 11.1 (continued)

Indonesia's commitment	Gender equality strategy	Agency responsible
Guidelines for gender mainstreaming and social inclusion in climate change projects	Ensure that climate change policies, programs, and projects are inclusive and consider the needs and roles of various social groups, particularly vulnerable groups such as women, persons with disabilities, indigenous peoples, and marginalized communities. These guidelines concentrate on gender and social analysis with a view to understanding the impacts of climate change on various groups and providing means to integrate gender considerations in project planning, implementation, and monitoring	Fiscal policy agency ministry of finance of the republic of Indonesia
Law No. 24 of 2007 on disaster management	Demands the participation of vulnerable groups, including women and children, in disaster risk reduction and recovery	National disaster management agency (BNPB)
Regulation of the head of BNPB number 13 of 2014	Regulates gender mainstreaming in disaster management so that disaster policies and programs take into account gender considerations and address the specific needs of women and vulnerable groups	National disaster management agency (BNPB)
Indonesia's medium-term development plan 2020–2024	Prioritizes gender equality in reducing disaster risk, adapting to climate change, and building resilience	National development planning agency (Bappenas)

Source Adapted from Indonesia UNFPA. (2023) with modifications

As an example, group of women farmers in Sayung Demak formed a cooperative and received training on climate-smart agricultural practices, including water conservation techniques, drought-resistant rice varieties, and organic farming methods. The cooperative shared their knowledge with other farmers in the village, leading to increased rice yields and improved food security. They also advocated for government support for sustainable agriculture practices.

Despite the severity of Indonesia's climate crisis, there's cause for optimism. A landmark 2021 court ruling held government officials accountable for environmental violations, demanding action on air and carbon pollution. Feminist activists are leading the charge for equitable solutions, advocating for increased women's participation in decision-making, secure land rights, economic empowerment, and regional cooperation on climate-related gender justice. Supporting these initiatives can build a sustainable and prosperous future for Indonesia.

Indonesia's government, NGOs, universities, communities, and the private sector are actively engaging women, girls, and people with disabilities in climate action.

Training, mentorship, and inclusive decision-making processes are proving effective in building community resilience and increasing the success of climate initiatives (Mulyasari & Shaw, 2013). These efforts are also boosting the number of skilled women in renewable energy and promoting inclusivity in business. This focus on women's leadership and development is integral to Indonesia's Nationally Determined Contribution (NDC) and its Long-term Strategy for Low Carbon and Climate Resilience 2050 (LTS-LCCR).

The Ministry of Environment and Forestry is taking steps to reduce the impacts of climate change by facilitating partnership and collaboration among all stakeholders in an endeavor to improve all activity capacity to adapt and mitigate changes by empowering communities through executing the Climate Village Program (Proklim) which is regulated in the Minister of Environment Regulation Number 19 of 2012 concerning the Climate Village Program. Proklim is a program at the local level that is carried out on a national scale. The Ministry of Environment and Forestry initiated this program as a method of environment conservation and management and utilization of potential within the local community, Proklim aims at reducing the impacts of climate change by strengthening the local population to contribute to all adaptation and mitigation actions being taken. Proklim implementation across various regions of Indonesia is also expected to provide benefits in the form of local scale contributions towards the achievement of the objective of reducing national greenhouse gas (GHG) emissions. Proklim also utilizes the potential of development at the regional level and its expression in terms of information and data that can be used as input to be processed during policy, program, activity and strategy formulation aimed at offsetting the impacts of climate change.

Proklim's strategy for implementation is on capacity development through the adoption of trainings, workshops, and awareness raising as a means of enhancing local capacities for climate adaptation and mitigation. Climate-resilient economic activities such as organic farming, renewable energy enterprises, and waste management techniques are promoted to improve green livelihoods (Usman & Balsalobre-Lorente, 2022). Disaster risk reduction involves the creation of early warning systems, community-level preparedness planning, and climate-resilient infrastructure. Public–private partnerships are promoted, with government agencies, private sector companies, and civil society organizations uniting in an effort to scale up climate action (Sufri & Lassa, 2024).

Women play a significantly crucial role in the success and implementation of Proklim initiatives. They enhance sustainable agriculture through the application of climate-smart agriculture, organic agriculture, and production of food for communities. In the promotion of renewable energy, women mobilize and utilize alternatives such as biogas and solar energy and therefore reduce dependence on fossil fuel (Setyowati et al., 2025; Utami & Indriyany, 2024). Women also engage in waste management operations involving recycling, composting, and plastic minimization towards a world with reduced plastics. Grassroots leaders and women activists are campaigning and demanding gender-responsive climate policies as a platform for ensuring that the voice of women is included in decision-making (Flavell, 2023).

To illustrate this further, the following section presents one of the case studies in the North Coast of Central Java from Semarang to Sayung Demak on how women play a role in disaster management efforts due to tidal flooding. This study highlights the role of women in disaster mitigation and response. The research applies the tools from the Kikkawa (1999) framework and women's DRR activities demonstrating how women contribute to effective disaster risk efforts.

11.4 Case Studies and Best Practices from Semarang and Sayung Demak, Indonesia

This chapter showcases successful programs that have built resilience among women and girls in Indonesia. These programs provide valuable insights into the effectiveness of gender-responsive approaches to climate change adaptation. The focus of this research area is the north coast of Semarang–Demak, considering that this area is an area affected by tidal floods that occur every year. The north coast of Semarang–Demak stretches along the coast of West Semarang, North Semarang, Genuk to Sayung Demak (as seen in Fig. 11.2).

The coastal area stretching from the coast of Semarang City to the coast of Sayung Demak is one of the areas prone to tidal flooding which has occurred for more than the last two decades (Mussadun et al., 2016; Hidayat et al., 2020; Maharlika et al.,

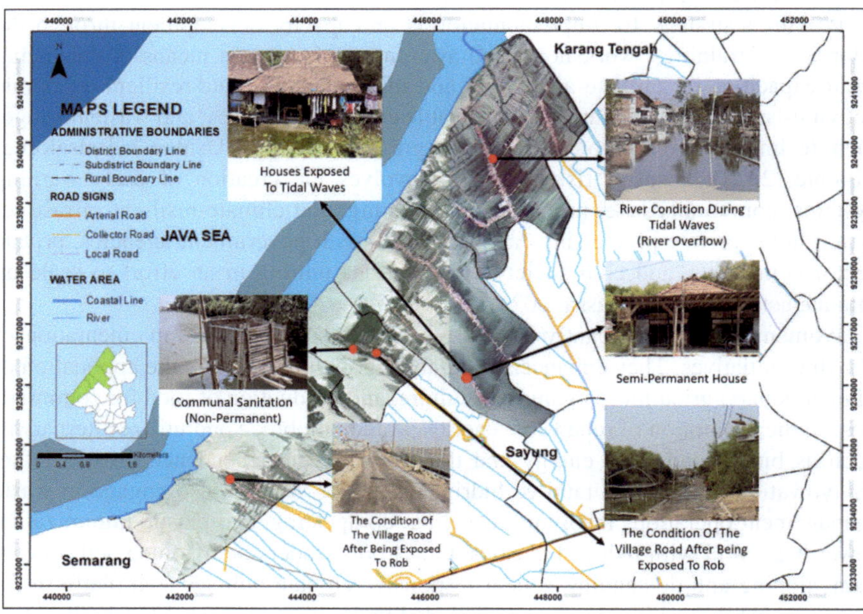

Fig. 11.2 North coast of Semarang–Demak. *Source* Sarasadi: 2021

2020). This has a social impact felt by the community, such as a shift in gender roles in carrying out daily activities (Bott et al., 2021). The public sector, which was initially dominated by men, is now largely replaced by women. The social impact felt by the community, such as a shift in gender roles in carrying out daily activities. The public sector, which was initially dominated by men, is now largely replaced by women. Coastal men who were initially predominantly working in the fisheries sector have now changed professions and even become unemployed due to the decline in fish production due to tidal flooding (Utami & Indriyany, 2024). This condition causes women to take part in taking control of the public sector.

In line with Pratiwi (2017) that coastal women also participate in earning a living as an additional means to sustain their lives. Coastal women work a lot in the fish processing industry such as preserving and packaging fish for export. The work done by women can improve the family economy, but on the other hand there is a difference in terms of wages received by women which are smaller than those of men. The results of research conducted by Utami & Indriyany (2024) in the Rob area of North Pekalongan District emphasized that the shift in gender roles in coastal communities in responding to rob floods gave rise to an imbalance in the roles between men and women, thus causing gender injustice. Gender injustice causes discrimination, because it is a double burden for coastal women. Gender injustice includes marginalization, subordination, stereotypes or negative labelling and a double burden for women (Fakih, 2006). The marginalization referred to here is the economic impoverishment of women. Women are considered to have a lower degree than men who earn a living for their families. So, women will get lower wages in doing work or helping men earn a living. Existing subordination such as the assumption that women do not need to have a higher education because they will only work at home. Negative labelling or stereotypes such as women have a weak physique compared to men. Women also have the possibility of having a double burden in allocating time in carrying out the roles of production, reproduction, and management of society and politics.

Based on the formulation of the Ministry of Women's Protection and Child Protection (Kemen PPA) regarding the National Action Plan for Gender and Climate Change (RAN GPI) in order to achieve gender equality in dealing with climate change in Indonesia, it aims to increase access to knowledge and information on disasters, so that it can increase women's participation in disaster mitigation. One of these objectives is implemented through the strategic role of women in tidal flood disaster mitigation. Various efforts to increase women's capacity in mitigating tidal floods include:

- Increasing women's awareness in understanding environmental situations and various threats of tidal floods.
- Providing an understanding of vulnerability and the ability to measure women's capacity.
- Providing an understanding of the ability to assess the risks that will be faced as individuals, families and communities.

Fig. 11.3 Community outreach and education program in a rural Indonesian village. *Credit* Author

- Providing an understanding of planning and action capabilities in tidal flood mitigation efforts.
- Improving women's capacity in monitoring, evaluating and ensuring tidal flood mitigation efforts

Various efforts to increase the capacity of women's roles in tidal flood mitigation efforts are implemented in various activities including, socialization and forum group discussions (FGD) on tidal flood mitigation management, forming disaster preparedness groups that contribute to the planning process, decision making, physical and material assistance (Andrista et al., 2025). This activity aims to increase preparedness through socialization and forum group discussions (FGD) attended by women from the Family Welfare Development (PKK) organization group (Fig. 11.3).

Based on the results of the discussion activities, most women experience obstacles in understanding and implementing tidal flood mitigation. This condition is indicated by their lack of understanding regarding the evacuation process. After the socialization of efforts to increase women's capacity in preparedness for tidal floods that have been understood include: (1) Making an evacuation plan through building communication with both family and the environment; (2) Preparing various important components such as documents, medical equipment and emergency needs, prioritizing elderly and disabled family members; (3) Preparing alternative communication tools as an anticipatory effort in the event of a loss of communication networks.

The involvement of women is very important as an initial foundation in adaptation planning, one of which is in tidal flood situations. According to Ritohardoyo et al. (2014), there are several forms of adaptation carried out by women in dealing with tidal floods, including:

A. *Economic Adaptation*

The decline in family income levels often forces women to seek alternative livelihoods, but women are often also the main breadwinners. This condition causes women to have a double burden, namely as breadwinners and taking care of domestic work. Women are the most sensitive to the impacts of tidal flooding. As in Timbulsloko Village, Sayung District, most of the ponds are no longer productive due to the

Fig. 11.4 Distribution of household needs affected by rob floods in Sayung District. *Credit* Puspita Bahari women fisherman community documentation

expansion of the sea level rise area. Various alternative efforts have been made by the people of Sayung District, including women who work as industrial workers, open stalls, and process seafood which are carried out with great limitations when tidal flooding occurs. Limited access out of the village as felt by women in Timbulsloko Village who sell processed salted fish and presto fish out of the village using a boat and continue on foot to leave the village.

B. *Social Adaptation*

Social adaptation is an effort that can be made by women in tidal flood mitigation including the social life of the community (Harwitasari & van Ast, 2011). As with women along the north coast of Demak Regency consisting of groups of mothers who work together to distribute basic food assistance, in addition to opening public kitchens. This activity aims to make it easier for local people to meet household needs (Ritohardoyo et al., 2014) (Fig. 11.4).

Other activities carried out by the Puspita Bahari Women Fishermen Community in 2024 include the distribution of 732 cloth sanitary napkins as an effort to increase efficiency and reduce waste, distribution of chili seeds, eggplant, ornamental plants, planting media. Based on Masnuah's explanation as the founder of the Puspita Bahari Women Fishermen Community, this activity is a step of solidarity to support women affected by rob floods, so that the needs of the community after the disaster can be met.

C. *Physical Adaptation*

Physical adaptation is an effort to overcome environmental changes (Harwitasari & van Ast, 2011), which cannot be separated from the economic capabilities of each individual or group. Seeing the current conditions such as in settlements along the northern coast of Demak Regency, women are trying to carry out physical adaptation by planting mangroves as an effort to prevent rising sea levels, raising household furniture as an initial step in evacuating household assets, if their house has a dirt floor, they will sprinkle scouring ash so that the water is absorbed by the ash (Listyowati & Siscawati, 2023).

To effectively address gender inequality and advocate for women's rights, it is crucial to embrace resilience-based approaches that confront these disparities and promote women's rights by using the concept of resilience that holds significant potential for development practitioners and policymakers, enabling them to enhance their efforts in supporting gender equality and advancing women's rights (Smyth & Sweetman, 2015). By these conditions, it is important to build the nexus to gender-responsive solutions, which could involve highlights of what women and girls can do to build knowledge on climate-resilient initiatives or tap resources that enable their role in decision-making for inclusive policy action.

The fate of Indonesia, a nation grappling with the escalating impacts of climate change, hangs in the balance. But the narrative is not predetermined. This not only illuminated the profound and disproportionate challenges faced by Indonesian women and girls in the face of climate change but has also showcased the extraordinary resilience and transformative potential they possess. Their strength, their knowledge, and their leadership are not merely assets; they are essential to building a sustainable and equitable future.

The path forward demands a fundamental shift in mindset and action. It requires a concerted effort from governments, international organizations, NGOs, and communities to prioritize gender equity in all climate change adaptation and mitigation strategies. This means dismantling systemic barriers that perpetuate inequality, ensuring women's equal access to resources, education, and decision-making processes. It means investing in programs that empower women to lead sustainable development initiatives, recognizing their invaluable contributions to climate resilience. It means listening to their voices, amplifying their experiences, and centering their needs in the design and implementation of climate action plans.

The challenge is immense, but the opportunity is even greater. By empowering Indonesian women and girls, we unlock not only their potential but also the potential of their communities and their nation. Their resilience, their ingenuity, and their unwavering commitment to a better future are the keys to navigating the climate crisis and building a more just and sustainable world. Let their strength inspire us to act, to innovate, and to create a future where climate resilience and gender equity are inextricably linked, not just as ideals but as lived realities. The time for decisive action is not tomorrow; it is now.

11.5 Conclusion

This chapter has illuminated the stark reality of climate change in Indonesia, revealing a crisis deeply intertwined with existing social and gender inequalities. The impacts are not evenly distributed; rather, they fall disproportionately on the shoulders of women and girls, exacerbating existing vulnerabilities and creating a complex web of interconnected challenges. From the relentless burden of securing food and water resources in the face of increasingly erratic weather patterns to the heightened risks they face during natural disasters, Indonesian women are on the frontlines of this crisis, often bearing the brunt of its devastating consequences. Their limited access to resources, land ownership, education, and decision-making processes further amplifies their vulnerability, leaving them marginalized in the very processes that determine their future.

The evidence presented throughout this chapter paints a clear picture: the warming temperatures, rising sea levels, and increasingly frequent and intense extreme weather events are not abstract threats; they are tangible realities shaping the lives of millions. The vulnerability of Indonesia, ranked third globally in climate change risk, underscores the urgency of addressing this crisis with immediate and decisive action (ADB, 2021). The projected changes in rainfall patterns, the intensification of droughts and floods, and the ever-present threat of coastal erosion all contribute to a precarious future for Indonesian communities, particularly those most vulnerable.

However, amidst the challenges, this chapter also offers a beacon of hope. The examination of successful programs and initiatives demonstrates the transformative power of gender-responsive approaches to climate change adaptation. By empowering women and girls with the resources, knowledge, and agency to participate fully in decision-making processes and resource management, we can unlock their immense potential as agents of change. These initiatives showcase the effectiveness of community-based solutions, highlighting the crucial role of local knowledge and leadership in building resilience. When women are empowered, their communities become more resilient, fostering sustainable practices that benefit everyone.

The conclusion, therefore, is not simply a summary of findings but a call to action. It is a demand for a fundamental shift in how we approach climate change adaptation and mitigation strategies. Policymakers, NGOs, international organizations, and community leaders must prioritize gender equity in their efforts to combat climate change. This requires not just acknowledging the disproportionate impact on women but actively integrating gender considerations into all aspects of planning, implementation, and evaluation. It demands the dismantling of systemic barriers that limit women's access to resources and decision-making power. It necessitates investing in education and training programs that equip women with the skills and knowledge to adapt to climate change and lead sustainable development initiatives. Only through such comprehensive and transformative action can we hope to build a more resilient, equitable, and sustainable future for all Indonesians, ensuring that women and girls are not just survivors but active participants in shaping their own destinies and the future of their communities. The time for decisive action is now.

References

Abedin, M. A., Habiba, U., & Shaw, R. (2013). Gender and climate change: Impacts and coping mechanisms of women and special vulnerable groups. In *Climate change adaptation actions in Bangladesh* (pp. 165–184). Tokyo: Springer Japan.

ACOG. (2018). Climate change and women's health: Position statement. https://www.acog.org/clinical-information/policy-and-position-statements/position-statements/2021/addressing-climate-change

ADB. (2021). *ADB climate risk country profile: Indonesia.* Manila: Asian Development Bank. Retrieved from https://www.adb.org/publications/climate-risk-country-profile-indonesia

Ahmad, N. (2011). Gender and climate change: Myth vs. reality. World Bank Blogs. https://blogs.worldbank.org/endpovertyinsouthasia/gender-and-climate-change-myth-vs-reality

Andrista, S., Utami, N. P., Hukom, V., Nielsen, M., & Nielsen, R. (2025). Responses to climate change: Perceptions and adaptation among small-scale farmers in Indonesia. *Journal of Environmental Management, 377*, Article 124593. https://doi.org/10.1016/j.jenvman.2025.124593

Benevolenza, M. A., & DeRigne, L. (2019). The impact of climate change and natural disasters on vulnerable populations: A systematic review of literature. *Journal of Human Behavior in the Social Environment, 29*(2), 266–281. https://doi.org/10.1080/10911359.2018.1527739

Bott, L. M., Schöne, T., Illigner, J., Haghighi, M. H., Gisevius, K., & Braun, B. (2021). Land subsidence in Jakarta and Semarang Bay-The relationship between physical processes, risk perception, and household adaptation. *Ocean Dan Coastal Management, 211*, Article 105775. https://doi.org/10.1016/j.ocecoaman.2021.105775

Calvin, K., Dasgupta, D., Krinner, G., Mukherji, A., Thorne, P. W., Trisos, C., Romero, J., Aldunce, P., Barret, K., Blanco, G., Cheung, W. W. L., Connors, S. L., Denton, F., Diongue-Niang, A., Dodman, D., Garschagen, M., Geden, O., Hayward, B., Jones, C., Jotzo, F., Krug, T., Lasco, R., Lee, J., Ha, M. (2023). *IPCC, 2023: Climate Change 2023: Synthesis report, summary for policymakers*. Contribution of Working Groups I, II and III to the Sixth Assessment Report of the Intergovernmental Panel on Climate Change. https://doi.org/10.59327/IPCC/AR6-9789291691647.001

Caruso, R., Petrarca, I., & Ricciuti, R. (2016). Climate change, rice crops, and violence in Indonesia. *Journal of Peace Research, 53*(1), 66–83. https://doi.org/10.1177/0022343315616061

Davison, C. (2022). The country trailblazing the fight against disasters. BBC Future.

Dietz, T., Shwom, R. L., & Whitley, C. T. (2020). Climate change and society. *Annual Review of Sociology, 46*(1), 135–158. https://doi.org/10.1146/annurev-soc-121919-054614

Doherty, T. J., & Clayton, S. (2011). The psychological impacts of global climate change. *American Psychologist, 66*(4), 265.

Egaputra, A. A., Ismunarti, D. H., & Pranowo, W. S. (2022). Inventarisasi Kejadian Banjir Rob Kota Semarang Periode 2012–2020. *Indonesian Journal of Oceanography, 4*(2), 29–40. https://doi.org/10.14710/ijoce.v4i2.13240

Ejem, A. A., & Ben-Enukora, C. A. (2025). Gendered impacts of 2022 floods on livelihoods and health vulnerability of rural communities in select Southern states in Nigeria. *Discover Social Science and Health, 5*(1), 67. https://doi.org/10.1007/s44155-025-00211-7

Flavell, J. (2023). *Mainstreaming gender in global climate governance: Women and gender constituency in the UNFCCC* (p. 199). Taylor & Francis.

Fahrudin, A., Patrianti, T., & Yusuf, H. (2022). Climate change, disaster, and social work in Indonesia. *The palgrave handbook of global social problems* (pp. 1–14). Springer International Publishing.

Fakih, M. (2006). *Gender analysis and social transformation.* Yogyakarta: Student Library.

Fredricks, K., Dinh, H., Kusi, M., Yogal, C., Karmacharya, B. M., Burke, T. F., & Nelson, B. D. (2017). Community health workers and disasters: Lessons learned from the 2015 earthquake in Nepal. *Prehospital and Disaster Medicine, 32*(6), 604–609.

Harwitasari, D., & van Ast, J. A. (2011). Climate change adaptation in practice: People's responses to tidal flooding in Semarang, Indonesia. *Journal of Flood Risk Management, 4*(3), 216–233. https://doi.org/10.1111/j.1753-318X.2011.01104.x

Hidayat, T. A., Helmi, M., Widada, S., Satriadi, A., Setiyono, H., Ismanto, A., & Yusuf, M. (2020). Pengolahan Data Satelit Sentinel-1 Dan Pasut Untuk Mengkaji Area Genangan Akibat Banjir

Pasang Di Kecamatan Sayung, Kabupaten Demak. *Indonesian Journal of Oceanography, 2*(4), 306–312. https://doi.org/10.14710/ijoce.v2i4.8405

Indonesia, U. N. F. P. A. (2023). The impact of climate change on gender-based violence in Indonesia.

IPCC. (2021). *Climate change 2021: The physical science basis. contribution of working group I to the sixth assessment report of the intergovernmental panel on climate change.* Cambridge University Press. In Press.

Jha, M. K., & Dev, M. (2024). Impacts of Climate Change. In *Smart Internet of Things for Environment and Healthcare* (pp. 139–159). Cham: Springer Nature Switzerland.

Kementerian Pemberdayaan Perempuan dan Perlindungan Anak Republik Indonesia. (2015). Pedoman Umum Adaptasi Perubahan Iklim yang Responsif gender. Jakarta: Kementerian Pemberdayaan Perempuan dan Perlindungan Anak Republik Indonesia.

Khandelwal, S., Mehra, M., & Parwez, S. (2025). Climate change and its impact on women in South Asia: evidence from a gender lens. In *Health and Climate Change* (pp. 289–311). Elsevier. https://doi.org/10.1016/B978-0-443-29240-8.00013-4

Kikkawa, T. (1999). *Risk communication: Aiming at mutual understanding and better decision making* (p. 197). Tokyo, Japan: Fukumura Press. (in Japanese)

Leiserowitz, A., Rosenthal, S., Verner, M., Lee, S., Ballew, M., Carman, J., Goldberg, M., Marlon, J., Paramita, E., Chamim, M., & Daggett, M. (2023). Climate change in the Indonesian Mind. Yale University, New Haven, CT, Yale Program on Climate Change Communication.

Listyowati, & Siscawati, M. (2023). Perempuan dan bencana: kerentanan, respon, dan inisiasi perempuan dalam menghadapi bencana. *VIDYA WERTTA: Media Komunikasi Universitas Hindu Indonesia, 6*(2), 158–172. https://doi.org/10.32795/vw.v6i2.4975

Margono, B. A., Potapov, P. V., Turubanova, S., Stolle, F., & Hansen, M. C. (2014). Primary forest cover loss in Indonesia over 2000–2012. *Nature Climate Change, 4*(8), 730–735.

Minister of Environment Regulation Number 19 of 2012 concerning the Climate Village Program

Mulyasari, F., & Shaw, R. (2013). Role of women as risk communicators to enhance disaster resilience of Bandung, Indonesia. *Natural Hazards, 69*(3), 2137–2160. https://doi.org/10.1007/s11069-013-0798-4

Oktari, R. S., Kamaruzzaman, S., Fatimahsyam, F., Sofia, S., & Sari, D. K. (2021). Gender mainstreaming in a disaster-resilient village programme in Aceh Province, Indonesia: Towards disaster preparedness enhancement via an equal opportunity policy. *International Journal of Disaster Risk Reduction, 52*, Article 101974. https://doi.org/10.1016/j.ijdrr.2020.101974

Pratiwi, N. A. H., Rahmawati, Y. D., & Setiono, I. (2017). Gender equality in climate change adaptation: A case of Cirebon, Indonesia. *The Indonesian Journal of Planning and Development*, *2*(2), 74. https://doi.org/10.14710/ijpd.2.2.74-86

Republic of Indonesia Law Number 24 of 2007 on Disaster Management.

Ritohardoyo, S., Sudrajat, & Kurniawan, A. (2014). *Aspek sosial banjir genangan (ROB) di kawasan pesisir*. Yogyakarta: Gadjah Mada University Press. Retrieved from https://balaiyanpus.jogjaprov.go.id/opac/detail-opac?id=270494

Rothschild, J., & Haase, E. (2023a). The mental health of women and climate change: Direct neuropsychiatric impacts and associated psychological concerns. *International Journal of Gynecology & Obstetrics, 160*(2), 405–413. https://doi.org/10.1002/ijgo.14479

Rothschild, J., & Haase, E. (2023b). Women's mental health and climate change Part II: Socioeconomic stresses of climate change and eco-anxiety for women and their children. *International Journal of Gynecology & Obstetrics, 160*(2), 414–420. https://doi.org/10.1002/ijgo.14514

Sarasadi, A., & Rudiarto, I. (2021). Kerentanaan dan Strategi Adaptasi Masyarakat Terhadap Bencana Rob di Kawasan Pesisir Kecamatan Sayung Kabupaten Demak. *Teknik PWK (Perencanaan Wilayah Kota), 10*(2), 91–102.

Setyowati, R., Rusdiyana, E., Widiyanto, W., Sugihardjo, S., Lestari, E., & Arwindianti, M. L. (2025). Women's contribution to climate change action: A gender analysis. In *BIO Web of Conferences* (Vol. 155, p. 06012). EDP Sciences. https://doi.org/10.1051/bioconf/202515506012

Sivakumar, B. (2011). Global climate change and its impacts on water resources planning and management: Assessment and challenges. *Stochastic Environmental Research and Risk Assessment, 25*, 583–600.

Smyth, I., & Hai, V. M. (2012). *The disaster crunch model: Guidelines for a gendered approach.* Oxford: Oxfam. Retrieved from http://hdl.handle.net/10546/247511

Smyth, I., & Sweetman, C. (2015). Introduction: Gender and Resilience. *Gender & Development, 23*(3), 405–414. https://doi.org/10.1080/13552074.2015.1113769

Stone, K., Blinn, N., & Spencer, R. (2022). Mental health impacts of climate change on women: A scoping review. *Current Environmental Health Reports, 9*(2), 228–243. https://doi.org/10.1007/s40572-022-00346-8

Sufri, S., & Lassa, J. A. (2024). Integration of disaster risk reduction and climate change adaptation in Aceh: Progress and challenges after 20 Years of Indian Ocean Tsunamis. *International Journal of Disaster Risk Reduction, 113*, 1–15. https://doi.org/10.1016/j.ijdrr.2024.104894

Suroso, D. S. A., Hilman, D., Setiawan, B., Pradono, Fitriyanto, M. S., Puspitasari, N., & Hastari, M. A. (2021). Climate change adaptation in Indonesia: Reviews on adaptation governance, metrics and financing. Strengthen national climate policy implementation: Comparative empirical learning & creating linkage to climate finance (SNAPFI) National Study Report.

Tickamyer, A. R., & Kusujiarti, S. (2020). Riskscapes of gender, disaster and climate change in Indonesia. *Cambridge Journal of Regions, Economy and Society, 13*(2), 233–251.

Usman, M., & Balsalobre-Lorente, D. (2022). Environmental concern in the era of industrialization: Can financial development, renewable energy and natural resources alleviate some load? *Energy Policy, 162*, Article 112780. https://doi.org/10.1016/j.enpol.2022.112780

Utami, W. K., & Indriyany, I. A. (2024). Gender Responsive Climate Change Adaptation and Mitigation Analysis through Flamboyan Climate Group in Tangerang City. *ijd-demos, 6*(3). https://doi.org/10.37950/ijd.v6i3.508

Widjanarko, M., & Indaryani, M. (2015). Women nurturing community during natural disaster: A study of the ecology of Kudus Rahtawu Muria mountain village. *Jurnal Perempuan, 20*(3), 207–211.

Wright, M. L., Drake, D., Link, D. G., & Berg, J. A. (2023). Climate change and the adverse impact on the health and well-being of women and girls from the women's health expert panel of the American academy of nursing. *Nursing Outlook, 71*(2), Article 101919. https://doi.org/10.1016/j.outlook.2023.101919

Chapter 12
Climate Crisis and Disaster Management in Japan: Scrutinizing from the Perspective of Gender Difference and Inclusion

Nina Takashino and Minakshi Keeni

Abstract Research on gender equality in disaster prevention and resilience remains underdeveloped, particularly in the context of Japan—a country frequently impacted by natural disasters. This chapter delves into climate crises and disaster reduction through the lens of gender equality and inclusion, analyzing the shifts in disaster management practices post the Great Hanshin Earthquake and the Great East Japan Earthquake. Utilizing databases such as Web of Science and ScienceDirect, this study systematically reviews scholarly articles to elucidate the roles of women in disaster prevention, mitigation, and post-disaster reconstruction. It highlights the evolving understanding of gender dynamics in disaster scenarios and examines the achievements and challenges of integrating gender perspectives in Japan's disaster management frameworks.

Keywords Gender and disaster management · Climate crisis · Disaster risk reduction (DRR) · Japan · Gender inclusion

12.1 Introduction

The escalating severity of climate crises globally highlights the urgent need for robust and effective disaster management strategies. As environmental catastrophes increase in frequency and intensity, safeguarding vulnerable populations and mitigating extensive damage has become a critical priority. Data from the International Disaster Database (EM-DAT) reveal a concerning trend: Economic losses attributed

N. Takashino (✉)
College of Policy Science, Ritsumeikan University, Ibaraki, Japan
e-mail: ninat@fc.ritsumei.ac.jp

M. Keeni
Graduate School of Agricultural Science, Tohoku University, Sendai, Japan

Md. A. Abedin et al. (eds.), *Climate Crisis in South and East Asia*, Sustainability Solutions, https://doi.org/10.1007/978-3-032-01012-4_12

to extreme weather events have surged sevenfold since the 1970s (World Meteorological Organization [WMO], 2021). This rise in disaster-related losses highlights the inadequacy of traditional disaster response frameworks, which often fail to account for the differential impacts on marginalized groups, particularly women and gender minorities (Chowdhury et al., 2022; Enarson & Fordham, 2001).

In this context, integrating gender perspectives into disaster management is increasingly recognized as essential for enhancing community resilience. Disasters disproportionately affect women due to pre-existing social, economic, and cultural inequalities (Amaratunga, 2023; Gaillard et al., 2017). Women often face limited access to resources, decision-making power, and critical information, exacerbating their vulnerability during crises (Fothergill, 1998; Sohrabizadeh et al., 2016). For instance, studies on the 2004 Indian Ocean tsunami revealed that women accounted for 70% of fatalities, partly due to sociocultural norms that restricted their mobility and survival skills (Enarson, 2000; Oxfam, 2005). Similarly, post-disaster recovery efforts frequently overlook women's specific needs, such as safe shelter, reproductive health services, and protection from gender-based violence (Daalen et al., 2022a, 2022b; Fisher, 2010).

Despite growing awareness of these disparities, gender-inclusive disaster management remains underdeveloped in many regions, including Japan—a country prone to earthquakes, tsunamis, and typhoons. While Japan has made strides in disaster risk reduction (DRR) through frameworks like the Sendai Framework for Disaster Risk Reduction 2015–2030 (United Nations Office for Disaster Risk Reduction [UNDRR], 2015), gender considerations are often superficially integrated into policy implementation (Priebe, 2024; Saito, 2014). For example, women's participation in disaster planning remains limited, and evacuation protocols rarely address the needs of caregivers or elderly women (Petraroli & Baars, 2022; Tanaka et al., 2019).

This chapter examines Japan's disaster management through the lens of gender equality and inclusion, analyzing shifts in policy and practice following catastrophic events like the 1995 Great Hanshin Earthquake and the 2011 Great East Japan Earthquake. By systematically reviewing scholarly literature from databases such as Web of Science and ScienceDirect, this study elucidates the evolving role of women in disaster prevention, mitigation, and reconstruction. It also evaluates the achievements and persistent challenges in Japan's efforts to mainstream gender perspectives in DRR, offering lessons for global disaster resilience strategies.

12.2 Gender Dynamics in Disaster Situations

12.2.1 Overview of Gender Impacts in Disasters Globally

The impacts of disasters are felt differently across genders globally, with women often bearing a disproportionately greater burden due to existing social, economic, and cultural inequalities. Disasters exacerbate vulnerabilities for women, primarily due to their limited access to resources and exclusion from decision-making processes. In many cultures, women are less likely to own land or have access to information and technology, which hampers their ability to recover. For instance, after the 2004 Indian Ocean tsunami, the mortality rate for women soared as many lacked essential survival skills such as swimming or climbing, which could have saved their lives (Enarson, 2000; Oxfam, 2005). Economically, women suffer more from the aftermath of disasters. Following Hurricane Katrina in 2005, studies indicated that women were significantly impacted, with a large gap in earnings and employment recovery compared to men (Fothergill & Peek, 2004). Additionally, disasters can heighten health and safety risks for women, including sexual violence and a lack of access to healthcare and reproductive services, as evidenced by the increase in gender-based violence following the 2010 Haiti earthquake (Kolbe et al., 2010). Women's roles as primary caregivers also limit their mobility during disasters, which was tragically highlighted during the 2011 Great East Japan Earthquake and Tsunami, where many women perished while trying to protect their families (Cabinet Office, Government of Japan, 2012). Furthermore, the long-term recovery phase often reveals gender-specific challenges; for example, after the 2013 Typhoon Haiyan in the Philippines, women faced substantial barriers to recovery due to a lack of legal rights to land and property, critical for securing housing and livelihoods in the rebuilding process (Oxfam, 2014).

12.2.2 Literature Review on Gender and Disaster Management

In recent years, the issue of gender in disasters has been actively discussed. Table 12.1 offers a comprehensive overview of pivotal research in the area of gender and disaster management, chronicling two decades of scholarly contributions. Fothergill (1998) addresses "The Neglect of Gender in Disaster Work," emphasizing the critical yet often overlooked impact of disasters on gender dynamics. The research specifically highlights the need for gender-sensitive approaches in disaster responses by focusing on how the pre-existing lower status of women and their limited power can lead to severe consequences during disasters. Similarly, Enarson and Morrow (1998), in "Why Gender? Why Women? An Introduction to Women and Disaster," provide foundational insights into the unique vulnerabilities and capacities of women in disaster contexts, advocating for more inclusive disaster management practices. Expanding this perspective, Gaillard et al. (2017), in their study "Beyond Men and Women: A Critical Perspective on Gender and Disaster," explore a more nuanced understanding of gender by including diverse gender identities and their experiences during disasters. The research cautions that discussing gender differences in disaster management in a dualistic manner will result in a lack of adequate support for sexual minorities. Further advancing this discourse, Hemachandra et al. (2018) discuss the "Role of Women in Disaster Risk Governance," highlighting the significant contributions women can make to disaster preparedness and response if involved in government structures. Collectively, these studies reflect a progressive shift towards recognizing and integrating gender-specific needs and contributions in disaster management.

When considering disaster and gender, several crucial issues arise. First, there are differential impacts during the disaster response phase (Fig. 12.1). Disasters disproportionately affect women and girls due to pre-existing gender inequalities. For example, of the 230,000 people killed in the 2004 Indian Ocean Tsunami, 70% were women (Okai, 2022). Women often have less access to resources, information, and decision-making power, making them more vulnerable during and after disasters. Additionally, from a caregiving perspective, women typically bear the primary responsibility for caring for children, the elderly, and the sick, which can hinder their ability to evacuate, seek safety, and access aid. After a disaster, the risk of violence could increase (Daalen et al., 2022a, 2022b; Fisher, 2010). Disasters can heighten the risk of gender-based violence, including sexual exploitation and trafficking. Furthermore, women are often more reliant on informal or low-paid employment, making them more susceptible to economic shocks caused by disasters.

Disaster preparedness, response, and recovery efforts often fail to adequately address the specific needs and vulnerabilities of women and girls. The lack of gender-sensitive approaches, such as insufficient sex-disaggregated data, exclusion from decision-making, and inadequate provision of gender-specific needs, can lead to increased gender inequality in disaster outcomes (Sams, 2019). The absence of sex-disaggregated data hinders the understanding of the differential impacts of disasters on women and men. Women are often excluded from decision-making processes related to disaster risk reduction and response (Alcobé & Harty, 2023). Additionally, relief efforts may not sufficiently address the specific needs of women and girls, such as access to safe spaces, menstrual hygiene products, and childcare.

Gender-inclusive approaches are essential for addressing gender inequality in disaster scenarios. These approaches should include collecting and analyzing sex-disaggregated data to understand the differential impacts of disasters, ensuring women's full and meaningful participation in all phases of disaster risk reduction—from preparedness to recovery. Integrating measures to prevent and respond to gender-based violence in all disaster contexts is crucial. Supporting women's economic empowerment through disaster risk reduction and recovery initiatives is

Table 12.1 Research on gender and disaster management in general

	Authors	Years	Contents
(a)	Forthergill	1998	The neglect of gender in disaster work
(b)	Enarson & Morrow	1998	An introduction to women and disaster
(c)	Gaillard et al.	2017	Beyond men and women
(d)	Hemachandra et al.	2018	Role of women in disaster risk governance

Source Created by Authors

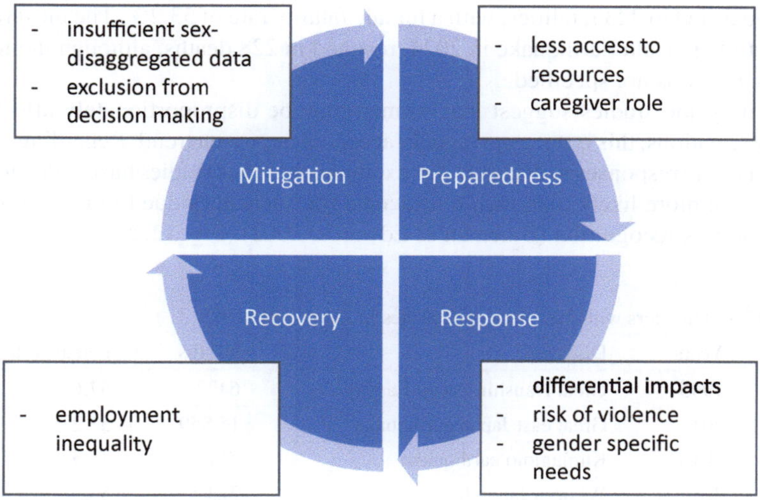

Fig. 12.1 Gender differences in disaster phases. *Source* Created by authors

also essential. By addressing these issues and adopting gender-inclusive strategies, we can significantly improve disaster preparedness, response, and recovery for all members of society.

12.3 Experiences and Challenges in Japan

12.3.1 Historical Context and Gendered Experiences in Major Disasters

Due to its geographical location and geological activity, Japan has experienced several significant disasters over the past 30 years. Table 12.2 provides an overview of significant disasters since 1995, each resulting in 200 or more fatalities and highlighting the death toll and the percentage of female victims. The Great Hanshin-Awaji Earthquake in 1995, a powerful tremor that struck the Kobe region, caused widespread devastation and over 6000 fatalities, with women comprising 57.6% of the deceased, emphasizing a significant gender disparity. Similarly, the 2011 Great East Japan Earthquake and Tsunami, a magnitude 9.0 earthquake followed by a devastating tsunami, struck northeastern Japan, leading to 15,859 deaths—53.2% of whom were female victims, highlighting the disproportionate impact on women during large-scale disasters. This disaster also triggered the Fukushima Daiichi nuclear crisis, with profound environmental and human consequences. The 2016 Kumamoto Earthquake resulted in 211 deaths, with a female fatality rate of 45.5%. The 2018 Western Japan Floods, triggered by torrential rains causing widespread flooding and landslides, led to 223 fatalities, with a female fatality rate of 53.2%. The most recent, the Noto Peninsula Earthquake in 2024, resulted in 228 deaths, although the female fatality rate was not specified.

While some studies suggest that women may be disproportionately affected in certain situations, this is not a universally accepted or proven trend. Regarding gender differences in responses to disasters, for example, previous studies have indicated that women are more likely than men to evacuate for hurricanes due to their caregiving roles and risk recognition (Bateman & Edwards, 2002) (Fig. 12.2).

Table 12.2 Disasters with 200 or more fatalities in Japan since 1995

	Years	Disasters	Deaths	Female rates
(a)	1995	Great Hanshin-Awaji Earthquake	6437	57.6
(b)	2011	Great east Japan earthquake	15,859	53.2
(c)	2016	Kumamoto earthquake	211	45.5
(d)	2018	Western Japan floods	223	53.2
(e)	2024	Noto Peninsula earthquake	228	

Sourc Compiled by Authors, (a) and (b) Gender Equality Bureau Cabinet Office, (2012), (c) Mainichi Shimbun, 2012, (d) 大原, 一郎. (2018), (e) Doi, (2024)

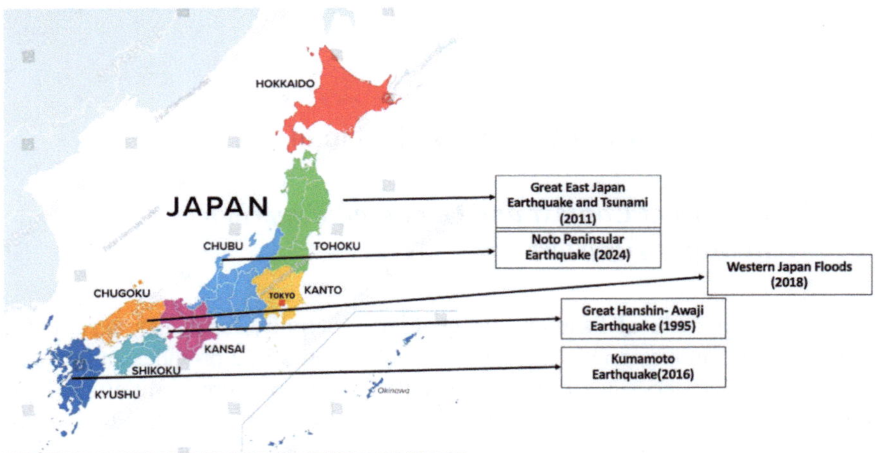

Fig. 12.2 Location of disasters in Japan. *Source* Created by Authors

12.3.2 Socio-Economic and Cultural Factors Influencing Gender Disparities

In Japan, the complex interplay of gender with other identity factors such as age, socioeconomic status, and ethnicity significantly compounds vulnerabilities in disaster scenarios. For example, elderly women are disproportionately affected due to both age and gender biases. Tanida's, 1996 study on the 1995 Kobe Earthquake found that elderly women were not only more likely to suffer from prolonged displacement but also exhibited higher mortality rates than their male counterparts, largely due to factors like social isolation and reduced physical mobility. This impedes their ability to access emergency services and support. Similarly, women from lower socioeconomic backgrounds often have less access to crucial resources for effective disaster preparedness and recovery, limiting their ability to secure safe housing and emergency supplies. Economic dependence can further hinder women's ability to evacuate or access resources during and after a disaster. This issue is further complicated for ethnic minorities, such as the Ainu in Hokkaido, whose unique cultural practices and community structures are deeply intertwined with specific geographic locations. When disasters strike, these communities not only face physical devastation but also significant cultural dislocation, as outlined in studies focusing on indigenous responses to natural disasters. Additionally, traditional gender roles often place women in positions of caregiving responsibilities, which can limit their access to information and physical mobility, further exacerbating their vulnerability in disasters. The specific type of disaster can significantly impact the differential effects on women and men, with certain disasters like tsunamis posing unique challenges for women. These intersecting identities—age, socioeconomic status, and ethnicity—shape how individuals experience and recover from disasters. The case of the Ainu, as noted in recent ethnographic research, illustrates the profound impact of natural disasters on both physical infrastructures and intangible cultural heritage, necessitating a disaster management strategy that is as culturally aware as it is responsive.

Gender differences in disaster management are prominently influenced by traditional roles, workforce participation, and policy frameworks. Women, traditionally viewed as caregivers, are often relegated to the responsibility of household safety but are underrepresented in decision-making processes within both community and professional settings (Freiner & Charlton, 2007). This disparity extends to the workforce where, despite increasing female participation, significant gaps remain in leadership positions within disaster-related fields such as emergency services and governmental agencies (Tanaka et al., 2019). Moreover, evacuation and relief efforts frequently overlook the specific needs of women, such as privacy, security, and access to sanitary products, which can adversely impact their safety and health during disasters (Hirani, 2024). Although Japan has integrated gender perspectives into disaster risk reduction, actual policy implementation often lacks depth in addressing these specific needs (Priebe, 2024).

This ongoing challenge and the shift toward more inclusive disaster management strategies are documented in Table 12.3, which serves as a comprehensive overview,

Table 12.3 Research on gender and disaster management in Japan

	Author	Year	Contents
(a)	Yasunari et al.	2011	Educational programme for pregnant women
(b)	Koikari	2013	Gender and post-3.11 Nationalism
(c)	Saito	2014	Gender perspectives in disaster management
(d)	Kobayashi et al.	2020	Perinatal women and public healthcare providers
(e)	Yoshihama	2021	Drivers of gender health disparities
(f)	Obi-Aso et al.	2022	Appraising the Sendai framework
(g)	Petraroli & Baars	2022	Disaster vulnerabilities and gendered discourses

Source Created by Authors

showcasing scholarly work spanning more than a decade. It starts with Yasunari et al. (2011), who evaluated educational programs designed for pregnant women, highlighting the specific needs and vulnerabilities of this group in disaster scenarios. Koikari, (2013), discussed how Japan used disasters as an opportunity to reinforce traditional gender roles and mobilize women for national purposes while simultaneously appearing to empower them. Saito, in 2014, expanded the scope to include broader gender perspectives within disaster management, marking a shift towards more inclusive methodologies. Subsequent studies by Kobayashi et al. (2020) emphasized the need for specialized attention to perinatal women's mental healthcare in postdisaster settings, drawing lessons from the Great East Japan Earthquake. Yoshihama (2021) identified the factors driving gender health disparities, illustrating the diverse impacts of disasters across different genders. That same year, Obi-Aso et al. (2021) critically assessed the Sendai Framework, probing its efficacy in integrating gender considerations. Most recently, in, Petraroli and Baars, (2022) explored the nuances of disaster vulnerabilities and gendered discourse, shedding light on the evolving roles and challenges that different genders face in disaster contexts.

12.3.3 Current Strategies and Future Directions for Gender-Inclusive Disaster Management

Disaster preparedness should integrate gender considerations throughout all phases, from planning and preparedness to response and recovery, to ensure equal safety and well-being for all. Women's economic and social empowerment is crucial for enhancing their resilience during and after disasters. Collecting and analyzing data that specifically considers how disasters affect women and men differently is essential for effective and equitable disaster response.

Japan has hosted several significant international conferences on disaster risk reduction, as shown in Table 12.4. The First World Conference on Natural Disasters, convened in Yokohama in 1994, established the foundational "Yokohama Strategy," emphasizing disaster prevention, preparedness, and mitigation as vital for a resilient society. This conference set the stage for enhanced global cooperation in addressing disaster risks. Following this, the Second World Conference on Disaster Reduction in 2005 in Kobe adopted the "Hyogo Framework for Action 2005–2015," which promoted a global plan for reducing disaster risks, emphasizing the critical role of community-based approaches and risk assessment. Building upon these developments, the Third UN World Conference on Disaster Risk Reduction in 2015 in Sendai culminated in the adoption of the "Sendai Framework for Disaster Risk Reduction 2015–2030." This comprehensive framework focuses on reducing disaster risks and building resilience by setting out clear targets and priorities for action that include preventing new and reducing existing disaster risks through the implementation of integrated and inclusive measures. These conferences collectively highlight the chronological progression and expanding scope of international disaster risk reduction efforts hosted in Japan, marking significant shifts from localized initiatives to global strategic implementations.

International conferences on disaster risk reduction have played a crucial role in shaping global approaches to managing and mitigating disaster risks. These conferences bring together governments, international organizations, and civil society to share knowledge, best practices, and develop collaborative solutions to address the growing challenges posed by disasters worldwide. Not only have these gatherings fostered international cooperation and promoted innovation, but they have also had a profound impact on how countries like Japan and others approach building more resilient communities.

In particular, the research by Obi-aso et al. (2021) focused on analyzing the Sendai Disaster Prevention Council's initiatives from a gender perspective, offering critical insights into the integration of gender considerations into disaster risk reduction policies and practices. Although there have been advancements in recognizing the importance of gender dimensions in disaster response and preparedness, significant gaps remain in practice. These studies reveal that women's participation in disaster planning and decision-making processes is still limited, despite their often greater vulnerability to disaster impacts. This underrepresentation can lead to preparedness

Table 12.4 International conferences on disaster risk reduction in Japan

	Years	Conference name	Place	Keywords
(a)	1994	First world conference on Natural disasters	Yokohama	"Yokohama Strategy" for disaster prevention, preparedness, and mitigation
(b)	2005	Second world conferences on disasters reduction	Kobe	Community-based approaches and risk assessment
(c)	2015	Third UN World Conference on disaster risk reduction	Sendai	"Sendai framenwork for disaster risk reduction 2015–2023"

Source Compiled by Authors, (a) United Nations, (1994), (b) United Nations Inter- Agency Secretariat of the International Strategy for Disaster Reduction, (2005), (c) United Nations office for Disaster Risk Reduction, (2015)

strategies that do not fully address or even recognize the specific needs and strengths of women in disaster situations.

The evaluations recommend adopting more inclusive policy frameworks that actively involve women and other marginalized groups right from the planning stages through to implementation. This approach ensures that disaster risk reduction strategies are both effective and equitable, ultimately leading to better-prepared societies that can respond to and recover from disasters more efficiently (Figs. 12.3 and 12.4).

Fig. 12.3 Emergence of women leaders from the Great East Japan Earthquake and Tsunami. *Source* UNDRR, (2021)

Fig. 12.4 Workshop held by Women's eye for 40 young women leaders from Tohoku. *Source* Women's eye, (2017)

After the Great East Japan Earthquake in 2011, the Japanese government and NGOs implemented targeted recovery programs for vulnerable populations, particularly women. One notable initiative was the formation of the "Women's Eye" organization, which supported women entrepreneurs in the Tohoku region, helping them rebuild their businesses and improve their economic status, thus acknowledging the unique economic setbacks women faced post-disaster (Women's Eye, 2020). Similarly, following the 1995 Kobe Earthquake, the Hyogo Prefecture established the "Gender Equality Center," which provided a range of programs from psychological counseling to vocational training, recognizing the disproportionate impact of the earthquake on women (Shaw & Goda, 2004).

Looking ahead, future disaster management strategies in Japan could leverage mobile technology to deliver gender-sensitive disaster alerts and educational content. Apps could provide tailored information on safe routes during evacuations and locations of gender-sensitive shelters that cater to women's health and safety needs. Moreover, utilizing artificial intelligence and big data analytics could predict and analyze disaster impacts by gender to enhance preparedness and response strategies, potentially mapping areas where women and children are most at risk and ensuring targeted delivery of resources (Wirjo et al., 2022). Development of policy simulation models might also assess the effectiveness of gender-inclusive policies before their implementation, helping policymakers visualize the potential impact of disaster response strategies on different gender groups, thus leading to more informed decision-making (UN Women East and South Africa Regional Office & Women's Resilience to Disasters team Geneva, 2021).

12.4 Conclusion

This chapter has provided a comprehensive review of the intersection between gender differences and disaster management in Japan. Our findings reveal that while some progress has been made, significant gaps in gender-inclusive disaster preparedness and response persist. Women, often disproportionately affected by disasters, continue to face challenges exacerbated by pre-existing social, economic, and cultural inequalities. However, it is crucial to recognize that the impact of disasters on women varies significantly depending on the specific context, the type of disaster, and the cultural and social factors at play, making generalizations difficult (Fig. 12.5).

The review highlighted the critical role that women can play in disaster risk governance when adequately empowered and included in decision-making processes. Moreover, reliable data on the differential impacts of disasters on women and men can be difficult to obtain, which complicates efforts to develop targeted interventions. This highlights the need for robust, gender-sensitive policies that not only recognize the unique vulnerabilities and capacities of women but also integrate these insights into mainstream disaster risk management frameworks.

Future research should focus on developing more refined methods for integrating gender perspectives into disaster risk reduction strategies, including creating and

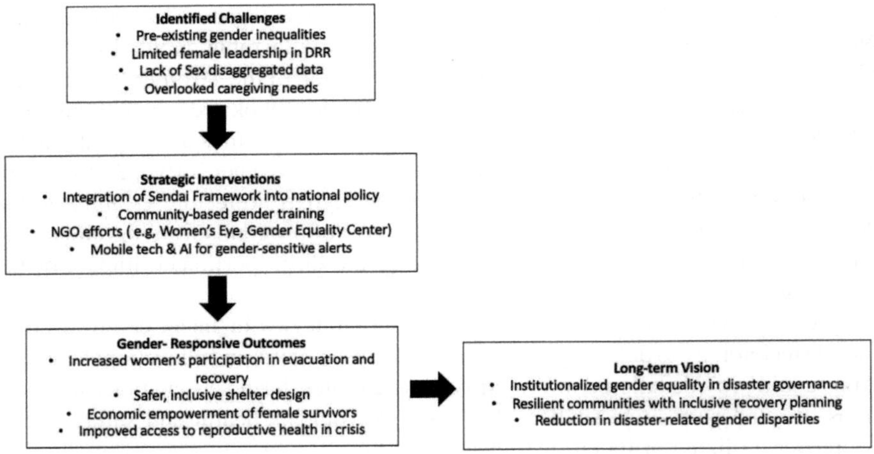

Fig. 12.5 Analytical flow chart for pathway to gender-inclusive disaster management. *Source* Created by Authors

testing models that address the specific needs of women and other vulnerable populations during disasters to improve their resilience and capacity to respond to and recover from disaster events. Additionally, longitudinal studies are needed to track the effectiveness of gender-sensitive interventions over time, providing a clearer picture of their impact and areas for improvement.

Through a continued focus on these areas, disaster management strategies can evolve to be more inclusive, ultimately leading to safer, more resilient communities that can better withstand and recover from the challenges posed by disasters.

References

Alcobé, F., & Harty, E. (2023). Understanding the gender imbalance at the international climate negotiations. *International Institute for Environment and Development*. https://www.iied.org/understanding-gender-imbalance-international-climate-negotiations (Accessed on Jan 25, 2025)

Amaratunga, D. (2023). Progress in the Development of Globally Accessible Early Warning Systems: The Gender Dimension of Disaster Risk Reduction. *UN Chronicle*. Accessed, 12.

Asano, S. (2021). Kokunai ni okeru jendā shiten no bōsai seisaku no tōtatsu-ten to kadai [Gender Sensitive Disaster Management Policies in Japan: Achievements and Challenges]. *Kokyoseisakushirin, 9*, 54–72. https://doi.org/10.15002/00024268

Bateman, J. M., & Edwards, B. (2002). Gender and evacuation: A closer look at why women are more likely to evacuate for hurricanes. *Natural Hazards Review, 3*(3), 107–117.

Cabinet Office, Government of Japan, Gender Equality Bureau. (2012). *Women and disasters: From the experience of the Great East Japan Earthquake*. https://www.gender.go.jp/english_contents/about_danjo/whitepaper/pdf/ewp2012.pdf

Chowdhury, T. J., Arbon, P., Kako, M., Muller, R., Steenkamp, M., & Gebbie, K. (2022). Understanding the experiences of women in disasters: Lessons for emergency management planning.

TheAustralian Journal of Emergency Management, 37(1), 72–77. https://doi.org/10.47389/37.1.72

Daalen, K. R., Kallesøe, S. S., Davey, F., Dada, S., Jung, L., Singh, L., Issa, R., Emilian, C. A., Kuhn, I., Keygnaert, I., & Nilsson, M. (2022). Extreme events and gender-based violence: a mixed-methods systematic review. *The Lancet Planetary Health, 6*(6), e504–e523

Daalen, K. R., Kallesøe, S. S., Davey, F., Dada, S., Jung, L., Singh, L., Issa, R., Emilian, C. A., Kuhn, I., Keygnaert, I., & Nilsson, M. (2022). Extreme events and gender-based violence: A mixed-methods systematic review. *The Lancet Planetary Health, 6*(6), e504–e523. https://doi.org/10.1016/S2542-5196(22)00088-2

Doi, Y. (2024). Death toll of Noto quake to hit 282, third highest since Heisei Era. *The Asahi Shimbun*. https://www.asahi.com/ajw/articles/15311450

Enarson, E., & Morrow, B. H. (1998). Why gender? Why women? An introduction to women and disaster. *The gendered terrain of disaster: Through women's eyes*, pp. 1–8.

Enarson, E. P. (2000). *Gender and natural disasters*. ILO.

Enarson, E., & Fordham, M. (2001). From women's needs to women's rights in disasters. *Global Environmental Change Part B: Environmental Hazards, 3*(3), 133–136. https://doi.org/10.3763/ehaz.2001.0314

Fisher, S. (2010). Violence against women and natural disasters: Findings from post-tsunami Sri Lanka. *Violence against Women, 16*(8), 902–918. https://doi.org/10.1177/1077801210377649

Fothergill, A. (1998). The neglect of gender in disaster work: An overview of the literature. *The gendered terrain of disaster: Through women's eyes*, pp. 11–25.

Fothergill, A., & Peek, L. A. (2004). Poverty and disasters in the United States: A review of recent sociological findings. *Natural Hazards, 32*, 89–110.

Freiner, N., & Charlton, S. E. (2007). Gender, political participation and the environment in Japan.

Gaillard, J. C., Sanz, K., Balgos, B. C., Dalisay, S. N. M., Gorman-Murray, A., Smith, F., & Toelupe, V. A. (2017). Beyond men and women: A critical perspective on gender and disaster. *Disasters, 41*(3), 429–447. https://doi.org/10.1111/disa.12209

Gender Equality Bureau Cabinet Office. (2012). 平成24年版男女共同参画白書 [2012 White Paper on Gender Equality]. https://www.gender.go.jp/about_danjo/whitepaper/h24/zentai/html/honpen/b1_s00_01.html

Hemachandra, K., Amaratunga, D., & Haigh, R. (2018). Role of women in disaster risk governance. *Procedia Engineering, 212*, 1187–1194.

Hirani, S. A. A. (2024). Barriers to women's menstrual hygiene practices during recurrent disasters and displacement: A qualitative study. *International Journal of Environmental Research and Public Health, 21*(2), 153.

Kobayashi, N., Nemoto, H., Seto, M., Sato, S., Kikuchi, S., Honda, N., & Tomita, H. (2020). Experiences of perinatal women and public healthcare providers in a community affected by the great east Japan earthquake and tsunami: Concerns that must be considered for the mental healthcare of perinatal women in postdisaster settings. *International Journal of Disaster Risk Reduction, 51*, 101767.

Koikari, M. (2013). Training women for disasters: Gender, crisis management (kiki kanri) and post-3.11 nationalism in Japan. *The Asia-Pacific Journal, 11*(26), 1.

Kolbe, A. R., Hutson, R. A., Shannon, H., Trzcinski, E., Miles, B., Levitz, N., Puccio, M., James, L., Noel, J. R., & Muggah, R. (2010). Mortality, crime and access to basic needs before and after the Haiti earthquake: A random survey of Port-au-Prince households. *Medicine, Conflict and Survival, 26*(4), 281–297.

Mainichi Shimbun. (2017). 森友学園問題で財務省、7カ月遅れの説明 [MoritomoGakuen issue: Ministry of Finance provides explanation after 7-month delay]. Mainichi Shimbun. https://mainichi.jp/articles/20170412/k00/00m/040/115000c

Obi-Aso, N. E., Ewurum, N. I., & Ifeanacho, K. P. (2021). Appraising the Sendai Framework: place of women in disaster risk mitigation. *British Journal of Environmental Sciences, 9*(4), 41–53. Ohara, M. (2015). Josei gijutsu-sha kara no message josei gijutsu-sha kara mita bōsai gensai

[Message from Female Engineers Disaster Prevention and Mitigation from a Female Engineer's Perspective]. *Civil engineering for life, 70*(3), 86–90.

Okai, A. (2022). Women are hit hardest in disasters, so why are responses too often gender-blind? Undp. Org. https://www.undp.org/blog/women-are-hithardest-disasters-so-why-are-responsestoo-often-gender-blind (Accessed on Jan 25, 2025).

Oxfam. (2014). *Beyond Safe Land: Why security of land tenure is crucial for the Philippines' post-Haiyan recovery.* Oxfam GB.

Oxfam. (2005). *The tsunami's impact on women.* Oxfam Briefing Note. ISBN 978-1-84814-684-6. Retrieved from: https://oxfamilibrary.openrepository.com/bitstream/handle/10546/115038/bn-tsunami-impact-on-women-250305-en.pdf (Accessed on Jan 25, 2025).

Petraroli, I., & Baars, R. (2022). To be a woman in Japan: Disaster vulnerabilities and gendered discourses in disaster preparedness in Japan. *International Journal of Disaster Risk Reduction, 70,* Article 102767. https://doi.org/10.1016/j.ijdrr.2021.102767

Priebe, M. B. (2024). Gender-blind DRR blunts response and recovery. Pacific Forum International.

Saito, Y. (2014). Progress or repetition? Gender perspectives in disaster management in Japan. *Disaster Prevention and Management, 23*(2), 98–111. https://doi.org/10.1108/DPM-08-2013-0134

Sams, I. S. (2019). Impacts of climate change induced migration on gender: A qualitative study from the Southwest Coastal Region of Bangladesh. *International Journal of Social Science Studies, 7,* 57.

Shaw, R., & Goda, K. (2004). From disaster to sustainable civil society: The Kobe experience. *Disasters, 28*(1), 16–40.

Sohrabizadeh, S., Tourani, PhD, S., & Khankeh, H. R. (2016). Women and health consequences of natural disasters: challenge or opportunity? *Women & health, 56*(8), 977–993. https://doi.org/10.1080/03630242.2016.1176101

Tanaka, Y., Ishiwatari, M., & Nonoguchi, A. (2019). Disaster recovery from a gender and diversity perspective: Cases following megadisasters in Japan and Asian countries. Contributing Paper to GAR.

Tanida, N. (1996). What happened to elderly people in the great Hanshin earthquake. *BMJ, 313*(7065), 1133–1135.

United Nations. (1994). *Report of the World Conference on Natural Disaster Reduction, Yokohama, Japan, 23–27 May 1994.* United Nations. https://www.undrr.org/publication/report-world-conference-natural-disaster-reduction-yokohama-23-27-may-1994

United Nations Inter-Agency Secretariat of the International Strategy for Disaster Reduction. (2005). *World Conference on Disaster Reduction: 18–22 January 2005, Kobe, Hyogo, Japan: Proceedings of the Conference.* United Nations. https://www.unisdr.org/2005/wcdr/thematic-sessions/WCDR-proceedings-of-the-Conference.pdf

United Nations Office for Disaster Risk Reduction. (2015). Sendai Framework for Disaster Risk Reduction 2015–2030. Retrieved from: https://www.undrr.org/publication/sendai-framework-disaster-risk-reduction-2015-2030 (Accessed on May 27, 2025).

United Nations Office for Disaster Risk Reduction. (2021). How women leaders emerged from the Great East Japan Earthquake and Tsunami. *UNDRR.* https://www.undrr.org/news/how-women-leaders-emerged-great-east-japan-earthquake-and-tsunami

UN Women East and South Africa Regional Office, & Women's Resilience to Disasters team Geneva. (2021). Toolkit: Gender-responsive implementation of the Sendai Framework for Disaster Risk Reduction. United Nations Women. Retrieved from: https://wrd.unwomen.org/sites/default/files/2022-07/Merged%20ESARO%20SF%20tools.pdf (Accessed on January 25th, 2025).

Wirjo, A., Calizo, S. Jr., Vasquez, G. N., & San Andres, E. A. (2022). *Artificial Intelligence in Economic Policymaking.* APEC Policy Support Unit, Policy Brief No. 52. Retrieved from: https://www.apec.org/docs/default-source/publications/2022/11/artificial-intelligence-in-economic-policymaking/222_psu_artificial-intelligence-in-economic-policymaking.pdf (Accessed on Jan 26, 2025).

Women's Eye. (2017). Grassroots Academy Tohoku. *Women's Eye*. Retrieved May 28, 2025, from https://womenseye.net/en-g_tohoku

Women's Eye. (2020). Disaster Recovery and Women's Empowerment 2011–2020. Women's Eye. Retrieved from: https://womenseye.net/wp02/wp-content/uploads/2022/12/booklet_E02.pdf (Accessed on Jan 26, 2025).

World Meteorological Organization. (2021). *WMO Atlas of mortality and economic losses from weather, climate, and water extremes (1970–2019)*. Retrieved from https://library.wmo.int/doc_num.php?explnum_id=10769.

Yasunari, T., Nozawa, M., Nishio, R., Yamamoto, A., & Takami, Y. (2011). Development and evaluation of 'disaster preparedness' educational programme for pregnant women. *International Nursing Review, 58*(3), 335–340. https://doi.org/10.1111/j.1466-7657.2011.00919.x

Yoshihama, M. (2021). Visualizing drivers of gender health disparities: Ongoing participatory action research following the 2011 disaster in Japan. *Social Science & Medicine, 283*, 114133. https://doi.org/10.1016/j.socscimed.2021.114133

大原, 一郎. (2018). 水関連災害における死亡・被害の実態と課題 [Mortality and Casualties in Water-Related Disasters: Current Status and Issues]. 地域安全学会梗概集, 43, 103–104. Retrieved from https://isss.jp.net/isss-site/wp-content/uploads/2019/02/29%E5%A4%A7%E5%8E%9F.pdf

Chapter 13
Enhancing Disaster Management Through Gender-Sensitive Approaches: Insights from South Korea's Climate Crisis

Bonjun Koo

Abstract As climate-induced disasters become more frequent and severe, South Korea is facing unprecedented challenges in building an inclusive and resilient disaster management system. This chapter examines how gender shapes both vulnerability to, and capacity for, responding to climate-related disasters. Drawing from a wide range of national disaster policy documents, municipal case studies, and community-level initiatives, it explores how structural gender inequalities—such as women's underrepresentation in decision-making, limited access to economic resources, and disproportionate caregiving responsibilities—compound the risks faced by women and other socially marginalized groups. The chapter provides empirical insight into how these disparities manifest during climate-related events such as floods, heatwaves, and typhoons, which have affected both urban and rural populations in recent years. It highlights promising gender-responsive practices, including inclusive risk communication strategies, community-based evacuation planning, and the mobilization of women's groups in post-disaster recovery. It also addresses the limitations of current frameworks, where gender is often treated as a secondary concern rather than a central component of risk governance. By situating the Korean experience within broader international frameworks—such as the Sendai Framework for Disaster Risk Reduction, the Paris Agreement, and the Sustainable Development Goals (SDGs)—this chapter argues for the institutionalization of gender-sensitive disaster governance. It emphasizes the importance of intersectional approaches that account for class, age, ethnicity, and migration status, in addition to gender, in shaping vulnerability and resilience. Ultimately, this chapter contributes to the growing discourse on climate justice by offering practical, context-specific policy recommendations for integrating gender into South Korea's disaster risk reduction and climate adaptation strategies.

B. Koo (✉)
The Great Hanshin-Awaji Earthquake Memorial Disaster Reduction and Human Renovation Institution, Kobe, Japan
e-mail: bonjun@dri.ne.jp

Md. A. Abedin et al. (eds.), *Climate Crisis in South and East Asia*, Sustainability Solutions, https://doi.org/10.1007/978-3-032-01012-4_13

Keywords Gender-sensitive disaster management · Climate change adaptation · South Korea · Social vulnerability · Inclusive governance

13.1 Introduction

As climate change intensifies, the frequency, severity, and unpredictability of disasters continue to grow, prompting urgent reassessments of disaster risk reduction (DRR) and adaptation strategies worldwide. Although often portrayed as universal phenomena, disasters are anything but socially neutral. They intersect with deeply ingrained inequalities related to gender, class, age, race, and other social categories, resulting in uneven impacts and highly variable recovery trajectories. Women, in particular, frequently face disproportionate burdens in the wake of climate-related crises. This gendered vulnerability is not inherent or unchangeable; rather, it stems from systemic social inequalities, cultural norms, and institutional practices that marginalize women's voices, limit their access to resources, and constrain their decision-making power. Understanding and addressing these gendered dimensions of disaster vulnerability is essential for creating more inclusive, effective, and sustainable disaster management frameworks.

This chapter focuses on the South Korean context, drawing insights that resonate broadly across South and East Asia. South Korea is renowned for its economic modernization, high-tech infrastructure, and capacity to respond swiftly to emergencies. Yet despite these strengths, the country's existing disaster management systems often overlook the nuanced social factors that shape vulnerability and resilience. Gender is a particularly critical dimension. While South Korea's policy documents, adaptation plans, and national strategies occasionally acknowledge women's vulnerability to climate change, the integration of a gender lens into actual DRR implementation remains patchy and insufficient. By examining how cultural norms, economic inequalities, institutional gaps, and community-level practices influence women's experiences of climate-related disasters, this chapter illuminates pathways toward more equitable, gender-sensitive approaches that can enhance disaster preparedness, response, and recovery.

The analysis presented here draws upon a wide range of sources, including international frameworks, academic studies, policy documents, case studies, and community initiatives. Beginning with an exploration of the gendered nature of disasters, the chapter then turns to the South Korean context, unraveling how pre-existing gender inequalities shape vulnerability to climate-induced hazards such as typhoons, floods, and heatwaves. It critically examines current policy frameworks, highlighting both institutional strengths and persistent gaps, before discussing how gender-responsive policies and practices can be integrated into disaster management systems at multiple levels. The chapter also considers how cultural change, technological innovation, cross-sectoral partnerships, and stable funding mechanisms can support these efforts. Ultimately, the South Korean experience underscores that gender-sensitive DRR is not a peripheral concern, but rather a strategic priority for ensuring that no one is left behind in an era of escalating climate risks.

13.2 Reframing Disasters Through a Gender Lens

Disasters have long been conceptualized as "natural" events, often described in gender-neutral terms. Historically, DRR frameworks concentrated on hazard prediction, infrastructure protection, and evacuation logistics—reflecting assumptions that all members of a community experience disaster risk similarly. However, a wealth of scholarly research and policy interventions now emphasizes that the impacts of disasters, including climate-induced events, are significantly influenced by social and cultural factors (Denton, 2002; Enarson & Chakrabarti, 2009; UN Women, 2019). Gender emerges as one of the most crucial variables, shaping exposure to hazards, access to resources, decision-making power, and the capacity to recover after catastrophe.

13.2.1 Differential Mortality and Morbidity

In numerous contexts around the globe, women and girls experience more severe consequences from disasters than men. Studies have found that women's mortality rates in certain disasters can exceed men's due to a variety of social constraints, from limited mobility and weaker access to warnings, to cultural expectations that women remain in the home (IPCC, 2014; Neumayer & Plümper, 2007). The 2004 Indian Ocean tsunami, for instance, saw disproportionately high female mortality in some affected regions, partly because women were less likely to learn about or respond to early warnings. Likewise, floods in Bangladesh have often been more lethal for women who could not swim or were restricted from leaving the household compound without male permission (Juran & Trivedi, 2015). These patterns underscore the critical role that social norms, family structures, and public messaging play in shaping disaster outcomes, revealing that factors like traditional clothing, caregiving responsibilities, and limited self-rescue skills all amplify women's vulnerability.

In addition to these stark differences in death tolls, morbidity rates also reflect a gendered dimension. Even when women survive disaster events, they can encounter a host of health challenges as shelters or clinics become overcrowded and resources like food, water, and medicine run short (UN Women, 2019). For example, overcrowded evacuation camps may lack hygienic facilities for menstruating women or fail to provide adequate nutritional support for pregnant mothers. Over time, these overlooked health needs can evolve into long-term medical complications, inhibiting women's ability to engage productively in rebuilding efforts.

13.2.2 Extended Caregiving Responsibilities and Economic Burdens

Even when survival is not at stake, the aftermath of a disaster can impose disproportionate burdens on women. They often shoulder the increased caregiving responsibilities that arise when health, education, or household support systems collapse, taking on tasks such as preparing food in disrupted environments, caring for children who may be out of school, or assisting the elderly and disabled who may have lost access to healthcare facilities. Employment opportunities for women—often concentrated in informal sectors or small-scale businesses—may be disrupted for longer periods, leading to more protracted economic insecurity. Women working as domestic laborers, market vendors, or seasonal agricultural workers may have little recourse to formal insurance or government compensation programs (UNDP, 2016). Because these informal jobs are less recognized by traditional labor protections, women might lose assets or savings with no institutional safety net, reinforcing poverty cycles that impede resilience.

Furthermore, these amplified caregiving responsibilities can severely limit women's ability to participate in long-term recovery planning or community decision-making. When electricity grids fail, for instance, mothers may spend hours gathering water or cooking meals without modern appliances, limiting the time they have to attend reconstruction meetings, learn new skills, or seek external assistance. Such constraints not only hamper immediate recovery but can also restrict women's long-term socio-economic mobility, making each subsequent disaster even more catastrophic at the household level.

13.2.3 Gender-Based Violence in Crisis Contexts

Additionally, the threat of gender-based violence can escalate in evacuation centers, shelters, and displaced communities that lack adequate safeguards (Juran & Trivedi, 2015; UNDP, 2016). Poorly lit corridors, communal washrooms without privacy measures, and insufficient security personnel can create environments where harassment and assault become more likely. The ensuing trauma can have long-term effects, making it difficult for survivors to access employment opportunities, continue their education, or fully participate in recovery efforts. Moreover, fear of violence or harassment may dissuade women from relocating to safer areas, opting instead to remain in compromised households—an agonizing choice driven by a perceived trade-off between personal safety and community stigma.

The heightened stress of disasters can also exacerbate domestic violence in households already grappling with power imbalances. Where men lose livelihoods or experience intense social pressures, aggression can escalate, manifesting as emotional or physical abuse against women (UN Women, 2019). In these contexts, the persistent silence surrounding gender-based violence, combined with insufficient legal protections or overwhelmed police forces, leaves survivors especially vulnerable. Breaking this cycle demands explicit acknowledgment of the risks, coupled with targeted measures such as installing better lighting in shelters, setting aside safe, private sleeping areas, and training shelter staff to identify and respond to abuse.

13.2.4 Women as Agents of Resilience and Change

Recognizing these disparities does not mean framing women solely as victims. Indeed, women frequently display remarkable resilience, leadership, and adaptability in disaster contexts. They can serve as first responders in their communities, mobilizing informal support networks, safeguarding children and the elderly, and actively participating in rebuilding efforts (Enarson & Chakrabarti, 2009). Women's knowledge of local resources, social networks, and cultural norms can be invaluable in creating recovery strategies that are both effective and sustainable. However, these contributions are often invisible to policy frameworks or overshadowed by narratives that focus predominantly on physical infrastructure or male-dominated emergency services. By acknowledging and incorporating women's agency, knowledge, and leadership, disaster management can become more inclusive, robust, and capable of addressing the multifaceted challenges posed by climate change.

13.3 The South Korean Context: Modernization, Inequality, and Cultural Expectations

South Korea's rapid ascent as a global economic powerhouse has been accompanied by improvements in human development indicators and living standards. The nation's technological sophistication in areas like meteorology, early warning systems, and emergency response logistics often sets it apart as a success story in disaster management. Complex flood-control structures along major rivers, high-speed communication networks that disseminate emergency alerts within seconds, and well-coordinated rescue operations underscore the country's readiness to tackle climate-induced hazards like typhoons, heatwaves, and heavy downpours.

To provide a structured overview of the intersecting factors discussed in this section, Table 13.1 summarizes key dimensions of gendered vulnerability and corresponding institutional gaps in South Korea's disaster management context. This analytical framing complements the detailed qualitative analysis in subsequent subsections and enhances clarity for comparative or policy-oriented audiences.

Although anecdotal reports and select disaster records suggest that gendered patterns of vulnerability may exist in events such as the Sewol Ferry disaster or recent heatwaves, consistent and comprehensive gender-disaggregated data in South Korea remains limited. This absence hinders the development of evidence-based,

Table 13.1 Gendered vulnerabilities and gaps in South Korea's disaster context

Category	Gendered vulnerability/Issue	Institutional or policy gap
Mortality and morbidity	Women face higher mortality and health risks post-disaster	Limited gender-disaggregated risk data collection
Caregiving and household roles	Women carry disproportionate care burdens during recovery	Disaster policy does not integrate caregiving realities
Economic participation	Women's informal labor highly affected, lacking insurance or aid	DRR funding and compensation frameworks overlook informal sectors
Decision-making	Underrepresentation in DRR planning and policy committees	No mandatory quotas or participatory mandates
Violence and safety	Increased GBV risk in shelters and post-disaster settings	Inadequate safety protocols, monitoring, and support systems
Cultural constraints	Social norms limit women's mobility and agency during emergencies	Cultural assumptions persist unchallenged in planning frameworks
Institutional capacity	Gender lens absent from most risk assessments and staff training	Weak gender-sensitive institutional capacity and auditing mechanisms
Budgetary resources	Few resources allocated to women-focused DRR initiatives	Gender-responsive budgeting not institutionalized

Source Compiled by the author based on analysis of national policy documents and secondary literature

gender-sensitive disaster risk reduction policies. Therefore, institutionalizing such data collection is a crucial first step toward inclusive and equitable governance in climate-related disaster management.

A relevant example is a study by Chung, (2017), which analyzed disaster-induced mortality patterns using national data. According to the findings, during the Sewol Ferry disaster in 2014, female passengers—particularly in their teens and twenties—experienced significantly higher mortality rates than their male counterparts, despite comparable overall exposure. The study also revealed that among total fatalities, 59.65% were female. Notably, in the 20 s age group, the female fatality rate reached 70.16%, illustrating the gendered dimensions of disaster vulnerability even among young adult populations (Fig. 13.1).

However, this highly developed context masks persistent gender inequalities and deeply ingrained cultural norms that continue to influence vulnerability and resilience in ways that are not always readily apparent. South Korea's ranking on global gender-equality indices lags behind that of many other nations with similar economic profiles. For all its technological prowess, the country still struggles with traditional family structures that place a heavy emphasis on women's roles as primary caregivers for children and elderly relatives—roles that become particularly onerous when disasters disrupt the availability of basic services.

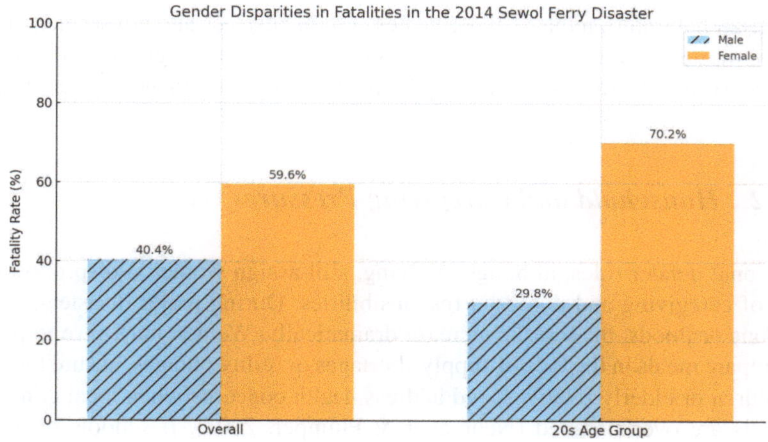

Fig. 13.1 Gender disparities in fatalities in the 2014 sewol ferry disaster a bar chart comparing male and female fatality rates by age group, based on Chung (2017). *Source* Compiled by the author based on Chung (2017)

Recent research that highlights the influence of social vulnerability indicators—including gender, income, and age—on risk perception in South Korea affirms that women are more likely to express concern about climate-related hazards and to engage in precautionary behaviors. Such findings reinforce the importance of incorporating gender-sensitive analysis into both risk communication and resource allocation strategies during all phases of disaster management.

13.3.1 Gender Gaps in Leadership and Economic Participation

The Korean labor market and political landscape remain marked by significant gender gaps. Women's workforce participation is lower than that of men, and a pronounced wage gap persists (OECD, 2020). Women in executive or parliamentary positions remain underrepresented, resulting in less visibility of women's perspectives in policy debates. This imbalance can influence budgetary allocations for DRR, the design and location of shelters, and the development of early warning systems that fail to consider the realities of women's daily responsibilities. Without direct input from women leaders, strategies often miss the crucial caregiving context that women manage, leading to incomplete or poorly suited interventions.

Additionally, socio-economic disparities can compound vulnerability. Many women work informally—running small market stalls or offering domestic services. When disasters strike, such informal businesses are easily destroyed, and owners rarely receive the same level of government assistance as formal enterprises. Given the patriarchal tradition that still permeates certain business and governmental institutions, women's entrepreneurial ventures may not be seen as legitimate or important enough to warrant targeted aid, leaving them further marginal in the recovery process.

13.3.2 Household and Caregiving Pressures

Traditional gender roles, although evolving, still assign women a disproportionate share of caregiving and domestic responsibilities. During major disasters, such as typhoons or floods, these tasks increase dramatically. Women may have to procure and prepare meals in the face of supply shortages or utility outages, ensure the safety of children or elderly relatives, and address health concerns when local clinics and hospitals are overburdened (Neumayer & Plümper, 2007). If schools or daycare centers close, the responsibility for childcare often defaults to mothers, reducing their ability to engage in paid work or participate in community decision-making forums. These daily constraints can fundamentally shape how a family weathers a crisis, with women bearing the brunt of logistical and emotional stress.

Over the long term, these caregiving demands can impede women's economic advancement, as they miss opportunities for skill development, vocational training, and networking activities essential to resilience. As a result, even if they survive a disaster, women often find themselves locked into precarious financial situations, inhibiting their capacity to rebuild effectively or insulate themselves against future hazards.

13.3.3 Societal Pressures and Stigma

Cultural expectations may encourage women to remain at home to care for children, the elderly, or disabled family members, even when it would be safer to evacuate. Women who do evacuate may face social stigma or concerns for their personal safety, particularly if shelters lack separate facilities for women or are perceived as risky places for female residents. Those remaining behind can become more vulnerable to hazards like floods, heatwaves, or landslides, especially if they lack the resources to fortify their homes or secure adequate cooling. Over time, these patterns reinforce a cycle of vulnerability, as women may lose assets, struggle to re-enter the workforce, or encounter difficulties accessing credit and insurance due to systemic discrimination. A single catastrophic event can thus have enduring effects on women's economic and social standing, entrenching gender inequalities more deeply.

13.3.4 Gender-Based Violence

Compounding these issues is the threat of gender-based violence, which can intensify in disaster situations (UN Women, 2019). Stress, resource scarcity, and reduced privacy in cramped settings can elevate the risk of domestic violence, sexual harassment, and assault. Although South Korea has made strides in recognizing and addressing gender-based violence, cultural silence and stigma may prevent survivors from seeking help (Juran & Trivedi, 2015). This situation is exacerbated during emergencies, when police and social services are stretched thin and may not prioritize or even detect emerging cases of violence. As a result, many women endure trauma without recourse, an injustice that deepens gender inequities and limits their capacity to recover fully.

13.4 Policy Frameworks and Institutional Gaps

South Korea's legal and institutional architecture for disaster management reflects a high level of technical competence, featuring a network of government agencies, scientific institutes, and local authorities working in concert to monitor hazards and coordinate responses. The Framework Act on the Management of Disasters and Safety provides an overarching legal structure, while the Korea Meteorological Administration and the Ministry of the Interior and Safety oversee operations ranging from early warnings to post-disaster rehabilitation (UNDRR, 2019). Large-scale exercises, real-time data collection systems, and modern emergency command centers demonstrate an impressive capacity to handle crisis situations at a logistical level.

However, the gender dimension often remains implicit rather than explicit within these frameworks. National climate adaptation plans and policy documents may sporadically acknowledge women's vulnerability, but they rarely translate recognition into concrete mandates, targeted programs, or budget allocations. Without gender-specific indicators or auditing mechanisms, progress remains difficult to track, allowing disparities to persist unchecked. As a result, much of the work to integrate gender considerations falls on local governments, NGOs, and women's groups, leading to fragmented, ad hoc approaches that lack systematic evaluation and scalability.

13.4.1 Capacity-Building Deficits

Institutional capacity-building is also limited. Although officials may receive specialized training in areas such as meteorological forecasting, infrastructural resilience, or logistics coordination, they often lack gender analysis or cultural competency training. This gap in institutional knowledge means that even well-intentioned policymakers may not know how to systematically incorporate a gender lens into risk assessments, evacuation planning, or shelter design. As a result, critical questions—such as whether women and men have equal access to shelters, whether childcare services are available, or whether privacy concerns have been addressed—may be overlooked, undermining the effectiveness of disaster responses. Over time, this oversight can solidify into a structural norm, where intangible social dimensions of vulnerability remain sidelined in favor of physical engineering solutions.

13.4.2 Budgetary Limitations

Moreover, gender-responsive budgeting is not yet institutionalized. This means that financial resources dedicated to disaster management rarely consider the different needs of women and men. For example, early warning campaigns might focus on broad messaging without ensuring that women's preferred modes of communication are utilized. Shelter funding might prioritize large, open dormitories that do not offer safe, private spaces for women. The lack of earmarked funds for community outreach or gender training can further perpetuate inequalities (Denton, 2002; UNDP, 2016). Where an initiative's success hinges on understanding diverse household contexts, ignoring these budgetary nuances can undercut well-intentioned programs.

13.4.3 Cultural Barriers in Institutional Spaces

Even when policies encourage greater female representation, cultural barriers may discourage women from speaking up in male-dominated meeting rooms, while bias in promotion processes may mean fewer women attain influential positions in disaster management agencies. Without institutional reforms that address these structural obstacles, gender remains a secondary consideration. Civil society organizations and international partners often step in to fill these gaps, offering toolkits, workshops, and pilot projects on gender and DRR. However, these efforts may be short-lived if reliant on external funding or leadership champions, and they often lack the formal authority to transform national strategies in a lasting way. Thus, women may remain confined to supportive or adjunctive roles in DRR, limiting their influence over large-scale funding decisions and policy formulations.

13.5 Integrating Gender-Sensitive Approaches: Practical Steps and Considerations

Despite these challenges, the integration of gender-sensitive approaches into South Korea's disaster management frameworks is both feasible and beneficial. A growing body of evidence confirms that gender-responsive DRR can enhance resilience, improve policy outcomes, and create a more equitable social landscape. Below are several practical steps and considerations that can guide the process of mainstreaming gender in DRR.

1. Gender-Responsive Risk Assessments

The first and most fundamental step involves gender-responsive risk assessments. Rather than treating the population as a homogenous group, policymakers and practitioners need to collect and analyze gender-disaggregated data on households, income streams, resource access, caregiving burdens, mobility patterns, and technology usage. These datasets reveal the different ways in which men and women might encounter disasters. For instance, if data show that single mothers in a certain district have limited access to smartphones or community alerts, officials can partner with local women's networks or NGOs to develop targeted communication strategies. Such assessments might also spotlight older women living alone in remote regions, highlighting the need for specialized outreach or physical assistance during evacuations.

Beyond quantitative data, qualitative methods such as interviews and participatory mapping provide a more nuanced glimpse into women's lived experiences. Community workshops can reveal hidden constraints—like families expecting female relatives to remain in the home at all times—that official surveys might fail to capture. Armed with these insights, authorities can prioritize infrastructure upgrades—like elevated roads, specialized shelters, or local water storage facilities—that align with the actual needs of women, thus enhancing the overall efficacy of DRR efforts.

2. Inclusive Decision-Making and Governance

Ensuring women's meaningful participation in decision-making processes is critical for designing DRR strategies that resonate with community realities. One viable approach is to implement quotas for women's representation on disaster committees, consultative councils, and planning boards. Such quotas help shift the balance of power and ensure that women's perspectives are not drowned out in male-dominated discussions (Enarson & Chakrabarti, 2009). Beyond quotas, capacity-building initiatives—such as leadership training, mentorship programs, and policy advocacy workshops—can bolster women's confidence and ability to influence high-level DRR policymaking.

Additionally, formal mandates for community consultations can create opportunities for women to voice their concerns and propose solutions in environments designed to encourage their participation. This could include holding public forums at times that accommodate women's caregiving schedules or providing childcare services during community meetings. When women contribute firsthand knowledge of household dynamics and local resource management, disaster strategies become more reflective of on-the-ground realities, ultimately enhancing their efficacy. Over time, this approach can foster greater trust between officials and the community, as women see their contributions mirrored in tangible policy outcomes.

3. Technological Innovation and Digital Solutions

South Korea's high rate of smartphone usage and robust IT infrastructure mean that early warning apps and digital communication platforms can be a powerful tool in reaching women effectively. However, the mere existence of such technology does not guarantee inclusivity. Developers must co-design these apps with input from women's associations and local communities to ensure features address real needs—such as step-by-step evacuation guidance, childcare checklists, and safety hotlines for survivors of domestic violence (Juran & Trivedi, 2015; UNDP, 2016). Integrating voice-based systems or language options might also prove essential for older women or immigrant populations with limited literacy.

Moreover, digital solutions must be accompanied by training programs to enhance digital literacy among women who may not be familiar with the latest technologies, particularly older adults or those in lower-income brackets. Community-based workshops could teach women how to navigate emergency apps, interpret meteorological data, or utilize group chat platforms to organize neighborhood responses. When women feel empowered by technology, they can serve as crucial communication hubs, relaying warnings to family members and neighbors who might otherwise miss critical updates. With consistent support, these digital tools can serve as transformative channels for real-time coordination and safety assurance.

4. Community-level Integration

On the local or community level, integrating gender considerations into preparedness plans can yield immediate and tangible benefits. Rural communities prone to flooding have seen improvements in household-level readiness when women-led cooperatives collaborate with local officials. These cooperatives can advise on infrastructure improvements—like raising floors or constructing water diversion channels—offer micro-insurance schemes to women entrepreneurs whose businesses are vulnerable to climate shocks, and coordinate neighborhood watch groups that look out for the needs of older women living alone. By giving women an active role in shaping these programs, local governments harness a wealth of specialized knowledge that might otherwise remain dormant.

Similarly, in urban neighborhoods grappling with extreme heat, setting up "cooling centers" staffed by volunteers trained in women's health and mobility issues can boost trust in local governance. These centers might include private breastfeeding areas, designated times for older women to avoid crowds, and access to medical consultations. Such initiatives also help reduce social isolation, encouraging women to come together, share resources, and support each other during crises. The communal bonds forged in these centers can endure long after the disaster subsides, leaving behind a network of proactive citizens who can swiftly mobilize for future challenges.

As illustrated in Fig. 13.2, the analytical framework linking gendered vulnerabilities to disaster policy interventions in South Korea begins with systemic inequalities and institutional gaps. These vulnerabilities, if unaddressed, tend to cascade into resilience and recovery inequities. However, the framework also highlights key policy entry points—opportunities where targeted intervention can promote inclusive and gender-sensitive disaster strategies that improve the effectiveness and equity of disaster risk reduction efforts.

This framework reinforces the need for systemic reform: policy and institutional change must be anchored in a clear understanding of the pathways through which social inequalities shape disaster outcomes.

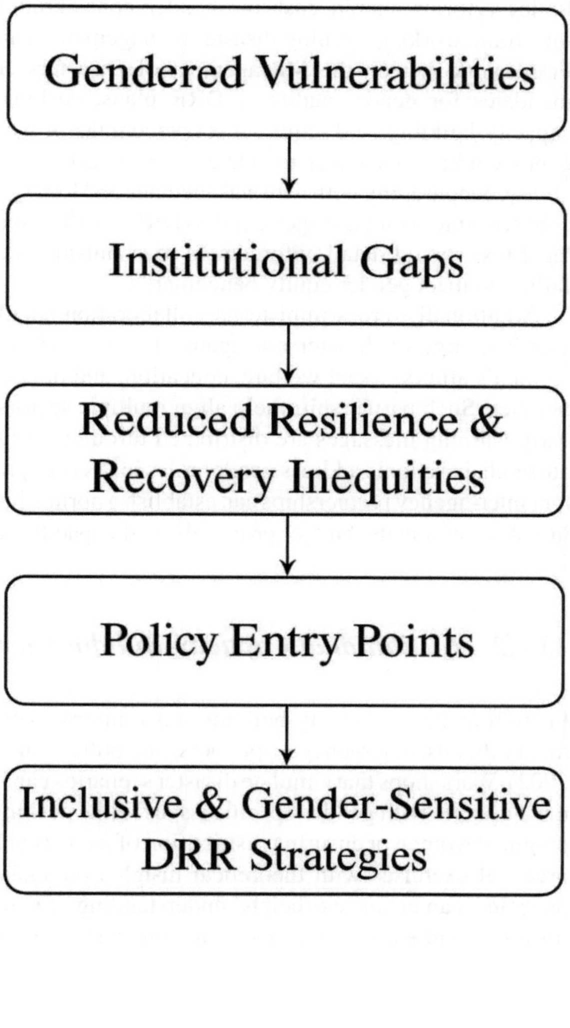

Fig. 13.2 Analytical framework linking gendered vulnerabilities to policy entry points in South Korea's disaster context

13.6 Long-Term Systemic Reforms: Policy, Culture, and Sustainability

While these practical measures demonstrate that gender-sensitive DRR is achievable, long-term sustainability requires deeper systemic reforms. Isolated successes or short-term pilot projects may not endure if the broader policy environment remains unresponsive to gender concerns. Systemic reforms must encompass policy changes, institutional restructuring, cultural transformation, and sustainable financing.

13.6.1 Policy and Legal Frameworks

Policy reforms should enshrine gender considerations into the legal and regulatory frameworks governing disaster management. This might involve amending the Framework Act on the Management of Disasters and Safety to include explicit mandates for gender audits of DRR plans, designate funds for women-focused capacity-building, and require a set percentage of women's representation in emergency committees. Clear mandates, benchmarks, and timelines can drive accountability, encouraging both national agencies and local governments to integrate gender concerns into their core agendas (UNDRR, 2019). To ensure these reforms take root, legal text should detail enforcement mechanisms, including penalties for agencies failing to meet gender equity benchmarks.

Additionally, cross-ministerial collaboration should be institutionalized. Agencies in charge of disaster management can collaborate with those responsible for women's affairs, social welfare, education, and public health to create holistic DRR policies. Such partnerships help align multiple sectors—ensuring, for example, that early warning messages are distributed through welfare offices, or that educational curricula in schools address gender roles in disaster preparedness. Over time, consistent inter-agency partnerships can establish a norm where gender is routinely included in risk assessments, budget proposals, and capacity-building initiatives.

13.6.2 Institutional Capacity-Building and Training

Institutionalizing capacity-building and training on gender-sensitive DRR for government officials, emergency responders, and policymakers is another key step (Denton, 2002). Workshops that simulate disaster scenarios can include sessions where trainees must grapple with gender-specific issues—like designing shelters that accommodate pregnant women or ensuring distribution of menstrual hygiene products. By blending practical exercises with theoretical insights on social vulnerability, these training programs can enhance officials' understanding of gender dynamics, leading to more informed and equitable decision-making across the disaster management cycle. To reinforce lessons, agencies might use checklists or guidelines specifically tailored to DRR, ensuring that every new project or policy is vetted through a gender lens before rollout.

13.6.3 Cultural Transformation and Public Engagement

Cultural transformation is arguably the most challenging yet most crucial dimension of fostering gender-sensitive DRR. Public awareness campaigns, educational reforms, and media initiatives can help shift social perceptions of women's and men's roles in disaster management. TV dramas or documentaries highlighting the leadership of women during floods or earthquakes can normalize and celebrate their contributions, thereby reducing the stigma attached to women who take on public roles. School programs can teach children about gender equality from an early age, encouraging boys and girls to collaborate in preparedness exercises and respect each other's capacities (Seager, 2009). These early interventions can gradually reshape generational attitudes, building a more supportive environment where women's leadership is accepted, even expected, in emergency contexts.

As societal attitudes evolve, it becomes more feasible to embed gender sensitivity into formal institutions. Traditional barriers—such as the assumption that emergency teams should be led by men—begin to erode, creating space for more women to ascend to leadership roles. Community leaders, religious figures, and youth organizations can also play pivotal roles in legitimizing women's participation and championing more inclusive policies at the local level. Through sustained messaging, face-to-face dialogues, and public recognition of women who excel in disaster response, communities can dismantle rigid gender stereotypes, enabling a more balanced distribution of responsibilities and authority in DRR processes.

13.6.4 Sustainable Funding and Cross-Sectoral Partnerships

Securing sustainable funding and forging long-term partnerships are essential to ensure that these efforts do not remain short-lived. Instead of relying on external grants or ad-hoc donations, stable budget lines should be incorporated into national and local disaster management appropriations. Engaging with the private sector, philanthropic foundations, and international bodies—like the United Nations, World Bank, or Asian Development Bank—can provide additional resources and technical expertise (UN Women, 2019). These cross-sectoral partnerships are invaluable for creating multi-stakeholder dialogues, sharing best practices, and standardizing monitoring and evaluation protocols that assess the effectiveness of gender-sensitive approaches over time.

In practice, this could involve formal agreements with technology companies to develop specialized apps or platforms that deliver real-time, gender-sensitive alerts. Financial institutions might offer microloans or insurance products tailored to women's informal businesses, thereby cushioning the economic blow of climate-induced disruptions. Over time, a synergy between public and private actors can bolster the entire DRR ecosystem, making it more inclusive and better equipped to handle recurring climate hazards.

13.7 Implications Beyond South Korea: Regional Relevance and Adaptation

Although this chapter focuses on South Korea, the insights have broader resonance across South and East Asia—regions that face a myriad of climate-induced challenges. From super-typhoons in the Philippines to recurrent flooding in Vietnam, to heatwaves in India, many countries grapple with gendered vulnerabilities comparable to those in Korea. Women's limited access to economic resources, their unpaid caregiving responsibilities, and their underrepresentation in political arenas are widespread phenomena that climate change exacerbates (IPCC, 2014; UNDRR, 2019).

13.7.1 Regional Knowledge-Sharing and Cooperation

Regional cooperation and knowledge-sharing platforms can help disseminate best practices and innovative solutions. Workshops, symposiums, and joint training programs—possibly hosted by entities like ASEAN, SAARC, or the Asian Disaster Preparedness Center—offer opportunities for policymakers, NGOs, community leaders, and academics from different countries to exchange experiences, refine tools, and create transnational networks. For instance, success stories from women-led cooperatives in South Korea's rural communities can serve as a model for smaller-scale coastal villages in Indonesia or agrarian communities in Cambodia.

Moreover, comparative research that highlights diverse cultural contexts can underscore the universality of certain gendered challenges, while also revealing local nuances that shape how policies are adopted. Governments in the region can adapt and refine these models, incorporating cultural considerations like religious norms, tribal affiliations, or linguistic differences into DRR strategies. By customizing solutions to fit local circumstances, states can more effectively reduce disaster-related losses and expedite community recovery.

13.7.2 Leveraging International Frameworks

The political momentum behind gender equality and climate action at the international level provides an additional supportive framework. The Sendai Framework for Disaster Risk Reduction, the Paris Agreement on climate change, and the Sustainable Development Goals (SDGs)—especially SDG 5 on gender equality and SDG 13 on climate action—all underscore the importance of inclusive and equitable climate strategies (UNDP, 2016; UN Women, 2019). By aligning national policies with these global standards, South Korea and its neighbors can leverage international recognition, funding opportunities, and technical guidance to accelerate progress.

In practical terms, this alignment might involve setting measurable targets—such as increasing the proportion of women in disaster management roles by a certain percentage within a given timeframe, or allocating a fixed percentage of DRR budgets to gender-specific training and community outreach. Reporting mechanisms tied to international agreements can then help track progress, holding governments accountable and encouraging continuous refinement of policies. If these goals are transparently reported to global bodies, governments may also attract additional support from donor agencies or philanthropic organizations interested in fostering gender-responsive solutions to climate threats (UNDRR, 2019).

13.8 Conclusion: Toward an Inclusive and Resilient Future

The intensification of climate-related disasters in South Korea and across South and East Asia calls for a fundamental rethinking of how DRR is conceptualized and implemented. Gender, once considered a "soft" issue or peripheral concern, emerges as a central axis of effective disaster management. Recognizing that men and women face different risks, possess different capabilities, and hold different responsibilities within their families and communities does not diminish the importance of technical solutions—it enriches and strengthens those solutions by ensuring they address the full range of human needs.

South Korea's experience illustrates that even in a highly developed country with advanced technology and infrastructure, social inequalities can undermine the efficacy of disaster response. By integrating gender-sensitive approaches—through risk assessments, inclusive decision-making, tailored communication strategies, institutional reforms, cultural shifts, and sustainable funding—disaster management can become more equitable, more effective, and ultimately more humane. Such transformations will not happen overnight; they require persistent effort, political will, and societal engagement. Confronting cultural taboos, challenging long-standing hierarchies, and embracing women's leadership as a source of strength and innovation are essential but achievable tasks.

The potential rewards are immense: by fully including women and considering gendered dynamics, DRR strategies can reduce human suffering, hasten recovery, and foster societies that not only survive disasters but emerge from them with stronger community bonds and more robust social structures. In an era of escalating climate challenges, societies cannot afford to overlook the needs, knowledge, and potential of half their populations. Addressing gendered vulnerabilities in disaster management is not just a matter of fairness—it is a strategic, forward-looking choice that will determine how effectively nations adapt to an uncertain climatic future.

By learning from the South Korean case and applying these lessons widely, policymakers, practitioners, and communities throughout South and East Asia can move toward a more inclusive and sustainable path—one where no one is left behind, and where resilience is built on the foundations of equity, empowerment, and shared responsibility.

References

Chung, G. (2017). 성인지적 관점을 고려한 재난안전관리 정책 수립의 필요성 [Necessity of disaster and safety management policy considering gender sensitivity]. *Journal of the Korean Society of Hazard Mitigation, 17*(4), 265–269. https://doi.org/10.9798/KOSHAM.2017.17.4.265

Denton, F. (2002). Climate change vulnerability, impacts, and adaptation: Why does gender matter? *Gender & Development, 10*(2), 10–20.

Enarson, E., & Chakrabarti, P. G. D. (Eds.). (2009). *Women, gender and disaster: Global issues and initiatives*. SAGE Publications.

IPCC (Intergovernmental Panel on Climate Change). (2014). *Fifth assessment report (AR5)*. Cambridge University Press.

Juran, L., & Trivedi, J. (2015). Women, gender norms, and natural disasters in Bangladesh. *Geographical Review, 105*(4), 601–611.

Neumayer, E., & Plümper, T. (2007). The gendered nature of natural disasters: The impact of catastrophic events on the gender gap in life expectancy, 1981–2002. *Annals of the Association of American Geographers, 97*(3), 551–566.

OECD. (2020). OECD Economic Surveys: Korea 2020. OECD Publishing.

Seager, J. (2009). *The penguin atlas of women in the world*. Penguin Books.

UNDP (United Nations Development Programme). (2016). Gender and Climate Change: Asia and the Pacific. UNDP.

UNDRR (United Nations Office for Disaster Risk Reduction). (2019). *Global assessment report on disaster risk reduction*. UNDRR.

UN Women. (2019). Gender Equality and the Sustainable Development Goals in Asia and the Pacific: Baseline and Pathways for Transformative Change by 2030. UN Women.

Chapter 14
Identifying Gender Perspectives on Green Roof Attributes Efficiency in Improving Mental Health

Shazmin Shareena Ab. Azis, Shastitharran Baskaran, Nur Hannani Ab Rahman, and Nursyuhaida Aziz

Abstract Green roof is believed to benefit community in terms of mental health and wellbeing. Previous studies claimed that green roof may influence woman more than man. However, there are limited studies examining the difference in gender perspective toward green roof efficiency. Thus, this study aims to assess the percentage of green roof efficiency in improving mental health according to gender perspectives. Questionnaire survey is conducted within green roof residents and visitors and analyzed using frequency analysis paired with independent sample *T*-tests. Results indicated that there is a significant difference between green roof efficiency in improving mental health between male and female, where females tend to experience better improvement from the green roof exposure, accessibility, distance, tree density, size, facility, security, and maintenance. This study supports decision-making of government in planning green building developments to improve mental health of community.

Keywords Green roof · Gender · Mental health · Green roof attribute

14.1 Introduction

The prevalence of mental health problems among the citizens has been rising recently, especially in the Central Business District (CBD) areas (Peen et al., 2010; Xu et al., 2023). Several common mental health symptoms include stress, depression, and anxiety. Various economic, environmental, and social factors are directly and indirectly influencing the increasing number of mental health problems which need to be addressed and improved for ensuring the sustainability of the city in terms of accommodating citizens with good health and wellbeing (Samsudin et al., 2024).

S. S. Ab. Azis (✉) · S. Baskaran · N. H. A. Rahman · N. Aziz
Faculty of Built Environment & Surveying, Real Estate, Universiti Teknologi Malaysia, Skudai, Johor, Malaysia
e-mail: shazmin@utm.my

Md. A. Abedin et al. (eds.), *Climate Crisis in South and East Asia*, Sustainability Solutions, https://doi.org/10.1007/978-3-032-01012-4_14

Among the various factors, lack of greenery exposure in daily lives of citizens was discovered to be one of the many factors that contribute to increasing mental health problems (Lederbogen et al., 2013; Liu et al., 2023).

Frequent green space visitation has been said to be a part of solutions that can be taken to improve the mental health of the citizens. Various researchers have emphasized and linked the positive relation on the green space visit with the improvement in mental health (Barton & Rogerson, 2017; Pasanen et al., 2023; Xian et al., 2024). The frequent visits to the green space in the residential and working area have improved one's mental health symptoms together with improving the overall mood of a person (Guizzo et al., 2018; Scopelliti et al., 2015). Green space is defined as a leisure place that is filled with vegetations that can be visited by the general publics which is commonly known as parks, gardens, and green roofs. The urbanization has reduced the availability of green spaces in urban areas and citizens tend to have limited options of green space for them to visit (White et al., 2011).

In the meantime, the prevalence of mental health and the severity of mental health vary according to the gender, where generally women tend to have higher probability in facing mental health symptoms compared to men's (Freeman, 2013). This arises the gap where if the severity of mental health is generally different across two genders, does it affect the green spaces efficiency toward the improvement of mental health between the genders too. To fill the gap, several researchers studied the tendency of green space in improving mental health across two genders. Koohsari et al., (2019) reported that the green space that is available around the respondents in Japan were more effective toward females compared to males. Apart from that, research that was conducted to determine the changes in depression symptoms in the availability of green space at a 0.5 km buffer reported that, the green space availability has more beneficial affect toward women than men in terms of mental health improvement (Sarkar et al., 2018). Apart from the general residential green spaces such as parks and gardens, green roof is also an infrastructure that has gained its popularity in the urban areas.

Green roof is a green infrastructure that is fitted into building to provide a green space to the public, which is also categorized as one of the sustainable options to implement green space in urban areas that are arising in number of buildings and decreasing green spaces (Gill et al., 2007; Ashley, 2011; Norton, 2015). Green roof provides many ecosystem services including urban climate control, filtration of stormwater runoff for flash flood mitigations, provides aesthetic values for the building and provides habitat for plant (Hoeben, 2021), including improving mental health of the citizens (Currie, 2008; Rezaei et al., 2021).

Previous studies have researched the benefits of green roof in improving mental health among the citizens across the world (O'Hara et al., 2022; Triguero-Mas et al., 2020; Williams et al., 2019). These studies analyzed the perceptions and state of mental health of the respondents before and after the green roof visitations to quantify the benefits gained. However, none of these studies explored the difference in perception between gender categories toward the attributes of green roof and its satisfaction level. The difference in the perceptions of green roof attributes on the mental health improvement benefits could provide a better understanding on the performance of green roof among male and females. Apart from that, the perception on green roof attributes that determines the satisfaction level of the green roof performance also could vary among the different genders. Thus, this study aims to identify the gender perspectives on green roof and its attributes in improving mental health. This study fills the gap in green roof knowledge where the gender perception on the green roof and its attributes effectiveness in terms of mental health improvement will be discovered.

14.2 Literature Review

14.2.1 Mental Health

The stability of mental health is one of the important aspects for citizens to be productive in their daily life (Saxena et al., 2015). According to the World Health Organizations (WHO), common mental health symptoms include depression, anxiety, stress, schizophrenia, and bipolar disorder. Mental health symptoms occur at different stages depending on the severity of the symptoms, and recent studies revealed that it occurs in almost all age groups, from young children to adults, and these studies also stated that depression is more likely to be experienced by women over men (Bhowmik, 2012). It is estimated that approximately a billion people are facing mental health symptoms with statistics showing that it is linearly increasing as days pass (Patel et al., 2023). Apart from that, the recent covid-19 pandemic also has worsened the mental health conditions of the citizens (Mohamed, 2022).

14.2.2 Green Space

Green space is known to be one of the many urban landscapes that is available to provide environmental and social benefits to the citizens and nature. Green spaces are widely known for their capabilities in improving people's health (Frank et al., 2019). Apart from that, the availability of accessible green spaces in an area attracts the citizens to visit and spend their leisure times in the green space due to its nature of a calming environment and previous researchers have reported there is a reduction in the citizens mental health due to the activities they carry out in the parks with green scenery and exposures (Cubukcu, 2013; Song, 2015).

14.2.3 Green Space Attributes

There are several attributes of green space that could enhance its capabilities in improving mental health according to various research findings including degree of exposure to green space, accessibility, tree density, size, facilities to perform leisure activities, distance, security, and maintenance at the green space. Exposure to the green roof is basically getting exposure to the greenery environment that the green roof provides to the community and the exposure to that environment has effects on the level of the efficiency of green roof in improving mental health (Lee and Lee, 2019; Bojic, 2018). Then the accessibility to green space is also one of the attributes that influences the visitation of citizens to the green spaces. Accessibility to the available green spaces in the urban area is the tendency of the citizens to visit the green roof without any restrictions together with no costs. Enabling green spaces that have got good accessibility is important to encourage the visitation of citizens (Biernacka & Kronenberg, 2019; Kiplagat et al., 2022). The accessibility of green space influences the mental health state of a person, and researchers concluded that lesser accessibility to the green area tends to result in a higher tendency to experience mental health problems (Guite et al., 2006; Lo, 2003).

Apart from that, tree density also is an important factor in the green space environment, which is the type of plants and trees that represent in the green space. Tree density is the measure of the amount of greenness in a particular green space. A green space with higher number of plants and vegetations naturally increases the density of greenness which is favorable to the citizens in terms of mental health improvement (Astell-Burt and Feng, 2019; Beyer et al., 2014). Furthermore, the facilities that are available in the green space also affect the probability of visiting citizens together with the enhancement in terms of mental health improvement. Facilities that are common in green spaces includes, jogging tracks, bench, playgrounds, evacuation sites, and parking lots (Kim et al., 2020; Pope et al., 2018). The availability of these facilities encourages citizens to spend more time in the green space and in the meantime have consideration to perform recreational activities, which is beneficial for the overall physical and mental health as well (Gililand, 2006; Kiplagat et al., 2022).

Moreover, size of the green roof is another aspect of the green spaces that area also influential in determining the quality of the green space. Bigger sizes of green spaces will have the ability to accommodate higher numbers of citizens while also allowing increase social communication opportunities between the visitors (Wood et al., 2017). Other than that, the distance to the green space is also an important attribute where when the green space is located near to the residential place and workplace it encourages the citizens to visit them more often during their tensed times (James, 2016; Reklaitiene et al., 2014). Lastly, the security and also maintenance of the green space is also an essential aspects that should be focused and sustained to ensure the quality of green space is always retained (Rosasco, 2019; Madureira et al., 2018). A green space that is safe preferably with a security guard always monitoring the activities in the green space together with proper scheduled maintenance such as cleaning and trimming the plants and vegetation will increase the aesthetics and retain the comfortability of the environment.

14.2.4 Mental Health Improvement Through Green Roof

Green roof is a green infrastructure that is constructed on buildings, where it partially or fully covers the roofs of buildings with the presence of various types of plants and vegetations (Yu et al., 2017). There are mainly two types of green roof which are known as intensive and extensive green roofs. The intensive green roof is placed inside the building where the citizens can access and enjoy the environment of the green roof. It has a higher substrate depth which can withstand bigger trees and various types of plantations. On the other hand, the extensive green roof is placed on the external site of the buildings normally on the roof which cannot be accessed by the public and due to its nature of lower substrate depth, only a few types of vegetation can be planted (Williams, 2010; Dunnett and Kingsbury, 2004).

Looking at the characteristics of both types of green roof, extensive green roof is constructed for its environmental benefits such as heat mitigation, urban stormwater runoff mitigation, and sound insulation where the intensive green roof is implemented for the environment benefits together with added social benefits such as providing mental health benefits as it can be accessed by the citizens (MacIvor, 2011; Jim, 2011). Green roof acts as a sustainable asset in increasing the amount of green space in the urban areas. Recently, apart from the many environmental benefits that green roof provides, it is also supportive in improving citizens mental health with its nature of aesthetics and scenery (Hartig et al., 2014; Oberndorfer et al., 2007). The visit to the green roof provides natural restorative effects to the citizens together with improving their mental health in terms of stress levels, anxiety, and depression. (White & Gatersleben, 2011). The continuous visit to the green roofs is associated with overall calmness toward the body and mind of the citizens (Loder, 2014).

14.3 Methods

A questionnaire survey between the green roof visitors was chosen as the data collection method. This is because the green roof visitors will have the experience on the benefits that green roof offer in terms of mental health improvement. Apart from that green roof visitors directly benefit from the aesthetics and green environment of the green roof which becomes the most appropriate respondent for this survey. The perception of green roof visitors on the levels of satisfaction toward green roof and its attribute will be surveyed. The research scope is chosen to be at Kuala Lumpur which is the urban city which is most favorite destination in the process of urbanization in Malaysia. Six buildings that have green roofs are selected as the research study place which is a combination of shopping malls, condominiums, and office buildings

(Rahman et al., 2013). As for the sampling method, the exact number of populations that visit the green roofs that are located at the six identified buildings is unknown. Thus, formula proposed by Cochran (1977) is used with an estimated variability value at 50% to identify the number of samples that are needed for research where the population size is not known.

14.3.1 Measuring Instrument

The scale that is used in the questionnaire is based on the items that are surveyed. The main purpose of the questionnaire survey is to identify the satisfaction levels of green roof visitors toward the improvement of their mental health and their perception of the green roof attributes efficiency in influencing the quality of green roof. Apart from the commonly used Likert scale in the questionnaire of social science studies, this study adapted the use of continuous rating scales. The continuous rating scale is a more appropriate scale to be used in the survey that relates to health matters such as the changes in mental health symptoms (Bradlow & Fitzsimons, 2001; Holte & Westgaard, 2012). The continuous rating uses the scale from 1 to 100 allowing more flexibility for the green roof visitors to express their satisfaction level and perception pertaining the green roof benefits in improving mental health. The questionnaire is divided into three sections where the first section contains the questions regarding the demographic profile of the respondents, second section is the survey of green roof satisfaction level on its effectiveness in improving mental health symptoms such as stress, anxiety, depression, and overall mood, and the last section consists of questions regarding the green roof attributes effectiveness in improving mental health.

14.3.2 Validity

The questionnaire was developed and a validity survey among the experts in the field of sustainable development, green buildings, and ecosystem services is undergone to identify whether the developed questionnaire is aligned with the aim of this research. A total of five panel experts were chosen to participate in the validity survey. Then, the content validity index was calculated from the responses of the panel experts (Davis, 1992; Lynn, 1986). Apart from that, the comments and suggestions given by the panel experts are also taken into consideration. The questionnaire was amended according to the comments and suggestions by the panel experts to enhance the overall quality of the questionnaire.

14.3.3 Reliability

Once the questionnaire survey is completed, the reliability of the relative responses of the questionnaires is identified using Cronbach alpha reliability test using the SPSS software. The test determines the internal consistency of the questionnaire results. (Sekaran, 2003) suggested that Cronbach's alpha value of more than 0.70 is advisable for the results of the research to achieve its internal consistency. Cronbach's alpha value for this survey was calculated to be at 0.985, which shows that the internal consistency in the questionnaire is good and reliable.

14.3.4 Statistical Analysis

The analysis of the questionnaire results was done using the SPSS software. Three sections of the questionnaire results data were imported into the software to identify the descriptive statistics together with carrying out the frequency for the green roof satisfaction level items together with the green roof attributes item. The green roof satisfaction was measured using four items and the benefits that green roof provides for each of items are quantified using the frequency analysis and identifying its mean value. Apart from that, the green roof has a total of 7 attributes which were identified in the literature review as green space attributes, which undergone the frequency analysis too. The results of the green roof satisfaction level in improving mental health together with its attributes were analyzed based on gender perception to understand the difference between males and females.

Then, to identify whether there is a significant change in the results between male and female satisfaction levels and perception independent samples *t*-test were carried out. This test is done to compare the average results of both gender groups in relation to the items in the questionnaire to determine whether their view on the green roof in improving mental health is same or different from each other.

14.4 Results

A total of 385 surveys were finalized to be included in the analysis after elimination of incomplete and missing value surveys. Table 14.1 shows the demographic profiles of the green roof visitors. The number of respondents also aligns with the sample size that was calculated according to Cochrane's formula. From the total number of respondents, females compromise 40% of the survey and the remaining was male respondents at 60%. Malay respondents dominated the survey with almost half of the total respondents followed by Chinese and Malay. Apart from that, most of the respondents hold a bachelor's degree and most of the respondents fall under the age group ranging from 26 to 45 years old.

Table 14.1 Demographic profiles of green roof visitors

Description		Frequency	Percent (%)
Gender	Female	155	40.3
	Male	230	59.7
Race	Chinese	105	27.3
	Indian	82	21.3
	Malay	198	51.4
Academic qualification	Bachelor's degree	243	63.1
	Diploma	137	35.6
	Doctor of philosophy	5	1.3
Age	18–25	15	3.9
	26–35	128	33.2
	36–45	136	35.3
	46–55	81	21.0
	55 or older	25	6.5
Place of visit	Lot 10 Kuala Lumpur	93	24.2
	Menara Binjai Ampang	61	15.8
	One Utama Secret Garden	84	21.8
	Riana Green East Kuala Lumpur	55	14.3
	The Maple Sentul	47	12.2
	The Saffron Sentul	45	11.7
	Total	385	100.0

14.4.1 Green Roof Satisfaction Level by Gender

The green roof performance in improving mental health among the green roof visitors was quantified using four different criteria of betterment in mental health such as stress reduction, lessened anxiety feelings, reduced depression symptoms, and mood improvement. Table 14.2 displays the green roof's potential in improving mental health among the visitors. Generally, females tend to have higher effect on improvement of their mental symptoms through green roof visitation than male.

The mean score on the improvement of mental health symptoms is higher for females on all four symptoms. Female respondents have higher satisfaction rate on their mental health symptoms than males at 27% for stress levels improvement, 31% improvement on their anxiety feelings 45% higher reduction in depression symptoms compared to male respondents. Apart from that, the overall mood of female respondents is 38% better compared to males. The levels of improvement in mental health of the citizens reflect the benefits that are provided by the green roof. Although females rated better scores in their mental health improvements, male respondents on the other hand also provided betterment rating as well. The results also support the findings of studies that identified the relationship between the quality of green

Table 14.2 Gender satisfaction level on mental health benefit by green roof

Symptoms	Gender	Nos	Mean	Difference in mean (%)	Average (male and female)	Std. deviation	Std. error mean
Stress	Female	155	82.15	27	73.51	13.892	1.116
	Male	230	64.87			20.806	1.372
Anxiety	Female	155	78.39	31	69.14	16.211	1.302
	Male	230	59.88			20.106	1.326
Depression	Female	155	74.42	45	62.83	18.642	1.497
	Male	230	51.25			21.204	1.398
Mood	Female	155	78.61	38	67.76	18.208	1.463
	Male	230	56.91			18.661	1.230

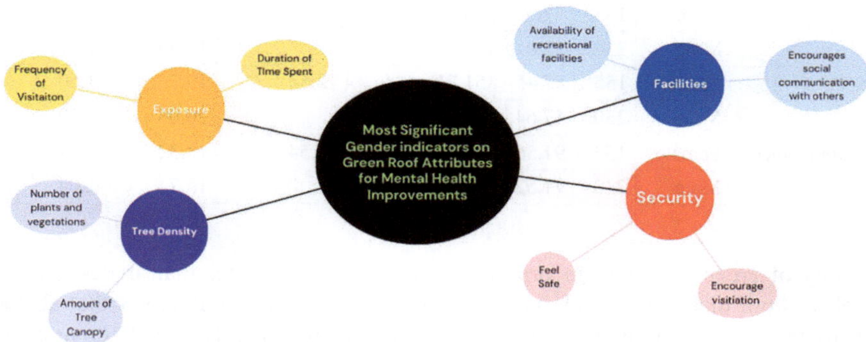

Fig. 14.1 Conceptual framework of the study

space and mental health betterment. Figure 14.1 shows the conceptual framework of the study where the attributes that has higher difference in mean proportion in the perception of male and females is discovered to be the most significant attributes that affects the mental health improvement between genders.

14.4.2 *Green Roof Attributes Perception by Gender*

Moving on to the results regarding the green roof attributes and what are the differences in gender perception toward each of the green roof attributes. Table 14.3 shows the results on the green roof that attributes perception by gender. From the findings, the females dominated the scores as well by having higher means scores on the perception of all the attributes. This result shows that females prefer better

Table 14.3 Gender perceptions on importance of green roof attributes

Attributes	Gender	N	Mean	Difference in mean (%)	Average mean score	Std. deviation	Std. error mean
Exposure	Female	155	79.71	46.54	67.05	16.12	1.29
	Male	230	54.40			22.74	1.50
Accessibility	Female	155	80.88	37.47	69.86	15.25	1.23
	Male	230	58.83			21.90	1.44
Distance	Female	155	82.55	37.72	71.24	16.20	1.30
	Male	230	59.94			17.99	1.19
Tree density	Female	155	81.94	53.09	67.73	17.20	1.38
	Male	230	53.52			20.13	1.33
Size	Female	155	82.06	42.05	69.92	20.28	1.63
	Male	230	57.77			20.16	1.33
Facility	Female	155	77.72	46.54	65.38	16.66	1.34
	Male	230	53.04			18.59	1.23
Security	Female	155	87.45	51.71	72.55	19.89	1.60
	Male	230	57.64			20.03	1.32
Maintenance	Female	155	91.56	28.02	81.54	13.32	1.07
	Male	230	71.52			15.74	1.04

quality of green roof with incorporation of all these attributes than males. Nevertheless, both male and female gave their perception and output shows that all the attributes presented to them is important to implement a quality and effective green roof improving mental health.

Among all the attributes, several attributes such as exposure tree density, facility, and security have relatively higher difference in perception compared to others which also have difference but these few are among the ones that resulted in higher differences. Females gave more importance to these attributes in the green roof while visiting apart from the other attributes. The security recorded the highest difference in the mean scores between male and female which portrays that females need the green roof to be safe for them to be comfortable visiting it for the purpose of mental health. Bahriny and Bell (2020) also discovered in their research that the safety features in the park together with other factors is more desired by women than men to attain the encouragement to visit a space which includes green roof as well.

14.4.3 Independent Samples T*-test*

The independent sample *t*-test was conducted to compare whether the difference between male and female on the green roof satisfaction level and green roof attributes perceptions is significant or otherwise. Tables 14.4 and 14.5 demonstrate the results of the test conducted. In terms of the green roof satisfaction level, there is statistically significant difference in between male and female for all the four mental health symptoms which quantifies p value less 0.05. The stress symptoms had significant differences at (t (382) = 9.777, $p = 0.000$), anxiety at (t (371) = 9.961, $p = 0.000$), depression at (t (383) = 11.031, $p = 0.000$), and overall mood at (t (383) = 11.302, $p = 0.000$). This results on the t-test demonstrates that the effect of green roof in improving the mental health between male and females differs significantly.

Apart from that, for the perception on the green roof attributes, the output also illustrates that there significantly difference in perception where female and male prefer different type of attributes and their perception toward a quality green roof that enhances mental health improvement is different from one another. Perception on the exposure attribute has difference at (t (382) = 12.779, $p = 0.000$), accessibility at (t (382) = 11.640, $p = 0.000$), distance at (t (382) = 12.582, $p = 0.000$), tree density at (t (383) = 14.383, $p = 0.000$), size at (t (383) = 11.569, $p = 0.000$), facility at (t (353) = 13.599, $p = 0.000$), security at (t (383) = 14.359, $p = 0.000$), and maintenance at (t (364) = 13.444, $p = 0.000$). The results prove that male and females have different desire when it comes to visiting green roof and both of them have difference in their preferred attributes that is needed to be included in the green roof for frequent visitation and higher time spending.

14.5 Discussions

The study is carried out to understand the difference in gender perception on the capabilities of green roof in improving mental health with the hypothesis that there is a difference in between male and females on the benefits in terms of mental health that is obtained from the green roof visitations. From the results of the independent samples t-test, the male and females do have different satisfaction levels on their mental health improvements upon green roof visitation. Females have better tendency on improving mental health through than males which indicates clear difference in perception between them. This results on the t-test demonstrates that the effect of green roof in improving the mental health between male and females differs significantly. The past studies concluded the effect of green spaces have more positive affect on women compared to men's which is also discovered in this study (Bos et al., 2016; Koohsari et al., 2019). Apart from that, male and females also have different hierarchy of importance when it comes to green roof attributes that is needed for the green roof to be at its highest and best use for providing mental health improvements. This t-test results also align with findings by Mulling et al. (2013) the output of the study figured out that green spaces with variation in attributes have different effects in terms of mental health improvement between male and females. The limitations of this study are that it is only done in the properties that has green roof in Kuala Lumpur. The results of the questionnaire may differ if the research is carried out in a different state or country. Nevertheless, the fact that proves there is significant difference in gender perception toward mental health improvement and validation

Table 14.4 Independent samples *t*-test on green roof satisfaction level between genders

				Levene's test for equality of variances		*t*-test for equality of means						
		Mean	Standard deviation	*F*	Sig	*T*	*df*	Sig. (2-tailed)	Mean difference	Std. error difference	95% confidence interval of the difference	
											Lower	Upper
Stress	Female	82.15	13.892	38.039	0.000*	9.777	382	0.000*	17.290	1.768	13.813	20.767
	Male	64.87	20.806									
Anxiety	Female	78.39	16.211	14.478	0.000*	9.961	371	0.000*	18.511	1.858	14.857	22.165
	Male	59.88	20.106									
Depression	Female	74.42	18.642	2.071	0.151	11.031	383	0.000*	23.172	2.101	19.042	27.302
	Male	51.25	21.204									
Overall mood	Female	78.61	18.208	0.287	0.592	11.302	383	0.000*	21.704	1.920	17.928	25.480
	Male	56.91	18.661									

Table 14.5 Independent samples *t*-test on green roof attributes perception between genders

				Levene's test for equality of variances		*t*-test for equality of means						
		Mean	Standard deviation	F	Sig	*t*	*df*	Sig. (2-tailed)	Mean difference	Std. error difference	95% confidence interval of the difference	
											Lower	Upper
Exposure	Female	79.71	16.12	26.702	0.000	12.779	382.008	0.000	25.314	1.981	21.419	29.209
	Male	54.40	22.74									
Accessibility	Female	80.88	15.25	31.660	0.000	11.640	382.569	0.000	22.043	1.894	18.319	25.766
	Male	58.83	21.90									
Distance	Female	82.55	16.20	1.852	0.174	12.582	383	0.000	22.608	1.797	19.075	26.141
	Male	59.94	17.99									
Tree density	Female	81.94	17.20	2.049	0.153	14.383	383	0.000	28.415	1.976	24.530	32.299
	Male	53.52	20.13									
Size	Female	82.06	20.28	0.016	0.898	11.569	383	0.000	24.295	2.100	20.166	28.424
	Male	57.77	20.16									
Facility	Female	77.72	16.66	4.599	0.033	13.599	353.461	0.000	24.682	1.815	21.112	28.252
	Male	53.04	18.59									
Security	Female	87.45	19.89	0.141	0.707	14.359	383	0.000	29.808	2.076	25.726	33.890
	Male	57.64	20.03									
Maintenance	Female	91.56	13.32	13.572	0.000	13.444	363.635	0.000	20.038	1.490	17.107	22.970
	Male	71.52	15.74									

on green roof visitation can improve mental health is sufficient as the contribution of this study. Future studies may implement and study more deeply on the perception on a larger scale of participants and bigger radius pf research area to obtain more variation and higher quality results.

14.6 Conclusion

In a nutshell, green roof has high capabilities in improving citizens mental health especially in the urban areas where high number mental health problems are recorded. In accordance with that, the acting parties that are involved and responsible in maintaining a sustainable city together with assuring the mental health and wellbeing of the citizens are being improved should consider implementing green roofs as way of effort in increasing the amount green spaces in the urban areas. Implementing green roofs not only benefits the owner of the building and the visitors economically but it also benefits them socially. Despite the discovery that both male and female have high satisfaction levels in terms of stress, depression, anxiety, and mood, there tend to be differences in their satisfaction levels. As for the green roof attributes, both male and female presented importance in implementing green roof with the consideration of all the discovered attributes. The safety feature was more preferred and needed by women than males and all the other attributes have significant different as well in terms of gender perceptions. The green roof implementers should consider all these attributes well studied and planned to provide a quality green roof for the citizens.

References

Bahriny, F., & Bell, S. (2020). Patterns of urban park use and their relationship to factors of quality: A case study of Tehran. *Iran. Sustainability, 12*(4), 1560.

Barton, J., & Rogerson, M. (2017). The importance of greenspace for mental health. *Bjpsych International, 14*(4), 79–81. https://doi.org/10.1192/s2056474000002051

Bhowmik, D., Kumar, S., Srivastava, S., Paswan, S., & Dutta, A. S. (2012). Depression-symptoms, causes, medications and therapies. *Pharma Innovation, 1*, 32–45.

Biernacka, M., & Kronenberg, J. (2019). Urban green space availability, accessibility and attractiveness, and the delivery of ecosystem services. *Cities and the Environment (CATE), 12*(1), 5.

Bojic, S. (2018). Organic agriculture contribution to the rural tourism development in the North of Montenegro. *Journal of Economics and Business, 1*(3), 222–233.

Bos, E. H., Van der Meulen, L., Wichers, M., & Jeronimus, B. F. (2016). A primrose path? Moderating effects of age and gender in the association between green space and mental health. *International Journal of Environmental Research and Public Health, 13*(5), 492.

Beyer, K. M., Kaltenbach, A., Szabo, A., Bogar, S., Nieto, F. J., & Malecki, K. M. (2014). Exposure to neighborhood green space and mental health: evidence from the survey of the health of Wisconsin. *International Journal of Environmental Research and Public Health, 11*(3), 3453–3472.

Bradlow, E., & Fitzsimons, G. (2001). Subscale distance and item clustering effects in self-administered surveys: A new metric. *Journal of Marketing Research—J MARKET RES-CHICAGO, 38*, 254–261. https://doi.org/10.1509/jmkr.38.2.254.18837

Cochran, W. G. (1977). Sampling Techniques (3rd ed.), Wiley, New York.

Cubukcu, E. (2013). Walking for sustainable living. *Procedia - Social and Behavioural Sciences, 85*. https://doi.org/10.1016/j.sbspro.2013.08.335

Currie, B. A., & Bass, B. (2008). Estimates of air pollution mitigation with green plants and green roofs using the UFORE model. *Urban Ecosyst, 11*(4), 409–422.

Davis, L. (1992). Instrument review: Getting the most from a panel of experts. *Applied Nursing Research—APPL NURSING RES., 5*, 194–197. https://doi.org/10.1016/S0897-1897(05)80008-4

Frank, L. D., Iroz-Elardo, N., MacLeod, K. E., & Hong, A. (2019). Pathways from built environment to health: A conceptual framework linking behaviour and exposure-based impacts. *Journal of Transport & Health, 12*, 319–335.

Freeman, D., & Freeman, J. (2013). *The stressed sex: Uncovering the truth about men, women, and mental health.* Oxford University Press (UK).

Gill, S. E., Handley, J. F., Ennos, A. R., & Pauleit, S. (2007). Adapting cities for climate change: The role of the green infrastructure. *Built Environment, 33*(1), 115–133.

Gilliland, J., Holmes, M., Irwin, J. D., & Tucker, P. (2006). Environmental equity is child's play: Mapping public provision of recreation opportunities in urban neighbourhoods. *Vulnerable Children and Youth Studies, 1*(3), 256–268.

Guite, H. F., Clark, C., & Ackrill, G. (2006). The impact of the physical and urban environment on mental well-being. *Public Health, 120*(12), 1117–1126.

Guizzo, A., Escoffier, N., Chan, J., & Puay Yok, T. (2018). Window view and the brain: Effects of floor level and green cover on the alpha and beta rhythms in a passive exposure EEG experiment. *International journal of environmental research and public health, 15*, 2358. https://doi.org/10.3390/ijerph15112358

Hartig, T., Mitchell, R., De Vries, S., & Frumkin, H. (2014). Nature and health. *Annual Review of Public Health, 35*(1), 207–228.

Holte, K., & Westgaard, RH. (2012). Exploring perceived tension as a response to psychosocial work stress. *Environment & Health.* Downloaded from www.sjweh.fi on Jan 13.

Hoeben, A., & Posch, A. (2021). Green roof ecosystem services in various urban development types: A case study in Graz, Austria. *Urban Forestry & Urban Greening, 62*, 127167. https://doi.org/10.1016/j.ufug.2021.127167

James, P., Hart, J., Banay, R., & Laden, F. (2016). Exposure to greenness and mortality in a nation-wide prospective cohort study of women. *Environmental Health Perspectives, 124*. https://doi.org/10.1289/ehp.1510363

Jim, C. Y., & Tsang, S. W. (2011). Ecological energetics of tropical intensive green roof. *Energy and Buildings, 43*, 2696–2704. https://doi.org/10.1016/j.enbuild.2011.06.018

Kim, H., Shoji, Y., Tsuge, T., Aikoh, T., & Kuriyama, K. (2020). Understanding services from ecosystem and facilities provided by urban green spaces: A use of partial profile choice experiment. *Forest Policy and Economics, 111*, Article 102086.

Kiplagat, A. K., Koech, J. K., Ng'etich, J. K., Lagat, M. J., Khazenzi, J. A., & Odhiambo, K. O. (2022). Urban green space characteristics, visitation patterns and influence of visitors' socio-economic attributes on visitation in Kisumu City and Eldoret Municipality, Kenya. *Trees, Forests and People, 7*, Article 100175.

Koohsari, M. J., McCormack, G. R., Nakaya, T., Shibata, A., Ishii, K., Yasunaga, A., Hanibunchi, T., & Oka, K. (2019). *Urban design and Japanese older adults' depressive symptoms. Cities, 87*, 166–173.

Lederbogen, F., Haddad, L., & Meyer-Lindenberg, A. (2013). Urban social stress–Risk factor for mental disorders The case of Schizophrenia. *Environmental Pollution, 183*, 2–6.

Liu, Z., Chen, X., Cui, H., Ma, Y., Gao, N., Li, X., Meng, X., Lin, H., Abuduo, H., Guo, L., & Liu, Q. (2023). Green space exposure on depression and anxiety outcomes: A meta-analysis. *Environmental research, 231*, 116303.

Lo, S., Yiu, E., & Lo, A. (2003). An analysis of attributes affecting urban open space design and their environmental implications. *Management of Environmental Quality: An International Journal, 14*. 604–614. https://doi.org/10.1108/14777830310495759

Loder, Angela. (2014). 'There's a meadow outside my workplace': A phenomenological exploration of aesthetics and green roofs in Chicago and Toronto. *Landscape and Urban Planning, 126*. https://doi.org/10.1016/j.landurbplan.2014.01.008

Lynn, M. (1986). Determination and quantification of content validity. *Nursing Research., 35*, 382–385. https://doi.org/10.1097/00006199-198611000-00017

MacIvor, J. S., & Lundholm, J. (2011). Performance evaluation of native plants suited to extensive green roof conditions in a maritime climate. *Ecological Engineering, 37*(3), 407–417.

Madureira, H., Nunes, F., Oliveira, J. V., & Madureira, T. (2018). Preferences for urban green space characteristics: A comparative study in three Portuguese cities. *Environments, 5*(2), 23.

Mohamed, N., Abidin, E., Rasdi, I., & Ismail, Z. S. (2022). The impact of COVID-19 pandemic on Malaysian police officers mental health: Depression, anxiety and stress. *International Journal of Infectious Diseases, 116*, S45. https://doi.org/10.1016/j.ijid.2021.12.108

Mullings, J. A., McCaw-Binns, A. M., Archer, C., & Wilks, R. (2013). Gender differences in the effects of urban neighborhood on depressive symptoms in Jamaica. *Revista Panamericana De Salud Pública, 34*, 385–392.

Oberndorfer, E., Lundholm, J., Bass, B., Coffman, R. R., Doshi, H., Dunnett, N., Gaffin, S., Kohler, M., Liu, K. K., & Rowe, B. (2007). Green roofs as urban ecosystems: ecological structures, functions, and services. *BioScience, 57*(10), 823–833.

O'Hara, A. C., Miller, A. C., Spinks, H., Seifert, A., Mills, T., & Tuininga, A. R. (2022). The sustainable prescription: Benefits of green roof implementation for urban hospitals. *Frontiers in Sustainable Cities, 4*, Article 798012.

Pasanen, T. P., White, M. P., Elliott, L. R., van den Bosch, M., Bratman, G. N., Ojala, A., Korpela, K., & Fleming, L. E. (2023). Urban green space and mental health among people living alone: The mediating roles of relational and collective restoration in an 18-country sample. *Environmental Research, 232*, 116324.

Patel, V., Saxena, S., Lund, C., Kohrt, B., Kieling, C., Sunkel, C., Kola, L., Chango, O., Charlson, F., O'Neill, K., & Herrman, H. (2023). Transforming mental health systems globally: principles and policy recommendations. *The Lancet, 402*(10402), 656–666.

Peen, J., Schoevers, R. A., Beekman, A. T., & Dekker, J. (2010). The status of urban-rural differences in psychiatric disorders. *Acta Psychiatrica Scand., 121*(2), 84–93.

Pope, D., Tisdall, R., Middleton, J., Verma, A., Van Ameijden, E., Birt, C., Macherianakis, A., & Bruce, N. G. (2018). Quality of and access to green space in relation to psychological distress: Results from a population-based cross-sectional study as part of the EURO-URHIS 2 project. *The European Journal of Public Health, 28*(1), 35–38.

Rahman, S. R. A., Ahmad, H., & Bin Rosley, M. S. (2013). Green Roof: Its awareness among professionals and potential in Malaysian market. *Procedia—Social and Behavioral Sciences, 85*. 443–453. https://doi.org/10.1016/j.sbspro.2013.08.373

Reklaitiene, R., Grazuleviciene, R., Dedele, A., Virviciute, D., Vensloviene, J., Tamosiunas, A., Baceviciene, M., Luksiene, D., Sapranaviciute-Zabazlajeva, L., Radisauskas, R., Bernotiene, G., & Nieuwenhuijsen, M. J. (2014). The relationship of green space, depressive symptoms and perceived general health in urban population. *Scandinavian Journal of Public Health, 42*(7), 669–676.

Rezaei, M., Emmanuel, N., Kim, J., & Kim, D. H. (2021). Analyzing the impact of green roof functions on the citizens' mental health in metropolitan cities. *Iranian Journal of Public Health, 50*(5), 900–907.

Rosasco, P., & Perini, K. (2019). Selection of (green) roof systems: A sustainability-based multi-criteria analysis. *Buildings, 9*, 134.https://doi.org/10.3390/buildings9050134

Samsudin, S., Ismail, R., Daud, S. N. M., & Yahya, S. (2024). The prevalence and underlying factors of mental health disorders in Malaysian youth. *Journal of Affective Disorders Reports, 15*, Article 100706.

Sarkar, C., Webster, C., & Gallacher, J. (2018). Residential greenness and prevalence of major depressive disorders: A cross-sectional, observational, associational study of 94 879 adult UK biobank participants. *The Lancet Planetary Health, 2*(4), e162–e173. https://doi.org/10.1016/S2542-5196(18)30051-2

Saxena, S., Funk, M. K., & Chisholm, D. (2015). Comprehensive mental health action plan 2013. *Eastern Mediterranean Health Journal, 461*.

Scopelliti, M., Lafortezza, R., Colangelo, G., Ferrini, F., Salbitano, F., Agrimi, M., Portoghesi, L., Semenzato, P., & Sanesi, G. (2015). Go greener, feel better? The positive effects of biodiversity on the well-being of individuals visiting urban and peri-urban green areas. *Landscape and Urban Planning. 134*. https://doi.org/10.1016/j.landurbplan.2014.10.022

Sekaran, U. (2003). *Research methods for business* (4th ed.). John Wiley & Sons, Inc.

Song, C., Ikei, H., Igarashi, M., Takagaki, M., & Miyazaki, Y. (2015). Physiological and psychological effects of a walk in urban parks in fall. *International Journal of Environmental Research and Public Health, 12*(11), 14216–14228.

Triguero-Mas, M., Anguelovski, I., Cirac-Claveras, J., Connolly, J., Vazquez, A., Urgell-Plaza, F., Cardona-Giralt, N., Sanyé-Mengual, E., Alonso, J., & Cole, H. (2020). Quality of life benefits of urban rooftop gardening for people with intellectual disabilities or mental health disorders. *Preventing Chronic Disease, 17*, E126.

White, E., & Gatersleben, B. (2011). Greenery on residential buildings: Does it affect preferences and perceptions of beauty? *Journal of Environmental Psychology, 31*, 89–98. https://doi.org/10.1016/j.jenvp.2010.11.002

Williams, K. J., Lee, K. E., Sargent, L., Johnson, K. A., Rayner, J., Farrell, C., Miller, E., & Williams, N. S. (2019). Appraising the psychological benefits of green roofs for city residents and workers. *Urban forestry & urban greening, 44*, 126399

Williams, N. S., Rayner, J. P., & Raynor, K. J. (2010). Green roofs for a wide brown land: Opportunities and barriers for rooftop greening in Australia. *Urban Forestry & Urban Greening, 9*(3), 245–251.

Wood, L., Hooper, P., Foster, S., & Bull, F. (2017). Public green spaces and positive mental health–investigating the relationship between access, quantity and types of parks and mental wellbeing. *Health & Place, 48*, 63–71.

Xian, Z., Nakaya, T., Liu, K., Zhao, B., Zhang, J., Zhang, J., Liu, Y., & Zhang, J. (2024). The effects of neighbourhood green spaces on mental health of disadvantaged groups: A systematic review. *Humanities and Social Sciences Communications, 11*(1), 1–19.

Xu, C., Miao, L., Turner, D., & Derubeis, R. (2023). Urbanicity and depression: A global meta-analysis. *Journal of Affective Disorders, 340*, 299–311.

Yu, W. D., Cheng, S. T., Miao, C. M., & Perng, G. Y. (2017). Green innovation of green roof technology-a case study: Umweltverträgliche Dachtechnologie-eine Fallstudie. *Materialwissenschaft und WeRkstofftechnik, 48*(5), 420–429.

Chapter 15
Climate and Disaster Risks and Resilience of Women Street Vendors in Metro Manila, Philippines

Emma Porio, Jessica Dator-Bercilla, and Noralene Uy

Abstract The study explores the experiences of exposure to environmental risks of women street vendors in Metro Manila, and their local knowledge systems on environmental pollution, health, and climate and disaster risks. It (1) characterizes the composition of street-based populations and their levels of exposure to air and noise pollution, (2) describes the resource-opportunity structure of the streets and their strategies of coping and responding to air and noise pollution, as well as to health and climate and disaster risks, and (3) identifies solution pathways for risk reduction from the perspective of highly exposed and vulnerable populations, including from the urban authorities. Through an ethnographic inquiry, the study used in-depth interviews, life histories, participant observation, direct and indirect observation, informal and unstructured interviews, and focus group discussions. The study reveals that the strategic practices of women street vendors are very much informed by their intimate knowledge of street and community power relations embedded in their perilous environment. This paper hopes to contribute to the literature on bridging, translating, and integrating local knowledge systems on air and noise pollution to environmental justice and climate disasters in cities. The study results provide valuable insights for the larger scientific community, development practitioners, and policymakers on air pollution risks among the most highly exposed and vulnerable street-based population in the cities of the global south.

Keywords Ethnography · Social vulnerability · Vulnerable populations · Urban poor · Environmental pollution · Climate change · Disaster risk · Local knowledge systems

E. Porio (✉) · J. Dator-Bercilla · N. Uy
Ateneo de Manila University, Manila, Philippines
e-mail: eporio@ateneo.edu

Md. A. Abedin et al. (eds.), *Climate Crisis in South and East Asia*, Sustainability Solutions, https://doi.org/10.1007/978-3-032-01012-4_15

15.1 Introduction

Metro Manila is one of the most densely populated urban centers in Southeast Asia, characterized by rapid urbanization and a high concentration of informal settlers and migrant workers from the urban-rural areas of the Philippines (Porio, 2024). About 40–60% of Metro Manila's working population belong to the informal sector belongs to the informal sector (ILO, 2009). Among the most vulnerable groups in the informal economy are women street vendors and their children, who rely on street-based vending for their daily income economies (Langbid-Roda & Collantes, 2024). These vendors operate in open spaces, often near busy roads and markets, where they are directly exposed to climate-related hazards such as worsening air pollution, prolonged heat waves, and seasonal flooding. Unlike formal sector workers, they lack social protection, secure vending spaces, and inclusion of their needs in urban planning and disaster risk reduction (DRR) efforts (Porio et al., 2019).

Women vendors are not passive victims of these risks but active agents in shaping their resilience strategies. Through lived experience, they develop adaptive mechanisms such as shifting vending locations based on environmental conditions, forming community support networks, and employing health-protective practices (Porio, 2021). However, these informal strategies remain largely unrecognized in official urban governance frameworks, leaving vendors without institutional support (Porio et al., 2019). This study explores how women vendors experience environmental hazards, construct resilience strategies, and negotiate institutional constraints within a congested urban landscape.

Most studies on climate and disaster resilience focus on formal sectors or rural communities, leaving informal workers, particularly women in urban settings, underrepresented. While some research exists on gendered vulnerability, very few studies have examined the local knowledge systems and adaptive strategies of street-based populations. This research aims to bridge that gap, providing a nuanced understanding of how gender, informality, and local knowledge intersect in urban resilience practices.

The study's key objectives are threefold, namely, (1) identify the primary environmental hazards affecting women street vendors in Metro Manila, (2) analyze their adaptive strategies in response to climate and disaster risks, and (3) examine how institutional constraints shape their resilience options at the micro-meso-macro levels. By documenting the experiences of these vendors, this research contributes to discussions on urban informality, gender-responsive climate adaptation, and the need for inclusive policymaking that recognizes the role of informal workers in city economies.

15.2 Significance of the Study

This research contributes to the growing body of literature on urban climate vulnerability, gender resilience, and informal economies. Documenting the experiences of women vendors offers insights for policymakers, urban planners, and climate and disaster risk reduction practitioners in designing inclusive and community-driven resilience strategies. Recognizing informal workers as key stakeholders in climate adaptation efforts can lead to more effective and socially just urban resilience policies and programs.

15.2.1 Climate Risks and Informal Workers in Urban Settings

Informal workers in rapidly urbanizing and highly congested cities are among the most climate-vulnerable populations (Porio et al., 2019; Ramin & Svoboda, 2009). Studies indicate that air pollution, heat stress, and extreme weather events disproportionately impact street-based economies, particularly in rapidly urbanizing Asian cities (World Bank, 2002). In Metro Manila, airborne pollutants from heavy traffic, combined with poor urban drainage systems, create a hazardous working environment for street vendors, transport workers, and market traders (DENR, 2016).

Women vendors face cumulative risks due to gendered social roles and economic precarity. Research shows that women in the informal economy are less likely to have access to climate change adaptation (CCA) and DRR resources, social protection programs, and secure workspaces (Recio et al., 2021; Santos & Reyes, 2020). Unlike male workers, many women vendors balance caregiving and other reproductive health duties with economic survival, further limiting their mobility and resilience options (Porio, 2014; Porio et al., 2019).

15.2.2 Local Knowledge Systems and Adaptation Strategies

Local knowledge systems are critical to understanding how vulnerable populations, especially women, adapt to environmental stressors. According to Lee et al. (2021), local knowledge on climate hazards and health risks helps communities create their solutions to various urban challenges. Women vendors in Metro Manila utilize their understanding of street environments, such as knowing the most flood-prone areas and how to adapt their vending stalls and livelihood practices to changing weather conditions. These adaptive strategies are often rooted in shared experiences and community knowledge passed down through generations (Porio et al., 2019). See Fuentes et al. (2024) also argue that these context-specific and place-based adaptive actions constitute local knowledge systems that promote transformative climate justice innovations.

Despite these strategies, women vendors often lack formal climate and disaster risk management support, and their local knowledge is not fully recognized in policy and program circles, like in the lowest political-administrative level of Barangay Disaster Risk Reduction and Management Council. This creates a gap between bottom-up resilience practices and top-down institutional CCA and DRR programs.

15.2.3 Gender, Informality, and Urban Resilience

Gender is a central factor in urban vulnerability. Studies show that informal workers, particularly women, often bear the brunt of environmental risks due to their social roles and economic responsibilities (Santos & Reyes, 2020). In many cases, these women vendors are primary caregivers, managing their households and livelihoods. This dual responsibility makes them particularly vulnerable to disaster impacts. However, as this study will show, these women also demonstrate significant resilience, largely through informal networks and community-based strategies.

15.3 Methodology

This study employs an ethnographic approach to capture the lived experiences, adaptive strategies, and institutional interactions of women sidewalk vendors in Metro Manila (Fig. 15.1). Ethnography is particularly suited for analyzing informal economies, as it provides rich, context-sensitive insights into how vulnerable populations construct resilience in real time.

This study employed a combination of semi-structured interviews, participant observation, focus group discussions, and policy analysis to capture the complex realities of women vendors' interactions with environmental risks.

In-depth interviews and life histories were conducted with 30 women sidewalk vendors across three major street markets in Metro Manila. These interviews sought to understand vendors' perceptions of climate hazards, their adaptive strategies, and

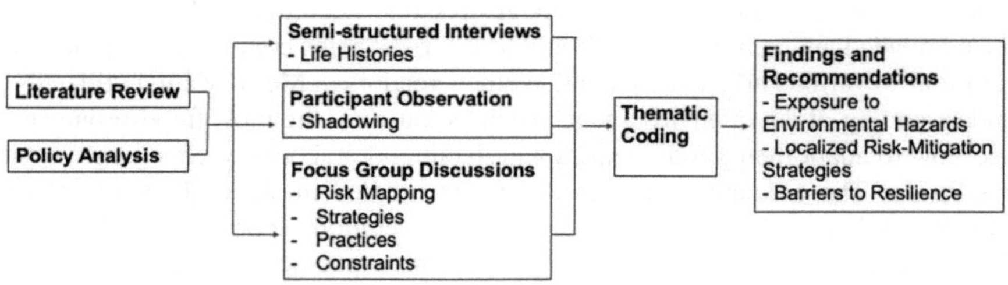

Fig. 15.1 Ethnographic approach employed in the study

their experiences with institutional policies. Life history interviews were particularly useful in tracing long-term resilience patterns, providing insights into how vendors have adapted to recurring climate-related events, such as extreme heat waves and monsoon flooding.

In addition to interviews, participant observation was carried out at vending sites over several months. This involved shadowing vendors throughout their daily routines, documenting their exposure to environmental stressors, and observing how they adjust their vending locations, manage health risks, and interact with customers, local authorities, and fellow vendors. By embedding the researcher in the vendors' everyday environment, this method allowed for a firsthand understanding of the spatial and social dynamics of informal vending.

To supplement individual interviews and observations, three focus group discussions (FGDs) were conducted with vendor communities. These discussions provided a collective perspective on resilience-building efforts, allowing vendors to share their risk-mitigation strategies, concerns, and suggestions for improving urban resilience policies. During the FGDs, vendors also participated in risk mapping exercises, identifying areas in the city that they perceived as high-risk zones for flooding, extreme heat, or government crackdowns. These group interactions helped reveal patterns of shared knowledge and collective action, further highlighting the importance of social networks in CCA and DRR.

In addition to collecting primary data, this study conducted policy and document analysis to assess how vendors are represented (or excluded) in urban CCA and DRR policies. Government reports, urban planning documents, and CCA and DRR frameworks were analyzed to determine whether existing policies recognize informal workers as stakeholders in urban resilience planning. This study evaluates gaps in policy implementation and enforcement by cross-referencing policy documents with vendors' lived experiences.

The collected qualitative data was analyzed using thematic coding, allowing for the identification of key patterns and recurring themes. The coding process was guided by three major analytical categories: (1) Exposure to environmental hazards, (2) localized risk-mitigation strategies, and (3) barriers to resilience. Under the first category, the study examined vendors' accounts of air pollution, extreme heat, and flooding, paying close attention to how these risks affected their health, income, and work routines. The second category focused on vendors' adaptive strategies, such as relocating to safer areas, relying on community support networks, and implementing informal health protection measures. The third category explored institutional constraints, including frequent evictions, lack of formal recognition in urban policies, and gendered vulnerabilities that shape vendors' ability to cope with environmental stressors.

15.3.1 Ethical Considerations

This research adhered to strict ethical standards to ensure the safety, confidentiality, and informed participation of all respondents. Before data collection, vendors were provided with detailed explanations about the study's objectives and their rights as participants. Informed consent was obtained verbally, considering the informal nature of the work and the participants' varying levels of literacy. Additionally, to protect the identities of vendors who may face risks due to law enforcement crackdowns, pseudonyms were used, and personally identifiable details were removed from the transcripts. Throughout the research process, care was taken to ensure that participation was voluntary, with vendors given the option to withdraw at any stage without consequences.

This study combines ethnographic methods, participatory research techniques, and policy analysis to provide a holistic understanding of women street vendors' resilience to climate, including institutional challenges. The integration of firsthand observations, lived experiences, and structural analysis ensures that the study captures not just individual strategies but also the broader socio-political context in which these adaptations occur.

15.4 Findings

The findings of the study are summarized in Table 15.1 and discussed below. These findings highlight the urgent need for gender sensitive policies that recognize informal workers as critical stakeholders in urban climate and disaster resilience planning.

15.4.1 Women Vendors' Exposure to Environmental Hazards

The findings confirm that women vendors in Metro Manila experience severe environmental hazards, including air and noise pollution, extreme heat, and flooding, which directly impact their health, livelihood, and well-being. Vendors operate in high-traffic areas, exposing them to continuous vehicular emissions that contribute to respiratory illnesses such as chronic cough, asthma, and bronchitis. Many participants reported frequent dizziness and shortness of breath, particularly during rush hours when air quality deteriorates.

The effects of extreme heat are also evident in vendors' daily struggles. With limited access to shade or cooling infrastructure, vendors report experiencing severe dehydration, headaches, and physical exhaustion, particularly during the summer months. Climate change has exacerbated these risks, with longer heat waves reducing vendors' ability to work safely for extended periods. Vendors must frequently relocate

Table 15.1 Summary of key findings

Area of inquiry	Experiences and strategic practices
Women vendors' exposure to environmental hazards	Women vendors in Metro Manila experience severe environmental hazards, including air and noise pollution, extreme heat, and flooding, which directly impact their health, livelihood, and well-being
Adaptive strategies	Spatial mobility—vendors relocate their stalls based on climate conditions, often moving to elevated or shaded areas to avoid flooding and heat stress Building strong community networks to share real-time information about weather conditions, enforcement activities, and potential risks Use of herbal remedies, makeshift masks, and hydration strategies to counteract pollution and heat-related illnesses
Local knowledge	Women vendors' local knowledge of street conditions and environmental hazards allows them to make quick, experience-based decisions about where and when to set up their vending stalls
Institutional constraints and contested urban spaces	Many women vendors operate in legally ambiguous spaces, making them vulnerable to frequent eviction, fines, and harassment from local authorities Exclusion of informal workers from formal CCA and DRR policies further weakens their ability to prepare and recover from climate hazards Many women vendors balance caregiving responsibilities with their work, making it harder for them to relocate or seek alternative employment

to shaded areas, but competition for these limited spaces adds to their economic precarity.

Flooding during the monsoon season presents an additional challenge. Women vendors in low-lying areas experience frequent inventory losses and income disruptions due to heavy rains and urban flooding. Some vendors are forced to evacuate temporarily, further affecting their economic stability. In extreme cases, stagnant floodwaters lead to waterborne illnesses, creating an additional health burden for vendors and their families.

15.4.2 Adaptive Strategies and Local Knowledge

Despite these risks, women vendors have developed diverse adaptive strategies to cope with climate and disaster risks. Their local knowledge of street conditions and environmental hazards allows them to make quick, experience-based decisions about where and when to set up their vending stalls (Fig. 15.2).

One of the most common adaptive strategies is spatial mobility—vendors relocate their stalls based on climate conditions, often moving to elevated or shaded areas to avoid flooding and heat stress. Vendors have also built strong community networks to share real-time information about weather conditions, enforcement activities, and potential risks. These informal support systems play a crucial role in enhancing their collective resilience, as vendors assist each other in securing safer vending spots,

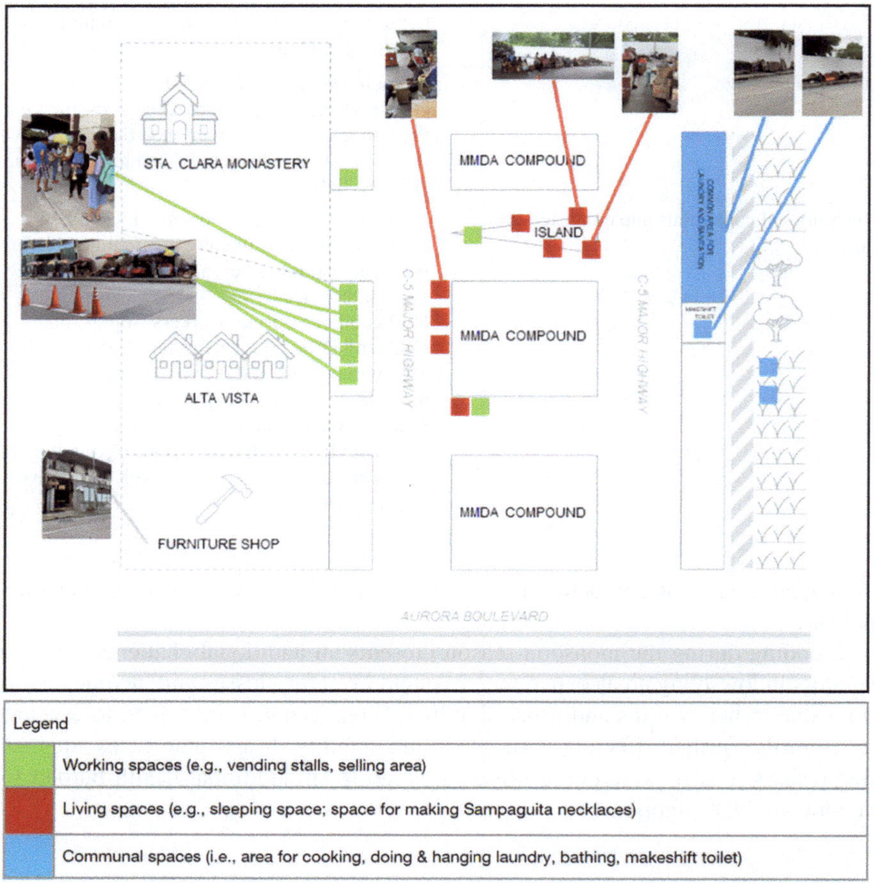

Fig. 15.2 Study area showing the working spaces, living spaces, and communal spaces of the women vendors in the study. *Source* Porio et al. (2019)

The above photos show two separate vendors' stalls of offertory eggs. The eggs are sold in 8s, and 6s, wrapped in colored cellophane or plastic corresponding to the intention that devotees can choose from. All color coding is the same for the other vendors' posters. The final code, 'Yellow' ends with the Filipino word, "Pangkalahatan" or "For all intentions"

Fig. 15.3 Stalls of women vendors showing specialized egg package offerings. *Source* Porio et al. (2019)

protecting their goods and inventory, and having access to temporary storage spaces during extreme weather events (Fig. 15.3).

Health protection is another key adaptation strategy. Vendors use herbal remedies, makeshift masks, and hydration strategies to counteract pollution and heat-related illnesses. However, these coping mechanisms remain largely informal and insufficient, underscoring the lack of formal support systems for informal workers exposed to climate hazards.

15.4.3 Institutional Constraints and Contested Urban Spaces

While women vendors demonstrate resourcefulness and resilience, their ability to adapt is significantly constrained by institutional policies and urban governance issues. Many vendors operate in legally ambiguous spaces, making them vulnerable to frequent eviction, fines, and harassment from local authorities. These periodic crackdowns disrupt vendors' economic stability and force them to relocate, often to areas with worse environmental conditions.

The exclusion of informal workers from formal CCA and DRR policies further weakens their ability to prepare and recover from climate hazards. Unlike formal sector workers, vendors do not receive heat advisories, pollution warnings, or access to emergency relief programs. The lack of institutional recognition means that vendors must rely solely on personal experience and social networks to mitigate risks.

Gendered vulnerabilities also shape vendors' experiences. Many women vendors balance caregiving responsibilities with their work, making it harder for them to relocate or seek alternative employment. Additionally, female vendors face greater harassment from law enforcement and local authorities, adding another layer of insecurity to their already precarious working conditions.

15.5 Discussion

15.5.1 Climate Vulnerability, Women Vendors, and Community Resilience

This study's findings reinforce that women street vendors in Metro Manila experience a unique set of climate vulnerabilities shaped by their occupation, gender, and socioeconomic status (Porio et al., 2019). Unlike formal workers who may have access to climate-controlled environments and employer-provided health protections, street vendors are fully exposed to environmental hazards with minimal institutional support. Their experiences align with previous research emphasizing that informal workers bear the brunt of climate change impacts due to their exclusion from urban resilience planning (Santos & Reyes, 2020).

The study also highlights the gendered nature of climate vulnerability, showing that women vendors face additional constraints related to caregiving, social norms, and environmental policies. Unlike male vendors, who may have more mobility and access to alternative vending spaces, women are often restricted to fixed locations due to safety concerns and family responsibilities. These findings contribute to broader discussions on gender and urban resilience, reinforcing the need for gender-sensitive climate adaptation policies.

15.5.2 Local Knowledge as an Untapped Resource

One of this study's most significant findings is the role of local knowledge in shaping vendors' adaptive strategies (Porio, 2014; Turnbull, 2008). Despite lacking formal education in CCA and DRR, vendors have developed highly effective, experience-based adaptation strategies that allow them to anticipate, prepare for, and respond to climate hazards using spatial adaptation, community networks, and health protection strategies. However, this knowledge remains unrecognized in formal urban planning and climate governance, resulting in policies that fail to account for the realities of informal workers.

This study argues that incorporating informal workers' knowledge into official CCA and DRR frameworks could significantly improve urban resilience policies. Rather than viewing informal vendors as policy challenges or nuisances, local governments should recognize them as active agents in climate and disaster resilience. By incorporating their insights into early warning systems, public health campaigns, and urban planning efforts, policymakers can create more inclusive, effective climate, and disaster resilience strategies.

15.5.3 Implications for Policy and Practice

Based on the above findings, the study recommends: (1) the integration of contextually driven adaptive strategies crafted by women street vendors into local CCA and DRR policies; (2) improved infrastructure that can reduce vendors exposure to extreme weather conditions (e.g., covered vending spaces, water access points, etc.); (3) leveraging vendor networks to disseminate real-time climate alerts, including advisories to improve climate information for community-based disaster response and preparedness; and (4) implementing fair, gender-responsive policies that protect street vendors from arbitrary displacement would enhance their long-term resilience. By addressing these institutional gaps, policymakers can create a more equitable urban CCA and DRR strategy that supports both informal workers and broader urban sustainability goals. Figure 15.4 describes a framework on moving the policy recommendations into practice through CCA and DRR and community-led and gender-responsive actions engaging national and local governments, community networks, academia, civil society, private sector, and women vendors (Porio, 2024).

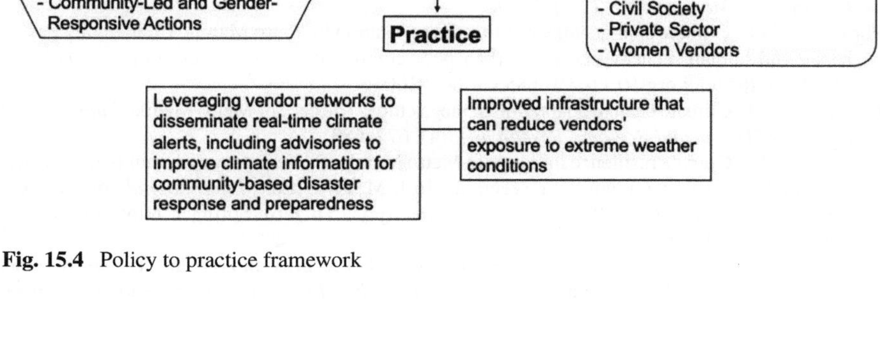

Fig. 15.4 Policy to practice framework

15.6 Conclusion

Women street vendors in Metro Manila are among the most climate-vulnerable urban populations, yet their street-based adaptation and resilience strategies remain largely unacknowledged in formal CCA and DRR policies (Porio et al., 2019). Their daily exposure to air and noise pollution, extreme heat, and flooding presents severe health and economic risks, which they navigate through experience-based local knowledge and informal community support systems (Porio, 2014). However, their ability to sustain these adaptive strategies is limited by institutional constraints, including policy exclusion, lack of disaster preparedness and relief access, and law enforcement threats (Porio, 2024; Porio et al., 2019).

This study reinforces the need for inclusive, gender-responsive urban resilience frameworks that recognize the role of informal workers in city economies (Langbid-Roda & Collantes, 2024). Rather than being treated as marginalized laborers, vendors should be seen as active catalysts of CCA and DRR, possessing valuable local knowledge that can contribute to broader, more contextualized resilience strategies (Porio, 2024). By integrating informal workers such as women street vendors into CCA and DRR programs, investing in climate-resilient urban infrastructure, and providing legal protections against displacement, policymakers can build more equitable and sustainable urban resilience systems that support both the informal workers and the urban population.

References

Department of Environment and Natural Resources (DENR). (2016). *Air quality monitoring in Metro Manila: Annual report*. Government of the Philippines.

International Labour Organisation (ILO). (2009). *Globalization and informal jobs in developing countries.*

Langbid-Roda, R. F., & Collantes, L. J. A. (2024). Overcoming barriers towards socio-economic empowerment: The case of women market vendors in Iligan City, Philippines. *Journal of Sustainable Equity and Social Research, 1*(Special Issue on Women). https://doi.org/10.5281/zenodo.13910472

Lee, A., Santos, M., & Reyes, P. (2021). Community-based adaptation strategies for climate change. *Climate Risk Management, 12*(3), 78–90.

Porio, E. (2014). Climate change vulnerability and adaptation in Metro Manila: Challenging governance and human security needs of urban poor communities. *Asian Journal of Science, 42*, 75–102. https://doi.org/10.1163/15685314-04201006

Porio, E. (2019). Climate change adaptation in mega cities: The case of Metro Manila. *International Journal of Urban and Regional Research, 43*(4), 672–689.

Porio, E. (2021). Climate resilience initiative in Metro Manila: Participatory community risk assessment and power in community interventions. In J. M. Fritz (Ed.), *International clinical sociology. Clinical Sociology: Research and Practice* (pp. 257–276). Springer. https://doi.org/10.1007/978-3-030-54584-0_14

Porio, E. (2024). Community resilience innovations in food security and risk governance systems. In G. Llanto & E. Basilio (Eds.), *Post-pandemic policy for food security*. National College of Public Administration and Governance R.P. De Guzman St., University of the Philippines Diliman, Quezon City.

Porio, E., Roque-Sarmiento, E., Salas, M. R. L., & Leynes, V. C. C. (2019). *From the skies to the streets: An ethnography of street-based populations in Metro Manila, Philippines and local knowledge systems regarding environmental pollution and climate disaster risks* (Technical Report). Coastal Cities at Risk in the Philippines.

Ramin, B., & Svoboda, T. (2009). Health of the homeless and climate change. *Journal of Urban Health, 86*(4), 654–664.

Recio, R. B., Gomez, J. E. A., Thai, H. M. H., & Nguyen, P. T. (2021). Street vending and co-production: Key lessons during the COVID-19 pandemic. *Space and Polity, 25*(3), 325–346. https://doi.org/10.1080/13562576.2022.2045931

Santos, M., & Reyes, P. (2020). Informal economies and climate vulnerability: A case study of market vendors. *International Journal of Disaster Risk Reduction, 34*, 566.

See, J., Fuentes, A. S., Porio, E., & Wilmsen, B. (2024). Conceptualising and enacting pathways to transformative climate justice: Examples from the Philippines. *Local Environment, 30*(7), 872–889. https://doi.org/10.1080/13549839.2024.2353040

World Bank. (2002). A preliminary desk review of urban poverty in the East Asia Region: With particular focus on Indonesia, The Philippines and Vietnam. In World Bank. World Bank. https://documents1.worldbank.org/curated/en/453321468749966784/314210v10EAP0w145220Sept0201public1.doc

Turnbull, D. (2008). Knowledge systems: Local knowledge. In H. Selin (Ed.), *Encyclopaedia of the history of science, technology, and medicine in non-western cultures*. Springer. https://doi.org/10.1007/978-1-4020-4425-0_8705

Chapter 16
Gender Transformative Approach in Climatic Disasters Management in Thailand

Shweta Sinha

Abstract Gender-sensitive disaster management strategies remain underdeveloped, especially in low- and middle-income countries. Significant gender disparities exist in disaster management, where women often face greater risks and have less representation in decision-making processes. Thailand ranks 74th on the Gender Inequality Index (GII) with a score of 0.310 in 2022 showing moderate progress toward gender parity in health, education and income. These disparities are pronounced, especially in the context of climate-induced disasters like floods. Gender-transformative disaster management aims to address these inequalities by empowering women, enhancing their leadership roles, and challenging systemic barriers that limit their engagement in disaster risk reduction (DRR). This paper attempts to identify challenges and approaches focusing on integrating gender-sensitive policies, improving access to resources, and gathering sex-disaggregated data to ensure that disaster response strategies are equitable and effective. As Thailand faces ongoing floods in 2024, adopting this approach is critical to achieving a more inclusive and resilient recovery, where women can contribute as key actors in shaping the future of disaster management.

Keywords Gender equality · Disaster risk reduction (DRR) · Climate resilience · Community engagement · Inclusive policy reform

16.1 Background

In 2024, Thailand was hit by widespread floods, flash floods, and landslides, affecting 30 provinces, including key areas like Chiang Rai, Chiang Mai, and Mae Hong Son. The disaster impacted 152 districts and 675 sub-districts, with around 141,400 households suffering the consequences (Thailand: Floods and Landslides—Aug 2024, 2024). Heavy rain, combined with multiple tropical storms, was a major factor in

S. Sinha (✉)
Pridi Banomyong International College, Thammasat University, Bangkok, Thailand
e-mail: shweta.s@pbic.tu.ac.th

Md. A. Abedin et al. (eds.), *Climate Crisis in South and East Asia*, Sustainability Solutions, https://doi.org/10.1007/978-3-032-01012-4_16

the flooding. Rainfall exceeded the average by 27%. According to the Federation of Thai Industries (FTI), 2024 flooding in the Northern Thailand has been most severe in the past 80 days with an economic damage of 27 billion baht. While Thailand has a history of devastating 2011 floods, the 2024 floods intensity led to sudden impact damaging crops, reducing yields, affecting livelihoods, damaging infrastructure leading to costly repairs and potential government debt. Flooding in tourist areas had slowed down tourist recovery and dependent businesses post-COVID, disrupting manufacturing in industrial zones impacting exports in electronics, automotive and textile sectors, increasing rebuilding cost, insurance claims and adding pressure on the financial institutions.

Monsoon and low-pressure conditions in neighboring countries caused flooding, especially in central and northern regions of Thailand. From the beginning of 2024, till September, 50 provinces experienced some degree of flooding, though post September 2024, the situation had returned to normal in 40 of these provinces and flood conditions only persisted in 10 provinces. Throughout the flooding period, demographic shifts take place because of migration to cities, health concerns due to waterborne diseases, displacement for temporary shelters, with some not returning to the flood prone areas adding more stress on the healthcare systems and government resources.

However, the effects of floods are not felt equally—women often face higher vulnerabilities during disasters. Unfortunately, many disaster management strategies fail to address these inequalities, leaving women disproportionately affected. Access to healthcare services, including maternal care and contraception, becomes limited during floods, further affecting women's physical health. Mental health challenges, such as anxiety, depression, and trauma, are also heightened by displacement, loss of livelihoods, and long recovery periods. The psychological stress of rebuilding homes and caring for family members during these disasters is immense. Moreover, women are at increased risk of gender-based violence in overcrowded, insecure temporary shelters where privacy and safety are lacking. Many studies indicate women are more vulnerable to the impacts of disaster, making it challenging and broadens the gap of gender inequality in the affected area.

The primary objective behind gender mainstreaming is to design and implement development projects, programs and policies that:

1. Do not reinforce existing gender inequalities (gender neutral).
2. Attempt to redress existing gender inequalities (gender sensitive).
3. Attempt to re-define women and men's gender roles and relations (gender positive/transformative).

The degree of integration of a gender perspective can be seen as a continuum (Table 16.1).

The Sendai Framework for Disaster Risk Reduction 2015–2030 emphasizes the importance of incorporating gender perspectives in disaster risk management and ensuring women's leadership in decision-making processes. It recognizes that addressing gender inequalities is crucial to reducing disaster vulnerabilities and achieving sustainable resilience. The framework advocates for gender-responsive

Table 16.1 Gender-related terms and concepts

Gender negative	Gender neutral	Gender sensitive	Gender positive	Gender transformative
Development objectives are pursued in ways that perpetuate gender disparities by relying on traditional gender roles, norms, and stereotypes	Gender is viewed as unrelated to development results, with existing gender norms, roles, and relationships remaining unchanged-neither improved nor deteriorated	Gender considerations are incorporated as tools to achieve predetermined development objectives, with attention given to norms, roles, and resource access only to the extent necessary for meeting project targets	Gender plays a crucial role in driving successful development outcomes, with transforming gender norms, roles, and resource access forming an essential part of the project's objectives	Placing gender at the core for fostering equality and driving meaningful development progress. This involves reshaping imbalanced gender dynamics to enable shared authority, equitable resource access, inclusive decision-making, and the advancement of women's empowerment

Source Author, adapted from UN-INSTRAW (now part of UN Women)

policies and the active participation of women's organizations in disaster preparedness, response, and recovery. Additionally, it stresses the importance of collecting sex-disaggregated data to tailor disaster risk reduction (DRR) strategies to the needs of both men and women (UNISDR, 2015). These provisions are intended to ensure that DRR policies are inclusive, effective, and equitable, helping to build a more resilient society.

16.2 Thailand—Gender Inequality

Thailand ranks 65th out of 146 countries in the 2024 Global Gender Gap index, rising from 74th in 2023 (UN, 2024). In terms of political leadership, women remain underrepresented. At the national level, women hold 19.2% of parliamentary seats, 10.4% of senatorial positions, and 22% of cabinet roles. At the local level, women comprise 7.1% of chief executives in provincial administrative organizations and 10.04% of mayors. This underrepresentation highlights the need for policies promoting women's participation in governance (OHCHR, 2024). The justice sectors also lag with only 9.54% women police officers, making it crucial for the government to expand opportunities for women to build capacity and engage in this sector. The labor force participation rate for women is 80.5%, compared to 94% for men. Additionally, women spend approximately 11% of their time on unpaid domestic and care work, while men spend about 3.3% (UN Women, 2024; World Bank, 2023). This disparity underscores the double burden women face in balancing paid employment and unpaid caregiving responsibilities.

In terms of education, government provides 12 years of free education in public schools, in addition to 3 years of free pre-primary schooling. Education is compulsory up to the ninth grade. Nearly all children in Thailand complete primary education, and 94% of girls complete lower secondary education, compared to 85% of boys. For upper secondary education, 73% of girls reach completion in contrast to 64% of boys, and for tertiary education, 48% of women complete at least two years, as opposed to 33% of men. Women, however, remain underrepresented in STEM fields. Access to education remains a challenge for rural, ethnic minority, low-income, migrant, stateless, and refugee women and girls. Girls who do not have birth registration and identity documents reportedly have difficulty enrolling in public schools. Completion rates for upper secondary school and beyond are noticeably lower for girls from low-income and rural areas, where persistent teacher shortages have resulted in lower quality of education.

In terms of health, adolescent birth rate has decreased from 27.41 per 1000 women aged 15–19 in 2020 to 23.5 in 2021. Furthermore, 88.2% of women of reproductive age have their family planning needs met with modern methods. However, challenges persist, such as 17% of women aged 20–24 being married before age 18, and 9.3% of women aged 15–49 reporting experiences of physical and/or sexual violence by an intimate partner in the past year (UN Women, 2024).

16.3 Gender-Sensitive Vulnerabilities with Climate Disasters

Many disasters indicate more women deaths as compared to men. There are multiple reasons that have been identified, putting women at higher risk or more vulnerable to disasters. They are also more likely to be responsible for caring for children and elderly people, which make it difficult for them to evacuate in a timely manner. In addition, women are often less likely to have access to education and information about disaster preparedness, which puts them at greater risk. After floods, women typically take on the primary responsibility for caring for children, the elderly, and the sick, which increases their workload and reinforces traditional gender roles, limiting their access to education, employment, and decision-making. In rural areas, girls are often kept out of school to handle household chores or care for younger siblings, while boys may return sooner post floods. This hinders girls' education and future opportunities, limiting progress toward gender equality in education and employment.

Climate disaster management is influenced by multiple gender-related factors that shape how individuals experience and respond to crises. Women, girls, and marginalized gender groups often face heightened risks and challenges due to existing

Table 16.2 Key gender-sensitive vulnerabilities with climate disasters

Exposure and vulnerability	Preparedness	Coping capacity
– Health needs – Type of assets (tangible or intangible) – Education and child labor – Labor (informal sector, agriculture) – Voice and agency (gender based violence (GBV) or early marriage)	– Risk perception – Access to information/early warning systems – Type of preparedness action – Evacuation behavior	– Access to finance/assets – Access to social protection – Labor (stability of income, number of sources of income) – Mobility and migration

Source Author

socio-economic inequalities, cultural norms, and policy gaps. Some of the key gender-sensitive vulnerabilities with climate disasters are as shown in Table 16.2.

16.4 Access to Resources and Services

Gender barriers exclude women from decision-making at household and institutional level, making them more vulnerable to climate change. Gender roles influence cropping choices which also influence access to productive and financial resources. Various climate risks like temperature increase, rainfall variability, sea-level rise impact lives and livelihoods with adverse effects to women.

16.5 Thailand Voluntary Commitments Toward Sendai Framework for DRR 2015–2030

Thailand's has commitment to the Sendai Framework for Disaster Risk Reduction 2015–2030 through various strategic initiatives and policy implementations. In 2015, the Thai government introduced a five-year national disaster prevention and mitigation plan that aligns with the Sendai Framework's priorities, focusing on comprehending disaster risk, enhancing risk governance, investing in disaster risk reduction, and improving preparedness measures (UNISDR, 2015). It further integrated health considerations into disaster risk management; Thailand played a pivotal role in formulating the "Bangkok Principles" in 2016. These principles aim to strengthen the collaboration between health sectors and disaster risk management, ensuring a comprehensive approach to future health emergencies (UNDRR, 2020). In its Voluntary National Report for the Midterm Review of the Implementation of the Sendai Framework, Thailand detailed its progress and ongoing efforts in disaster risk reduction. The report highlighted the country's initiatives in enhancing disaster resilience, improving early warning systems, and fostering community engagement in disaster preparedness (Department of Disaster Prevention and Mitigation, 2022).

16.6 Climate Resilience and Inclusive Policy Reform

Department of Disaster Prevention and Mitigation (DDPM) of Thailand worked in collaboration with the Department of Empowerment of Persons with Disabilities for providing effective risk communications to persons with disabilities and conducted gender inclusive community-based DRR projects with the Thai Red Cross according to the Midterm Review of the Implementation of the Sendai Framework for Disaster Risk Reduction 2015–2030: National Voluntary Report of Thailand (DDPM, 2022).

16.7 Case of Chiang Rai Province: Impact During Thailand Floods 2024

A gender impact survey was conducted online during Thailand floods 2024 in Chiang Rai Province between October to November 2024. Based on the 23 responses received, 60.9% of respondents were females and 39.1% were male. About 56.5% respondents were of the age between 20 and 39 years, 26.1% were of the age between 40 and 59 years, 13% were above 60 years and 4.3% were between the age of 13 and 19 years old. The majority of those affected were between 20 and 39 years old (Fig. 16.1), with a higher percentage of female respondents. The crisis has led to housing damage, with many homes destroyed or severely impacted, causing displacement. Public health is threatened due to disruptions in healthcare services, increasing the risk of disease outbreaks.

Population Affected and Displacement: In Chiang Rai floods, 56.5% of respondents were displaced and most impacted were between 20 and 39 years of age (Fig. 16.2). The remaining 26.1% were between 40 and 59 years old, 13% were 60 years old and above. And 4.3% were between the age of 13 and 19 years. Among the total displaced respondents, 60.9% were women and 39.1% were male. Though no fatalities were reported, the displacement caused significant disruption and financial strain, affecting entire households.

Impact on Gender: Gender-specific challenges emerged during the floods. Women, often primary caregivers, faced additional burdens in managing family needs and accessing resources. Nearly half of the respondents, about 43.5% observed inequalities in the distribution of aid between men and women (Fig. 16.2), reflecting the need for gender-sensitive disaster management strategies. Addressing these disparities is essential to ensure equitable access to relief and recovery resources.

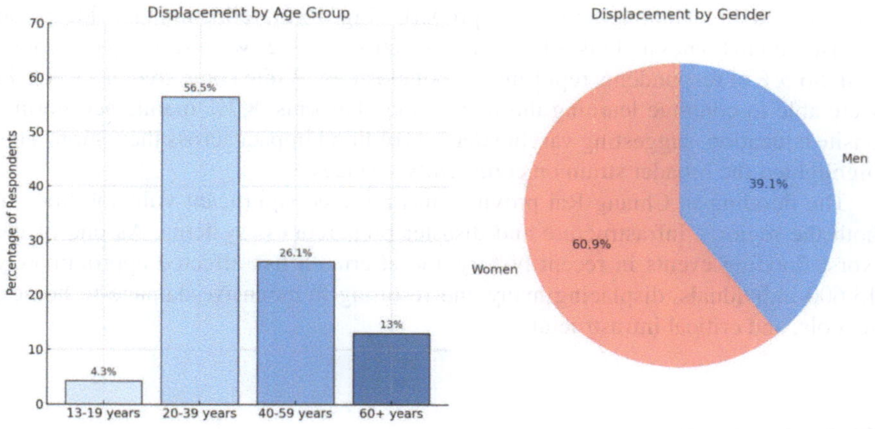

Fig. 16.1 Displacement by age and gender. *Source* Author, based on gender impact survey during Chiang Rai, Thailand floods 2024

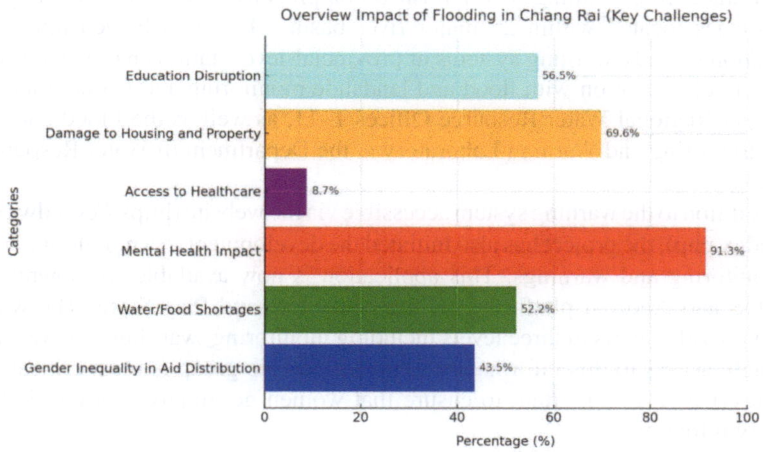

Fig. 16.2 Overview impact of flooding in Chiang Rai, Thailand 2024. *Source* Author, based on gender impact survey during Chiang Rai, Thailand floods 2024

Public Health Impacts: Flooding in Chiang Rai caused severe public health challenges. About 52.2% respondents faced water and food shortages (Fig. 16.2), while poor sanitation led to waterborne diseases. Mental health suffered, with 91.3% reporting increased stress. Access to healthcare was limited, with only 8.7% rating it as "good," and vulnerable groups, like the elderly, struggled to receive adequate support.

Impact on Housing and Community Services: Housing and community services suffered extensive damage during the floods. Approximately, 69.6% of respondents experienced direct damage to life and property (Fig. 16.2), reflecting the widespread devastation to homes and infrastructure. Education services were severely disrupted, with 56.5% of respondents reporting school closures. While some students (34.8%) were able to continue learning through online platforms, 8.7% maintained normal onsite education, suggesting varying degrees of flood impact across the region. This highlighted the broader strain on community services.

The flooding in Chiang Rai province has revealed significant vulnerabilities in both the region's infrastructure and disaster preparedness systems. As one of the worst flooding events in recent history, the aftermath has affected approximately 13,600 individuals, displacing many and resulting in extensive damage to homes, schools, and critical infrastructure.

16.8 Early Warning

The Department of Water Resources has implemented the Early Warning System (EWS) since 2005, aiming to establish a comprehensive warning system for at-risk villages located within 25 major river basins. This is achieved through the installation of early warning systems at provincial-level stations nationwide, which operate in coordination with flood and landslide monitoring and warning laboratories under Regional Water Resource Offices 1–11, as well as the Flood and Landslide Monitoring and Warning Laboratory at the Department of Water Resources in Bangkok.

In addition to the warning system accessible via the website (https://ews.dwr.go.th/ews/index.php), the project has also initiated the development of a mobile application for monitoring and warnings. This application is now available for download on both iOS and Android platforms via the App Store and Play Store. The warning system provides alerts in three levels including monitoring, warning and evacuation. Although, access to this information to the vulnerable group is most critical at the time of crisis. It is important to ensure that women get timely disaster alerts and recovery information.

16.9 Gender Responsive Policies in Thailand

The Constitution of Thailand B.E. 2560 (2017) guarantees equality between men and women and prohibits discrimination based on sex under the Gender Equality Act of 2015. Thailand has various legislative and institutional frameworks for promoting gender equality, including the Ministerial Regulation on the Promotion and Coordination of Women's Affairs, the Gender Equality Promotion Action Plan 2023–2027, and the Women's Development Action Plan 2023–2027. Thailand has also established a National Committee on the Policy and Strategy for the Advancement of Women under the Office of the Prime Minister, as well as the Committee for the Promotion of Gender Equality (CPGE) and the Committee on Unfair Gender Discrimination Complaints (CDUGD).

16.10 Discussion

Flooding in Thailand has a disproportionate effect on women, particularly in rural areas where they are the primary caregivers for children and the elderly. The consequences of flooding on women's lives are multifaceted and far-reaching. Many women in rural areas working in agriculture and small businesses are severely impacted by flooding leading to financial instability, increasing their caregiving responsibilities and limiting their access to healthcare and education. The loss of livelihoods and income exacerbates the burden on women, making them more vulnerable to poverty and exploitation.

The caregiving responsibilities of women are intensified during flood situations that includes providing emotional support and managing the household. This added burden can lead to physical and mental exhaustion, further compromising their well-being. Moreover, the risk of gender-based violence increases during disaster situations like flooding. The breakdown of social structures and the lack of access to support services create an environment conducive to violence and exploitation. Women are more vulnerable to abuse, harassment, and trafficking during these times.

Flooding also disrupts healthcare services, making it difficult for women to access medical care, including reproductive healthcare. The destruction of educational infrastructure and the displacement of families limit women's access to education, perpetuating the cycle of poverty and inequality. Furthermore, flooding affects our progress toward several Sustainable Development Goals (SDGs), including SDG 2 (Zero Hunger), SDG 5 (Gender Equality), SDG 11 (Sustainable Cities and Communities), and SDG 13 (Climate Action).

Thailand faces several sustainability challenges, particularly in terms of gender disparities. Creating a safe shelter for gender, access to disaster alerts, and economic recovery after disasters are some of the challenges that need to be addressed. For instance, setting up separate shelters or ensuring gender-sensitive facilities becomes difficult when basic needs are already hard to meet. Women in rural areas often lack access to technology and education, making it difficult for them to receive disaster alerts. Moreover, women in informal jobs, such as farming and small businesses, are often the most affected during disasters and lack protections like insurance or paid leave.

Table 16.3 Policy recommendation for gender equitable disaster management

	**Raise public awareness**—Launch campaigns to educate communities, local leaders, and policymakers about the importance of gender equity in disaster management. This can help change cultural attitudes and encourage supportive policies
	**Improve access to technology**—Invest in technology and infrastructure to help women access information tools like mobile phones and the internet. This ensures they get timely disaster alerts and recovery information. Enhance the existing early warning systems with improved technology and broader reach to ensure timely alerts for all residents, especially vulnerable populations
	**Implement gender-responsive disaster recovery frameworks**—Create recovery plans that focus on the specific impacts of disasters on women and prioritize their needs. This includes ensuring access to healthcare, social services, and job opportunities

Source Author

16.11 Conclusion and Recommendations

Floods in Thailand have had a significant impact on both the economy and women's well-being, particularly in rural areas. Women in these regions face heightened challenges, including job losses, limited access to essential hygiene products, and increased domestic responsibilities. Since many women rely on agriculture for their livelihoods, floods often result in substantial financial and food insecurity. Additionally, the lack of gender-sensitive shelters and safety measures increases their vulnerability during disasters. Addressing these issues requires the development of recovery plans that prioritize women's needs, improved access to disaster alerts through technology, and awareness campaigns to ensure equitable disaster management (Table 16.3).

Integrating gender equality into disaster risk reduction (DRR) can significantly enhance long-term resilience and sustainable recovery. By focusing on gender-specific vulnerabilities, empowering women's participation in disaster response, and ensuring equal access to resources and decision-making, Thailand can create a more inclusive and effective disaster management framework.

Acknowledgements The author would like to thank Pridi Banomyong International College, Thammasat University for supporting the research paper. The author declares no conflict of interest.

References

Thailand Department of Disaster Prevention and Mitigation. (2022). *Midterm review of the implementation of the Sendai Framework for disaster risk reduction 2015–2030: National voluntary report of Thailand*. United Nations Office for Disaster Risk Reduction. https://sendaiframework-mtr.undrr.org/publication/thailand-voluntary-national-report-mtr-sf

UN. (2024). *UN Working Group on discrimination against women and girls*. Bangkok. https://www.ohchr.org/sites/default/files/documents/issues/women/wgdawg/statements/2024-12-13-eom-thailand-wgdawg-en.pdf

UN Women. (2024). *Thailand—Country fact sheet*. https://data.unwomen.org/country/thailand

UNISDR. (2015). *Sendai Framework for disaster risk reduction 2015–2030*. United Nations Office for Disaster Risk Reduction. https://www.undrr.org/publication/sendai-framework-disaster-risk-reduction-2015-2030

United Nations Human Rights Office of the High Commissioner. (2024, December 13). *End of mission statement on Thailand by the UN Working Group on discrimination against women and girls*. https://www.ohchr.org/sites/default/files/documents/issues/women/wgdawg/statements/2024-12-13-eom-thailand-wgdawg-en.pdf

United Nations Office for Disaster Risk Reduction. (2020, September 24). *Helping implement the Bangkok Principles: Integrating health into disaster risk reduction*. UNDRR. https://www.undrr.org/publication/bangkok-principles

World Bank. (2023). *Thailand—Gender data portal*. https://genderdata.worldbank.org/en/economies/thailand

Chapter 17
Seeding the Sustainability: Enhanced Women's Role in Climate-Resilient Agriculture in Vietnam

Thi Tuyet Tran, Huy Ngoc Ha, Thi My Thi Tong, and Thuy Dung Dang

Abstract Rural women are particularly vulnerable to the impacts of climate change due to their dual responsibilities of managing household affairs and serving as the primary labor force in agricultural production. This burden is further exacerbated by labor migration, which shifts a significant portion of farm work onto women's shoulders. Despite their awareness of climate-adaptive agricultural practices, women face considerable barriers in accessing the necessary resources to transition to more sustainable production methods. Their continued engagement in agricultural livelihoods is often driven by the need to retain land ownership and ensure household food security. Therefore, targeted interventions are required to enhance women's capacity and motivation to adopt climate-resilient agricultural practices. This study was conducted to analyze the impacts of climate change on agricultural production, focusing on the roles and challenges of rural women in climate-resilient agriculture. The study used a combination of qualitative and quantitative methods, through two sociological surveys in Ha Tinh province, Vietnam. The results of the study suggested some solutions to improve women's awareness, access, and roles in agricultural adaptation to climate change. These should focus on improving technical skills, expanding access to essential resources, and increasing women's participation in decision-making processes related to climate change and agricultural adaptation strategies.

Keywords Women's role · Climate change adaptation · Climate-adaptive agriculture · Ha Tinh · Vietnam

---

T. T. Tran
Institute of Human Geography and Sustainable Development, Vietnam Academy of Social Sciences, Hanoi, Vietnam

H. N. Ha · T. M. T. Tong (✉)
Institute of Vietnam and World Economic, Vietnam Academy of Social Sciences, Hanoi, Vietnam
e-mail: Tongthimythi149@gmail.com

T. D. Dang
School of Languages and Tourism, Hanoi University of Industry, Hanoi, Vietnam

Md. A. Abedin et al. (eds.), *Climate Crisis in South and East Asia*, Sustainability Solutions, https://doi.org/10.1007/978-3-032-01012-4_17

17.1 Introduction

Agricultural production plays a crucial role in rural livelihoods in Vietnam, serving as a primary income source while ensuring household food security. However, labor migration has significantly reshaped the rural workforce, as a majority of young men and women have left their home provinces to seek employment or work abroad. Consequently, women have assumed the primary role in agricultural production. Alongside their domestic responsibilities, this dual burden leaves them with limited time for self-care and participation in capacity-building activities, making them particularly vulnerable to the impacts of climate change. Climate change is widely recognized as a force that exacerbates existing threats and deepens social inequalities (IPCC, 2022; MONRE, 2022; UNDP, 2023; Van & Mai, 2021).

Research findings illustrate that agriculture, as an economic sector largely dependent on natural resources, is among the most severely affected by climate change. Climate-adaptive agriculture plays a critical role in promoting sustainable development by implementing best agricultural practices, meeting environmental protection standards, and leveraging ecosystem functions to enhance competitiveness along the value chain and improve labor productivity. However, transitioning to climate-resilient agriculture requires substantial investment and a robust policy framework, whereby transforming thinking and human capacity is a prerequisite (Giller et al., 2021; Milyausha et al., 2020; Tran, 2023; Wang, 2022). Women engage in agricultural production mainly to sustain their families' agricultural livelihoods. However, they encounter numerous barriers in responding to climate change, including technological constraints, limited capacity, traditional norms, and financial challenges. Additionally, women face significant limitations in fully exercising their rights and accessing the necessary resources and knowledge to implement climate-adaptive agricultural practices successfully. Enhancing women's capacity to adapt to climate change in agricultural production requires a more tailored approach, focusing on capacity and responsibility, with active involvement and support from authorities at all levels and households (FAO et al., 2022; Hang & Huyen, 2025; UNDP, 2023).

This study examines the impacts of climate change on agricultural production and explores the roles and challenges faced by rural women in adapting to these changes. Drawing on secondary data from research, reports on climate change and gender, and sociological survey data, the study proposes solutions to enhance women's adaptive capacity in agriculture.

17.2 Research Methods

In the context of labor migration taking place strongly in rural areas, women play an increasingly important role in maintaining and developing agricultural production, especially in responding to the challenges of climate change. Women with their knowledge and practical experience have been changing their thinking and are increasingly proactive in implementing measures to adapt to climate change (Ngoc & Son, 2018; Phuong et al., 2018; The & Hai, 2019). Within the framework of the study, the concept of developing climate-resilient agriculture led by women is that they take actions and strategies to cope with the impacts of climate change; while taking advantage of the opportunities it brings in the agricultural sector. This may include the implementation of sustainable farming techniques, the rational use of crop varieties, effective land management, and the integration of production with the conservation of natural resources.

The study utilized a mixed-method approach, integrating secondary data analysis, in-depth interviews, focus group discussions, and household questionnaire surveys. Field surveys were conducted twice in October 2023 and February 2024 in Ha Tinh Province, Vietnam.

(i) In-depth interviews and focus group discussions: Two group discussions were conducted at the Communal People Committee of Mai Phu Commune and Loc Ha Town, Ha Tinh Province. Each discussion involved 10 participants.
(ii) Household questionnaire surveys: The study utilized a semi-structured household interview questionnaire to gather information on awareness, labor division, and adaptive capacity to climate change in agricultural production. A total of 180 questionnaires were administered using a simple random sampling method.

The research targets were individuals aged 18 years or older who were capable of engaging in household responsibilities and income-generating activities (Table 17.1).

Table 17.1 Demographics of the interviewed households

	Total		Male		Female	
	Number	Rate	Number	Rate	Number	Rate
Occupation						
Crop farming	105	58.3	30	28.6	75	71.4
Aquaculture	75	41.7	50	66.7	25	33.3
Educational level						
Non-literate	0	0	0	0	0	0
Primary education	4	2.2	0	0	4	4
Lower secondary education	95	50.6	25	31.3	71	71
Upper secondary education	85	47.2	55	68.7	30	30
Household characteristics						
Affluent	6	3.3	6	7.5	0	0
Upper middle income	77	42.8	20	25.0	57	57
Lower middle income	95	52.8	49	61.3	46	46
Near poor	2	1.1	0	0	2	2

17.3 Results and Discussions

17.3.1 Awareness of Climate Change

Group discussions in Mai Phu Commune and Loc Ha Town, Loc Ha District, Ha Tinh Province, reveal that women are directly and frequently impacted by environmental challenges, natural disasters, and climate change, particularly extreme weather events such as droughts, heatwaves, irregular rainfall, and floods. Poor women who have limited access to support resources are especially vulnerable to these typical climate effects. In response, the local authority has implemented diverse information, communication, and education initiatives to enhance awareness of climate change and adaptation measures among the community, with a focus on women. These efforts have contributed to improving awareness and strengthening the capacity to identify and respond to climate change impacts.

The results from the household survey questionnaire indicate that 100% of respondents are aware of erratic changes in weather patterns over the past 5 years. Most perceive an increase in temperature shifts, extreme heat, and droughts, with a consistent rate of approximately 85–87%. Likewise, irregularly occurring storms and tropical depressions, coupled with rising unseasonal rainfall, have disrupted crop calendars and agricultural production. Floods, inundation, and saltwater intrusion are recognized by fewer respondents, at 58% and 53%, respectively, while awareness of river and sea erosion stands at the lowest level, 36% (Fig. 17.1).

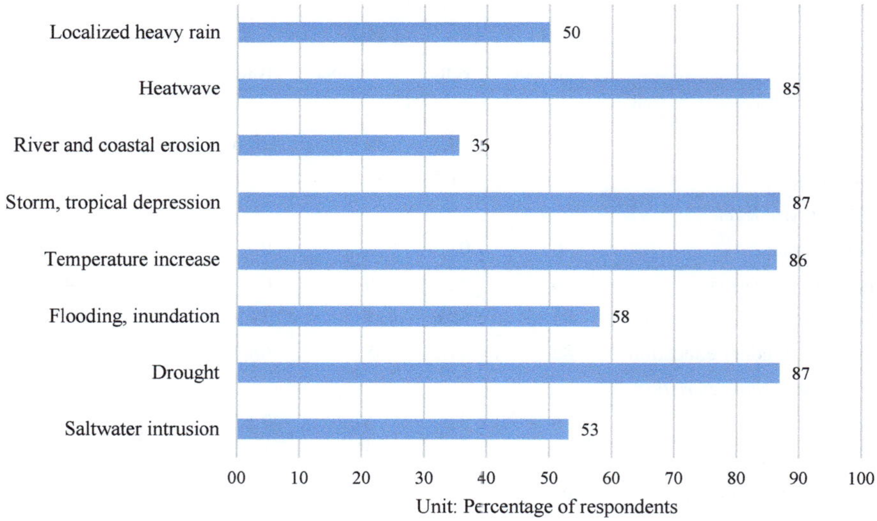

Fig. 17.1 Awareness level of climate change

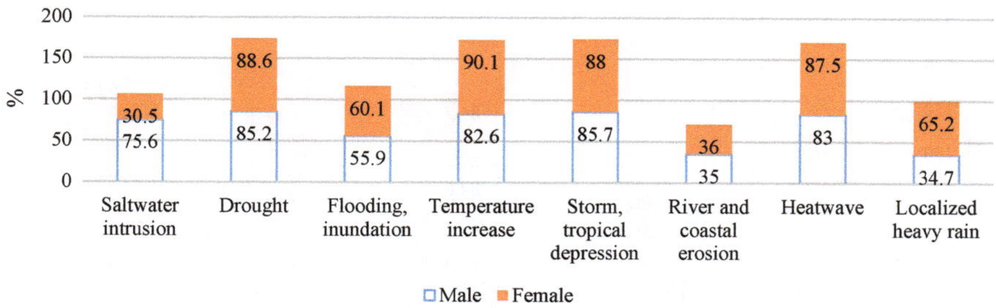

Fig. 17.2 Differences between male and female awareness of climate change

A comparison of awareness regarding typical climate change manifestations between men and women reveals that, among 7 key manifestations, women outperform men in recognizing 6/7 of them. Notably, for common identifiable changes such as temperature fluctuations, heat waves, floods, and especially localized heavy rain, the difference in awareness reaches nearly 31%. For the remaining manifestations, the disparity between men and women is negligible, below 3%. Conversely, men exhibit greater awareness of saline intrusion, with a difference exceeding 45%, largely attributable to the fact that most interviewed men work in coastal aquaculture areas, heightening their sensitivity to this issue (Fig. 17.2).

Awareness of climate change impacts on agricultural activities and daily life exhibits broadly comparable levels between men and women. Nearly 100% of respondents believe that climate change impacts human health, particularly through rising temperatures and heat waves, which contribute to deteriorating health. Women are considered more vulnerable to diseases than men. Over 70% of people realize that climate change affects household livelihoods, particularly agricultural activities reliant on natural capital, which is declining in quality, including water shortages and land resource degradation, resulting in reduced cultivable land area. About 35% of respondents are affected by the scarcity and declining quality of domestic water, despite the surveyed area being equipped with a proactive supply system designed to regulate both the quantity and quality of water usage (Fig. 17.3).

Results show that understanding of climate change manifestations and effects on daily life and agriculture is predominantly informed by personal experience, enhanced by data from management authorities and community sources. In-depth interviews with local leaders highlight the effectiveness of local communication policies and programs in bolstering public awareness and understanding of climate change and its consequences. These efforts have promoted proactive participation and enhanced the capacity of both the community and the locality to respond and adapt to climate change.

"*The achieved results stem from ongoing dissemination and propaganda efforts on environmental issues and the transition to green production. Various methods are tailored for different target groups: officials receive training through meetings and programs by the Department of Natural Resources and Environment, while the public is engaged through events like Environment Day and campaigns like legal awareness*

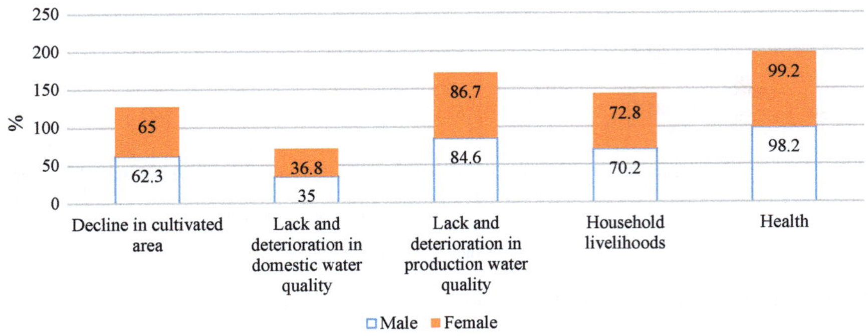

Fig. 17.3 Impacts of climate change on agricultural activities

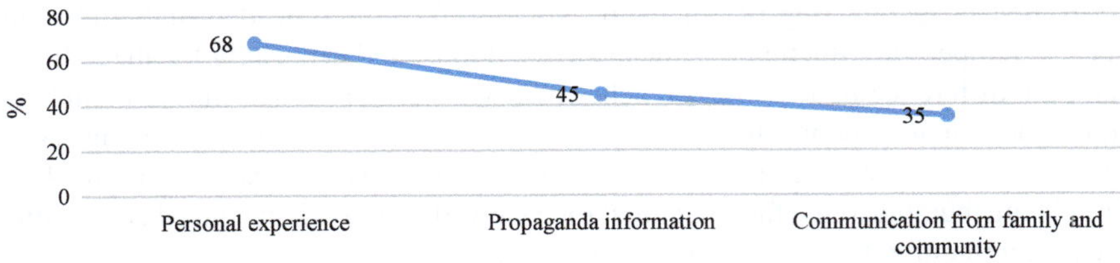

Fig. 17.4 Factors shaping awareness of climate change

and new rural building. These initiatives encourage sustainable livelihood models aligned with local conditions and national green growth strategies"—interview with a male officer in Mai Phu Commune, Loc Ha District, Ha Tinh Province (Fig. 17.4).

According to household survey results, 92% of respondents recognize the importance of transforming agricultural production methods and diversifying livelihoods to reduce risks and maintain livelihood security. However, constrained resources, particularly financial limitations and insufficient capacity to adopt climate-adaptive techniques, hinder the adaptation efforts. Certain cultivated areas still face challenges in managing water resources and coping with weather variability.

17.3.2 Women's Contribution to Climate Change Adaptation in Agriculture

In Vietnam, women account for 50.1% of the total population as of 2023. Women are disproportionately vulnerable to climate change, particularly in rural areas, where they often bear a heavier workload than men. Their responsibilities encompass household duties, healthcare, and family education, while simultaneously participating in production activities to support family income and sustain livelihoods.

Labor Distribution Role in Agricultural Activities

Research findings provide evidence that women participate in nearly all phases of agricultural production, with varying levels of involvement in crop farming and aquaculture. In crop farming, women play a decisive role in each stage of production, especially the seed selection stage (59%). Women are also responsible for crop care, selection of fertilizers and pesticides, timing of fertilization and spraying, and harvest. These activities require meticulousness, skill, knowledge, and experience in production. Men are mainly involved in land preparation and harvesting, particularly tillage, which demands significant physical strength.

For aquaculture, men play a key role, particularly for location decisions, production techniques, and breed selection. Women primarily engage in the care and harvesting phases, assisting men with care activities throughout the agricultural process (Table 17.2).

The increasing participation of women in farming reflects a shift in household labor division, particularly in the context of widespread labor migration in rural areas, where men tend to seek employment in industrial zones and urban areas. Consequently, women assume greater responsibility for production tasks to sustain and develop their livelihoods. This trend is evident in the high participation rate of rural women in agricultural extension training programs organized by local authorities, with nearly 80% of surveyed women attending training courses on farming techniques and serving as the primary labor force in agricultural livelihoods. In contrast, this shift in roles and responsibilities is not observed in aquaculture livelihoods. The production processes in aquaculture require physical strength, technical expertise, and market knowledge, making men more likely to take on key responsibilities than women.

Utilization of Climatic Information in Climate-Adaptive Agriculture

In the context of climate change and increasingly erratic natural disasters, people have gradually adjusted their habits and production time according to changes in climatic factors to minimize the damage caused by changes in climatic factors to crops and livestock. Due to the irregularity and changes in weather and climate, indigenous knowledge and the agricultural production experience of local people are no longer suitable. Instead, farmers mainly rely on forecasts and guidance information from commune-level authorities to time their sowing and planting. In addition, they also

Table 17.2 Labor distribution role in agricultural activities (unit: percentage (%))

Production stages	Crop farming			Aquaculture		
	Male	Female	Both	Male	Female	Both
Land preparation and production techniques	11.4	38.1	50.5	56.0	0	19
Seed and breed selection	4.8	59.0	36.2	20.0	6.7	48.3
Crop care, irrigation, and fertilization	2.9	33.3	63.8	46.7	20	8.3
Harvest	2.9	21.9	75.2	46.7	20	8.3

follow recommendations on farming methods and the use of fertilizers and pesticides from management agencies; thereby, partly limiting the damage caused by weather and pests.

However, the survey results show that there are differences in the way information is used between the two genders, with women mainly interested in weather to arrange their crop schedules and minimize damage caused by extreme weather events. On the contrary, men mainly use information to arrange their working time. The reason for the difference is that men participating in aquaculture activities are mainly hired workers, not have to decide on the time and method of cultivation, as well as the damage caused by natural disasters. The way weather information is used in production activities shows women's proactiveness in adapting to the unusual changes of climate factors in production activities, especially farming livelihoods (Fig. 17.5).

About 57% of respondents employ adaptive measures to minimize the impacts of climate change, which can be categorized into four groups as follows:

(i) Crop and breed variety adjustment by changing varieties, crops/livestock, and crop structure, which is the most chosen option, accounting for 77.9% of respondents, with a relatively similar distribution between men and women.
(ii) Water management by water tank repair and water-saving irrigation techniques was chosen by 58.2% of respondents. Men prefer this method 10% more frequently than women. This trend is particularly evident among workers engaged in aquaculture activities, where water quality during the farming process plays a crucial role in determining productivity and income.
(iii) Farming technique renovation, including five measures of crop conversion and management of fertilizers, pesticides, crop quantity, and organic matter. This technique was integrated into training sessions on crop cultivation, where participants were equipped with appropriate farming skills. Notably, over 57% of women choose this method, 13% higher than that of men. Interviews with local people highlight that they have increased the use of fertilizer and pesticides to sustain crop yields and mitigate the adverse effects of rising pest and

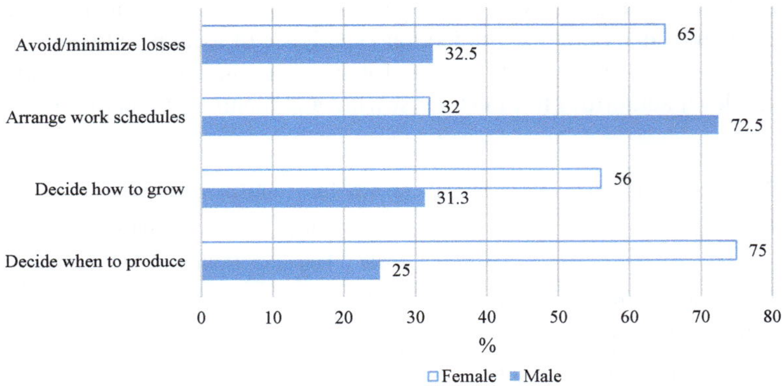

Fig. 17.5 Gender-sensitive utilization of climatic information

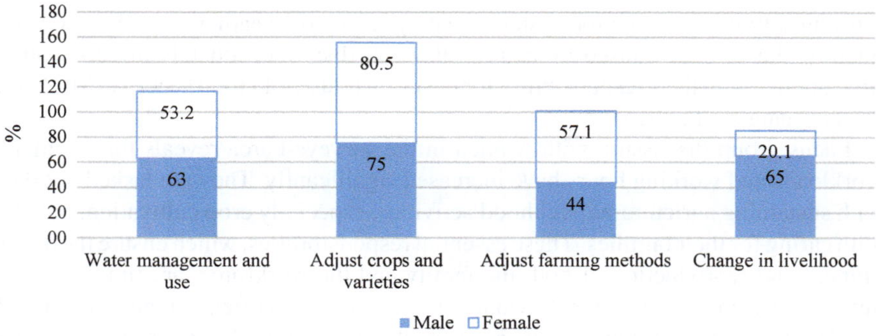

Fig. 17.6 Measures to adapt to climate change in production activities

disease occurrences induced by changing weather, corresponding to the selection rate from 49 to 50.5%. Meanwhile, environmentally friendly measures such as switching to organic fertilizers, adjusting the number of crops, or renovating cultivating methods receive less attention (less than 10% of selection).

A notable gender disparity is observed in the selection of livelihood adaptation strategies, with a predominant shift toward hired labor, accounting for nearly 20% of such transitions, where men surpass women by 45% (65.3% for men compared to 20.1% for women). This proves that men exhibit greater flexibility and adaptability through decisive, foundational choices aimed at ensuring the stability of household income generation (Fig. 17.6).

17.3.3 Challenges Faced by Women in Responding to Climate Change

Climate change remains a complex and unpredictable phenomenon, impacting all aspects of life and production by altering the fundamental components of the natural environment and diminishing the quality of development resources. It has emerged as one of the primary obstacles to the territorial development process. According to statistics, in the past 10 years in Vietnam, natural disasters have taken the lives of nearly 1000 people. The area of flooded deltas and coastal areas directly affects the lives and livelihoods of 20% of the population, and losses can be up to 10% of GDP each year. In 2020 alone, natural disasters caused death and left missing to 357 people missing; estimated economic damage was more than 37.4 trillion VND (NEU, 2023; VDMA, 2021).

Women are more vulnerable than men to the impacts of climate change, a disparity compounded by the diversity among different groups of women. Limited adaptive capacity has hindered women's access to climate change response interventions. In the context of labor migration, women often assume the primary role as laborers in farming activities, production systems heavily reliant on weather conditions. Consequently, the damage caused by natural disasters disproportionately affects women first, intensifying their responsibility for ensuring household food security (MONRE, 2022; Thang et al., 2015).

Focus group discussion with women in the surveyed area reveals that women's workloads and working hours have increased significantly. They are tasked not only with sustaining agricultural livelihood activities, especially crop cultivation, but also with caring for their families. These essential responsibilities, which ensure the development and reproduction of both the family and the workforce, are time-intensive yet typically unpaid. As a result, women's working hours often extend beyond 12 h per day, frequently spanning from 5 a.m. to 9 p.m. Such prolonged schedules leave women with less time for rest and self-care, delay their access to early warning information, and heighten their vulnerability to the impacts of climate change. Moreover, these extended working hours limit women's time and opportunities to engage in decision-making processes at all levels, including those related to climate change adaptation.

Additionally, the motivation to transition to production models adapted to climate change remains limited, primarily due to the low income generated from agricultural activities. Survey results reveal a significant disparity in income from production activities across different groups. There are 100% of aquaculture households have reported accumulated income from these activities. The remaining crop-based households have accumulated only 2% of their income from production activities, while the remainder is sufficient merely for subsistence or partially meets their need. Specifically, nearly 50% of respondents reported that cropping activities provide income sufficient for only a portion of their livelihood, requiring support from additional sources. Over 30% indicated that cropping activities yield no accumulated income. For households engaged in agricultural production, income is predominantly derived from migrant workers rather than farming. Crop-based livelihoods are primarily maintained to ensure "self-sufficiency" and to preserve production land (Fig. 17.7).

Skills required to transform agricultural production methods for climate change adaptation remain limited. Workers in the agricultural sector largely adhere to

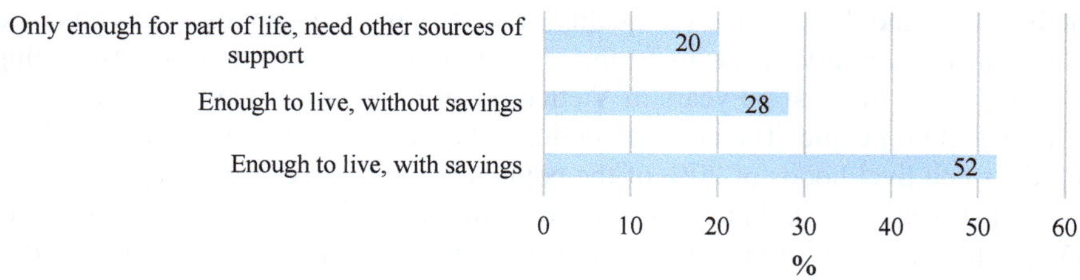

Fig. 17.7 Incomes from agricultural activities

traditional production practices, characterized by low qualifications and a persistent small-scale production mindset. The adoption of mechanization and scientific-technological advancements in production progresses slowly, with reliance predominantly on independent efforts, personal experience, and a largely unchanged perspective on large-scale or collective production. Vocational training outcomes remain modest, compounded by the fact that the majority of the workforce engaged in farming activities consists of elderly individuals, with minimal participation from younger workers. This demographic and skill gap creates significant challenges in transitioning to production methods that demand technical expertise, strict adherence to production processes, and soft skills necessary for rapid, market-responsive operations. Consequently, human resources struggle to adopt and implement scientific and technological advances in production, making it difficult to meet the escalating demands of transforming agricultural production models and methods. Simultaneously, the low quality of human resources and diminished labor productivity constrain income from agricultural and rural production activities, thereby hindering progress in the development of new rural areas.

According to the survey results, local production methods have begun to align with the guidance provided by local authorities. Approximately 48% of respondents reported cultivating crops following instructions from cooperatives or commune professional staff, while 36% combined these instructions with their own production experience. However, adherence is primarily limited to the timing of sowing. For other stages of the production process, individuals predominantly rely on personal experience or sensory observations made during visits to fields and production sites, rather than consistently following the guidance of professional staff (Fig. 17.8).

The agricultural labor force is increasingly characterized by trends of "feminization" and "aging." This phenomenon is widespread, with women frequently serving as household heads. Group discussions with residents reveal that young people in these households typically seek employment in industrial zones outside the province or participate in labor export programs, returning home only during holidays, New Year celebrations, or significant family events. Consequently, the household is often left with elderly individuals and school-aged children. As a result, middle-aged women and elderly individuals, many beyond traditional working age, constitute the primary workforce in farming. Unable to secure income from alternative livelihoods, these individuals tend to be conservative in adopting recommendations from local

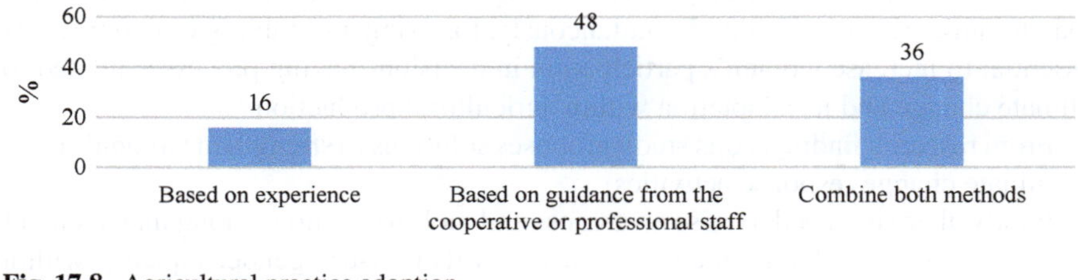

Fig. 17.8 Agricultural practice adoption

authorities, relying instead on experiential knowledge and peer learning. They undertake labor-intensive tasks such as tilling the land, sowing rice, fertilizing, pumping water, and spraying pesticides. The primary objectives of these farming activities are to "retain the land" and ensure self-sufficient food production. During the harvest season, labor shortages compel households to either exchange agricultural labor or hire workers at a wage of approximately 200.000 VND per day.

17.4 Conclusion and Moving Forward

Climate change is growing increasingly complex, impacting all actors and sectors, with agricultural production emerging as the most vulnerable sector. This vulnerability stems from its dependence on natural capital, interrelated factors such as land, water resources, and organisms, which fluctuate in response to climate and weather patterns shifts.

Research findings indicate that the role of women has been institutionalized in the legal policy system on climate change. Women have the right to decide on actions to adapt to climate change, including agricultural production activities. In practice, the labor force engaged in agricultural production activities is predominantly female, particularly in crop cultivation. Women not only bear the responsibility of sustaining agricultural livelihoods, participating in most stages of the production process, but also assume primary caregiving duties for their families, a burden intensified by labor migration. This dual role heightens their susceptibility to the adverse effects of climate change. Although women have a good awareness of the need for proactive solutions to adapt to climate change in their lives and maintain their livelihoods, they do not have much time to take care of themselves and improve their skills and knowledge about adapting to climate change in their production activities. Furthermore, farming livelihood activities in the study area are mainly self-sufficient and produced to "keep the land" due to limited post-harvest profits and the "aging" trend in the farming workforce, so they do not create motivation to transform farming models toward adapting to climate change, only for changing crop varieties and complying with sowing time, other production stages are mainly based on experience.

Consequently, there is a pressing need for integrated solutions to enhance women's capacity to access opportunities for transforming production methods, enabling effective adaptation to climate change, diversifying livelihoods, and reducing their workload to mitigate vulnerability. Simultaneously, fostering an enabling environment is essential to increase women's participation in decision-making processes related to climate change and its adaptation within agricultural production.

From research findings, this study proposes solutions to strengthen women's roles in climate change response activities.

Firstly, there is a need to raise awareness and mobilize action among management agencies to clearly define goals and measures that ensure gender equality within policy documents, addressing women's vulnerability, limited access to information, and lack of decision-making power in responding to climate change. The Vietnam

Women's Union should develop a comprehensive program to enhance the capacity of union officials, from central to grassroots levels, by improving their knowledge and skills in coordinating the implementation of climate change response policies.

To enhance women's awareness and access to opportunities in climate change adaptation activities, soft skills training should be provided for rural women, involving the Vietnam Women's Union and influential community. Emphasis should be placed on time management skills, enabling women to balance time between sustaining livelihoods and family care, while also increasing opportunities for self-care and effective adaptation to climate change. Comprehensive communication strategies are essential, leveraging multiple channels and delivering information in formats tailored to women's habits and educational levels. This approach acknowledges the distinct ways in which men and women in agriculture respond to climate change, as well as their differing levels of access to and utilization of climate information. Such efforts will bolster women's confidence in engaging in decision-making processes, thereby amplifying their roles and voices in programs and activities related to climate change adaptation.

Develop and effectively implement a project aimed at supporting women in entrepreneurship, integrate with mechanisms and policies on climate change adaptation and sustainable development to improve women's access to opportunities, from access to information to material, financial, human, and social resources, contributing to proactive and equal adaptation to climate change. At the same time, continue to coordinate with relevant units to research the gender impacts of climate change on sectors and regions, mainstream gender into climate-related policies; increase women's access to sustainable livelihoods and alternative livelihoods to be able to withstand natural disasters; at the same time promote women's empowerment in decision-making on climate change adaptation, mitigation, and disaster risk management at all levels and in all fields effectively.

Policies should prioritize support for the cooperative economy, focusing investments on core or "central" households. These households serve as key drivers, both attracting participation and generating momentum for establishing and strengthening market value chain linkages and cooperation. This approach enhances the capacity to negotiate and sell agricultural products through economic contracts, mitigating the prevalent issue of "good harvests, low prices" in agricultural markets.

Efforts should also encourage organizations and individuals, particularly private sector enterprises, to engage in research and technology transfer. Emphasis should be placed on developing tangible, hands-on models that guide farmers in acquiring knowledge and skills. These initiatives should be paired with incentive programs that inspire farmers to innovate and adopt farming methods suited to local conditions and the demands of climate change adaptation.

A proactive support strategy targeting women is essential to bolster agricultural production capacity in the face of climate change. This involves creating conditions for women to access critical skills and resources, thereby improving the quality of income derived from agricultural activities. Coordinated solutions are needed to elevate the role of the agricultural economy within the broader economic transformation toward proactive climate adaptation. In this context, women can emerge as a vital source of skilled labor and a catalyst for economic restructuring.

Given the trends of "feminization" and "aging" in agriculture, tailored policy mechanisms are required to consolidate land resources, encourage labor-rich households to expand cultivated areas, and reduce the extent of abandoned or uncultivated land. Mechanization should be promoted through policies that facilitate linkages or provide subsidies for households to invest in machinery, alongside efforts to enhance the quality of agricultural services. These measures aim to reduce labor intensity and boost income.

Innovative solutions must ensure that public services and resources are specifically designed to address the challenges of "feminization" and "aging" in agricultural production, particularly in farming. This includes addressing health, education, and family responsibilities. Simultaneously, equipping women with the knowledge and skills to take a proactive role in transforming production methods is critical. Training programs should prioritize quality and align with job creation for women, integrating with women's entrepreneurship initiatives to meet labor market demands and local needs. Policies should synchronize vocational training and capacity-building efforts with sustainable job creation, enhancing livelihoods and income through entrepreneurship support or the practical application of acquired knowledge in production.

References

FAO, IFAD, UN Women, & WFP. (2022). *Rural women's economic empowerment through climate-resilient agriculture: Experiences and ways forward from the joint programme on rural women's economic empowerment*. New York.

Giller, K. E., et al. (2021). The future of farming: Who will produce our food? *Food Security, 13*, 1073–1099. https://doi.org/10.1007/s12571-021-01184-6

Hang, B. T. M., & Huyen, N. T. T. (2025). Overview of the vulnerability assessment framework for rural women under climate change (In Vietnamese: Tổng quan khung phân tích tình trạng dễ bị tổn thương với biến đổi khí hậu của phụ nữ nông thôn). *Vietnam Journal of Agricultural Science, 23*(2), 267–277.

Intergovernmental Panel on Climate Change (IPCC). (2022). Climate change 2022: Impacts, adaptation, and vulnerability. In H.-O. Pörtner, D. C. Roberts, M. Tignor, E. S. Poloczanska, K. Mintenbeck, A. Alegría, M. Craig, S. Langsdorf, S. Löschke, V. Möller, A. Okem, & B. Rama (Eds.), *Contribution of Working Group II to the sixth assessment report of the Intergovernmental Panel on Climate Change*. Cambridge University Press (in press).

Milyausha, S. F., et al. (2020). Human resources in the context of the digitalization of agriculture. *BIO Web of Conferences, 27*, 00020. https://doi.org/10.1051/bioconf/20202700020

Ministry of Natural Resources and Environment (MONRE). (2022). *National adaptation plan for climate change 2021–2030, vision to 2050* (In Vietnamese: *Báo cáo Kế hoạch Quốc gia thích ứng với Biến đổi khí hậu giai đoạn 2021–2030, tầm nhìn đến năm 2050*). Hanoi.

National Economics University (NEU). (2023). *Proceedings of the International Conference on Climate Change Economics and Policy Implications for Vietnam* (In Vietnamese: *Kỷ yếu hội thảo khoa học quốc tế "Kinh tế học biến đổi khí hậu và gợi ý chính sách đối với Việt Nam")*. Vietnam National Economics University Press.

Ngoc, P. T. B., & Son, N. H. (2018). Agricultural livelihoods adapting to climate change: Experiences of Can Loc district, Ha Tinh province (In Vietnamese: Sinh kế nông nghiệp thích ứng với biến đổi khí hậu: Kinh nghiệm của huyện Can Lộc, tỉnh Hà Tĩnh). *Journal of Climate Change Science, 8*, 63–72.

Phuong, L. T. H., et al. (2018). Understanding smallholder farmers' capacity to respond to climate change in a coastal community in Central Vietnam. *Climate Development, 10*(8), 701–716.

Thang, N. T., et al. (2015). The impact of climate change on gender roles in agricultural production: A case study in Giao Lac, Giao Thuy, Nam Dinh (In Vietnamese: Ảnh hưởng của biến đổi khí hậu đến vai trò giới trong sản xuất nông nghiệp: Trường hợp nghiên cứu tại Giao Lạc, Giao Thủy, Nam Định). *Journal of Science and Development, 13*(3), 464–473.

The, B. D., & Hai, P. M. (2019). Economic evaluation of farming models towards climate change adaptation in the sandy area of Hai Lang district (In Vietnamese: Đánh giá kinh tế mô hình canh tác theo hướng thích ứng với biến đổi khí hậu ở vùng cát huyện Hải Lăng). *Hue University Journal of Science: Development Economics, 128*(5A), 5–15. https://doi.org/10.26459/hueuni-jed.v128i5A.5062

Tran, T. T. (2023). Green growth in agriculture in Vietnam: Awareness and strategy for development (In Vietnamese: Tăng trưởng xanh trong nông nghiệp tại Việt Nam: Nhận thức và chiến lược phát triển). *Review of Vietnam Academy of Social Sciences, 5*(217), 14–29. https://doi.org/10.56794/VSSR.5(217).14-29

United Nations Development Programme (UNDP). (2023). *Overview report on gender mainstreaming in the national adaptation plan for climate change*. Vietnam.

Van, H. T. H., & Mai, L. H. (2021). Adaptation policies to climate change in Vietnam's agriculture (In Vietnamese: Chính sách thích ứng với biến đổi khí hậu của nông nghiệp Việt Nam). *Vietnam Journal of Social Sciences, 8*.

Vietnam Disaster Management Authority (VDMA). (2021). *Summary report on natural disasters and disaster recovery in 2020* (In Vietnamese: *Báo cáo tổng hợp tình hình thiên tai và công tác khắc phục hậu quả năm 2020*). Hanoi.

Wang, S. (2022). The positive effect of green agriculture development on environmental optimization: Measurement and impact mechanism. *Frontiers in Environmental Science, 10*, 1035867. https://doi.org/10.3389/fenvs.2022.1035867

Part IV
Concluding Remarks

Chapter 18
Gender Inclusion in Climate Crisis: Policies and Institutional Approaches in Asia

Md. Esraz-Ul-Zannat, Swarna Bintay Kadir, Kazi Farzana Shumi, and Gulsan Ara Parvin

Abstract According to the World Meteorological Organization (WMO), climate change and extreme weather will hit Asia in an extremely hard way. The warming trend of Asia has nearly doubled since the 1961–1990 period and this warming is faster than the global average. Therefore, along with international organizations, Asian nations and Governments are initiating different policies, priorities, and action planning. Further, to combat climate crisis multi-sectoral approaches are adopted and implemented by different institutions and communities as well. Nonetheless, these policies and approaches do not always have an adequate and appropriate focus on the poor and marginalized segments of a community. It is claimed that climate change impacts men and women differently, largely due to their gender-differentiated relative powers, roles, and responsibilities at the household and community levels. Due to lack of effective and efficient gender inclusion in climate crisis management policies and initiatives, women still have less economic, political, and legal clout and are hence less able to cope with and are more exposed to the adverse effects of the changing climate. Therefore, it is pivotal to understand how and to what extent gender associated issues are dealt with and addressed in climate crisis policies, planning, and institutional approaches in different countries in Asia. This paper scrutinizes major national climate policies in Asian countries with gender-inclusive approaches, inclusion levels, and associated sectors. Success, challenges, and Gap

Md. Esraz-Ul-Zannat (✉)
Department of Urban and Regional Planning, Faculty of Civil Engineering, Khulna University of Engineering & Technology, Khulna, Bangladesh
e-mail: esraz@urp.kuet.ac.bd

S. B. Kadir
Institute of Disaster Management, Khulna University of Engineering & Technology, Khulna, Bangladesh
e-mail: swarna@idm.kuet.ac.bd

K. F. Shumi
Bangladesh University of Professional (BUP), Dhaka, Bangladesh

G. A. Parvin
College of Policy Science, Ritsumeikan University, Osaka, Japan

Md. A. Abedin et al. (eds.), *Climate Crisis in South and East Asia*, Sustainability Solutions, https://doi.org/10.1007/978-3-032-01012-4_18

in the inclusion of gender specific policies and further key recommendations to overcome the challenges are also highlighted in this paper.

Keywords Gender-inclusive policy · Climate crisis · Climate change · Disaster risk reduction · Asia

18.1 The Context: Gender, Climate Crisis, Vulnerability, and Financing

18.1.1 Asia and Its Climate Crisis

Asia, covering one-third of the Earth's surface, is home to 58% of the global population (UN-IOM, 2024). It is the largest and most climatically diverse continent, with geographical variations that range from the freezing polar regions in the north to the warm, humid tropics in the south, and mountain climates in between (IPCC, 2014). Asia holds the highest average elevation above sea level, and is the wettest, coldest, and most physically diverse of all continents (Dando, 1998; Oliver, 2008). The continent's climate is shaped by variations in solar energy, temperature, humidity, precipitation, atmospheric pressure, winds, and storms, leading to regional climatic differences. Despite this diversity, a unifying climatic feature across Asia is the monsoon effect, which influences the overall climate pattern (Dando, 1987).

Asia is highly vulnerable to the effects of climate change, facing extreme temperatures, flooding, droughts, cyclones, and rising sea levels (Roy, 2022). Despite recent economic growth, the region remains at significant risk from climate-related impacts (Sanderson & Islam, 2000). The World Meteorological Organization's 2023 report highlights a dramatic increase in warming trends across Asia, which have been almost double those of the global average from 1991 to 2023. Furthermore, rates of sea-level rise across Asia have exceeded the global mean from 1993 to 2023 (WMO, 2024). In case of most of the key climate change-related parameters, such as, rising temperature, glacier retreat, sea level rise Asia is having accelerating rates, which are imposing enormous challenges and threats in every sort of life including livelihood, society, economies, and ecosystems. The UNDP (2024) warns that climate change poses an "existential threat," exacerbating extreme weather events, deepening poverty, and posing risks to public health. As climate change affects both natural and managed ecosystems, the livelihoods and ecosystem services in Asia are projected to suffer substantial impacts (Wu et al., 2018).

Asia is especially prone to climate hazards, including floods, droughts, and cyclones, which account for over 90% of global climate-related disasters (Ancha, 2025). The frequency and intensity of these disasters have risen, causing significant loss of life and economic disruption, particularly among vulnerable communities (UN ESCAP, 2023b). The UN-ESCAP (2023a) reports that climate-induced disasters in Asia are now more severe and frequent than previously assessed, pushing millions into poverty, displacing communities, and threatening public health. Given these trends, implementing effective climate strategies is critical to supporting vulnerable populations and promoting climate justice in the region (Roy, 2022). However, studies indicate that the impacts of climate change are not gender-neutral, with women and marginalized groups disproportionately affected (Chidakwa et al., 2020; Tanou et al., 2022). Environmental institutions in the Global South, including those in South Asia, are increasingly acknowledging the importance of gender sensitivity in climate adaptation and mitigation efforts (Prakash et al., 2024). However, many policies still overlook gender biases, leading to the reinforcement of existing inequalities rather than addressing them. This chapter analyzes gender inclusion in climate crisis policies across five Asian regions: South Asia, Southeast Asia, East Asia, the Middle East, and Central Asia, excluding Russia and the Caucasus. This exclusion is due to significant differences in climate policy development and gender integration.

18.1.2 Gender Vulnerabilities in Changing Climate

Since climate change impacts are altered through the social, economic, and political characteristics of the community or society, the reality is that climate change poses a gender-specific set of challenges and risks that cause disproportionate hardships for women (Eastin, 2018). Studies prove that climate change associated impacts are not even for all and these impacts disproportionately affecting the lives and livelihoods of women and girls. It is also evident that climate change has more detrimental consequences for women than men, due to their unequal power and physical vulnerability (Anjum & Aziz, 2025; Sidun & Gibbons, 2024). Furthermore, as the marginalized sub-groups have minimal capacity to adapt to climate change variations, climate change and extreme weather events have laid bare major gender inequalities (Benjamin & Thomas, 2020; Memon, 2020). It is asserted that pre-existing societal and domestic roles of women increase their exposure to climate-related risks and that adversely impact their livelihood, financial stability, and physical and mental health (Tanou et al., 2022). In contrast, climate change exacerbates existing gender inequalities, leading to increased vulnerability for women and reinforcing gender disparities across various socioeconomic contexts (Eastin, 2018). Therefore, women in different socio-economic, political, and even institutional settings are having different extent and nature of vulnerabilities and disparities. In terms of both geographic exposure to environmental shocks and stressors and underlying systemic social inequalities Asia is particularly vulnerable to climate change impacts, which have intensified gender specific vulnerabilities and disparities in Asia, especially in the developing nations of the continent.

18.1.3 Gender Specific Climate Change Vulnerable Cases in Asia

In ASEAN region, despite substantial advances toward universal access to clean drinking water, women are more likely than men to die from consuming unsafe water sources across the region. In Indonesia women comprise 29% of deaths associated with unsafe water sources, compared to 25% of men; while the rates are 23% for women versus 18% for men in the Lao People's Democratic Republic (UN Women, 2024a).

In Pakistan, shelters and camps built for internally displaced people are considered as a place of high risks in gender-based violence (GBV) for women and girls. Generally, provisional shelter spaces are overly crowded and factors such as poor lighting and lack of adequate space for specific gender enhance GBV. Further due to transportation damage during climate induced disasters, GBV risks increase in Pakistan (Qayum et al., 2012; UN-IOM, 2024).

Shelters lacked privacy, heightened the risk of abuse, and women faced social stigma if displaced from their homes during a disaster. The women also experienced health risks because of inadequate hygiene and poor living arrangements such as a lack of adequate shelters, toilet facilities, and secure gender-segregated living arrangements if relocation was essential during a disaster (Fatema et al., 2023).

18.1.4 The Need for Gender-Sensitive Policies and Institutional Approaches

On the eve of International Women's Day 2025, United Nations Climate Change celebrates a new decade of action to advance gender equality in the face of climate change. Further United Nations Climate Change Conference that took place in Baku, Azerbaijan, at COP29, 2024, the Parties committed to extend the Enhanced Lima Work Program on Gender and Climate Change for another ten years. Considering the critical issues of gender-based vulnerabilities, challenges, and efforts at COP29, parties also pledged to develop a new gender action plan to be adopted at COP30, which will outline concrete measures for the effective implementation of gender-responsive climate action moving forward (UNFCCC, 2025). Apart from these, other global agendas, such as Sustainable Development Goals (SDGs), Paris Agreement on Climate Change, and the United Nations Framework Convention on Climate Change have recognized the critical role of women in combating climate change (Desai & Zhang, 2021). Furthermore, it was over a decade, realizing the importance of gender, health and climate change the World Health Organization (WHO) emphasized to offering mitigation strategies to address the issues (WHO, 2014).

Despite all these efforts of the International Organizations and Global Agendas in promoting a gender-lens in international frameworks and commitments to address climate change, transforming the proposals and targets into policy action has been inadequate. Besides, in the Asia and pacific region there is limited attention to gender-specific needs in the context of DRR, although various countries in the region have integrated gender considerations within their Nationally Determined Contributions (NDCs) (UN UNDRR, 2021; United Nations Environment Programme, 2016; Women, 2022a). Women's underrepresentation in governments and in DRR decision-making are enhancing this situation and that leads to deprioritizing and undervaluing gender-specific responses in climate governance (UN ESCAP, 2023b). Nevertheless, ADB (2020), warns that failure to adopt gender-responsive policies, strategies, and action plans may exacerbate pre-existing gender inequalities, which have led to disparities in resource access, services, information, training, and decision-making. Therefore, gender inclusion in climate change policies and action are foremost requirements. In fact, the implementation of gender-specific measures is crucial for increasing women's access to benefits, developing their resilience, and achieving sustainable development goals (ADB, 2020).

18.2 Climate Crisis Policies and Institutional Approaches

The degree of gender mainstreaming in climate change legislation varies across Asia's five sub-regions. National climate policies in these regions generally develop governing structures, define roles, distribute budgets, and outline frameworks for risk mitigation and adaptation. While many policies acknowledge women as a vulnerable group, few include explicit gender-based provisions. Based on a gender-transformative framework, these policies can be classified into five categories: gender-discriminative, gender-unaware, gender-sensitive, gender-responsive, and gender-transformative. The latter three incorporate gender either implicitly or explicitly. Gender-sensitive policies recognize differential needs; gender-responsive ones promote women's participation in decision-making; and gender-transformative policies aim to ensure equal leadership opportunities (UNICEF, 2024).

Asian countries frequently align national climate policies with global frameworks such as the Sendai Framework for Disaster Risk Reduction 2015–2030, the Gender Action Plan 2017, the Green Climate Fund Gender Policy, and the Sustainable Development Goals—particularly SDGs 5 and 13 (ADB, 2021; UN Women, 2022a). There also exist regional frameworks like ASEAN Gender Mainstreaming Strategic Framework 2022–2030 for Southeast Asia, SAARC Gender Policy Advocacy Framework 2010 for South Asia, UN ESCAP's Regional Gender Equality and Women's Empowerment Strategy for Southeast and South Asia. These frameworks aim to integrate gender equality across the respective regions and are parts of a broader push to operationalize gender commitments. However, national policy plays the most crucial roles as they have the scope of direct implementation and enforcement. Table 18.1 displays the list of national polices in Asia and their gender inclusion level. Many of these policies cover multiple sectors like disaster management, food and agriculture, environment or ecology, water, energy, infrastructure or livelihood, education, and health. Since the adoption of Sendai Framework of Action, Asian countries have made significant progress in mainstreaming gender into disaster risk reduction compared to the other sectors (ADB, 2021).

Gender inclusive policy development requires gender-disaggregated data collection, climate vulnerability assessment of all genders and stakeholders mapping. It is also essential to ensuring equal participation while policy making and incorporating gender action plans with targeted support such as gender budgeting, institutional triggering, and creating feedback loops. Middle East and central Asia suffer from severe data gaps on gender-climate intersections and patriarchal norms that limit women's participation while policy making. While South Asia and South-East Asia have ensured gender inclusive languages in climate policies, cultural barriers have limited the voices of women in the design stage. When it comes to putting these policies into practice, the main obstacles are a lack of funding at the grassroots level and resource allocation. Moreover, gaps are frequently observed in awareness building among policy makers, targeted actions, and monitoring gender responsive indicators (UN Women, 2022a, 2022b).

18.2.1 Climate Policies with Gender Integration

According to a study, the connection of climate change and gender policy mostly deals with gender differences rather than gender mainstreaming in Hindu Kush-Himalayan region (Deshar & Koirala, 2020). Despite prevailing gender and poverty challenges, countries under South Asian Association for Regional Cooperation (SAARC) have made strides toward gender-inclusive climate policy (Jafry, 2016). Nepal's Disaster Risk Reduction and Management Act 2017 emphasizes female representation on disaster councils and mandates the inclusion of the women's ministry. Moreover, Nepal's Climate Change Policy 2019 promotes gender equality as a guiding principle. Bangladesh updated its Climate Change and Gender Action Plan (ccGAP) 2024 addressing gender gaps in key sectors and promoting women's leadership, education, and access to climate-resilient resources (Bangladesh Climate Change Trust & UN Women Bangladesh, 2024). Other foundational policies like Bangladesh Climate Change Strategy and Action Plan 2009, National Adaptation Plan and Eighth Five Year Plan 2020–2025 address specific vulnerabilities and capacities of different genders and emphasize gender-responsive actions. Pakistan adopted its ccGAP in 2022 with similar gender focus as Bangladesh. India's National Action Plan on Climate Change 2008 includes missions with potential for women's empowerment, though often lacks dedicated financing for women-led initiatives (UNICEF, 2024). Sixteen states of the country have incorporated gender in their State Action Plans on Climate Change (SAPCCs), though often through existing development projects rather than transformational change. India's Environment Protection Act 1986 does not address gender explicitly, though its broad definition of the environment allows for gender-sensitive interpretation (ADB, 2021). Bhutan's Disaster Management Act 2013 mandates women's inclusion in disaster committees and prioritizes their needs during emergencies. Maldives, a frontrunner among small island states, launched a

Table 18.1 Major national climate policies in Asia with gender-inclusive approaches, inclusion levels, and associated sectors

Region	Country	Existing Policies and Guidelines	Gender Inclusion Level	Focusing Sectors								No. of Sectors
				Disaster	Agriculture	Water	Energy	Environment	Infrastructure	Education	Health	
Southeast Asia	Philippine	Philippine Disaster Risk Reduction and Management Act, 2010	Responsive	✓								1
		National Disaster Risk Reduction and Management Plan 2020-2030	Responsive	✓								1
		Philippine Climate Change Act of 2009	Sensitive	✓	✓	✓		✓	✓			5
		National Climate Change Action Plan 2011-2028	Sensitive	✓	✓	✓		✓	✓			5
	Indonesia	Disaster Management Act 2007	Sensitive	✓								1
	Cambodia	Cambodia Climate Change Strategic Plan 2014-2023	Sensitive		✓	✓		✓	✓	✓	✓	2
		Gender and Climate Change Action plan 2014-2018	Sensitive	✓	✓	✓	✓	✓		✓	✓	7
		Master Plan on Gender and Climate Change 2018	Responsive		✓	✓		✓	✓	✓	✓	6
		Guidelines for Mainstreaming Gender in Inclusive Disaster Management 2021	Transformative	✓								1
		Climate Change Action Plan for Energy Sector 2021-2023	Responsive				✓					1
		Gender mainstreaming and Strategic Framework in Agriculture 2016-2020	Transformative		✓							1
	Vietnam	Disaster Risk Management guidelines 2009	Responsive	✓								1
	Thailand	Climate Change Master Plan 2025–2050	Sensitive	✓	✓		✓	✓	✓	✓	✓	7
	Laos	Law on Catastrophe Management 2019	Sensitive	✓								1
South Asia	Bhutan	Bhutan Disaster Management Act 2013	Sensitive	✓								1
	Nepal	Nepal Disaster Risk Reduction and Management Act 2017	Responsive	✓								1
		Nepal Climate Change Policy 2019	Responsive		✓	✓	✓	✓			✓	5
	Bangladesh	Bangladesh Climate Change and Gender Action Plan (ccGAP) 2024	Responsive	✓	✓	✓		✓	✓	✓	✓	7
		Eighth Five Year Plan 2020–2025	Sensitive	✓	✓			✓	✓	✓	✓	6
		Bangladesh Climate Change Strategy and Action Plan 2009	Sensitive	✓	✓	✓			✓		✓	5
		National Adaptation Plan of Bangladesh 2023-2050	Sensitive	✓	✓	✓			✓	✓	✓	6
	India	National Action Plan on Climate Change 2008	Responsive		✓	✓	✓	✓	✓			5
		State Action Plans on Climate Change (SAPCCs)	Responsive	✓	✓	✓		✓		✓	✓	6
	Pakistan	Pakistan Climate Change and Gender Action Plan (ccGAP) 2022	Responsive	✓	✓	✓	✓	✓		✓	✓	7
East Asia	China	Modern Environmental Governance Guidelines 2020	Sensitive					✓				1
	Mongolia	Mongolian Law on Disaster Protection 2017	Sensitive	✓								1
		Environment Protection Law 2012	Sensitive					✓				1
		Environmental Impact Assessments Law 1998	Sensitive					✓				1
		Environmental Sector Gender Strategy 2014-2030	Responsive	✓		✓		✓				3
	Taiwan	Climate Change Response Act (2023)	Sensitive	✓			✓					2
		National Climate Change Adaptation Action Plan	Sensitive		✓	✓	✓	✓	✓		✓	6
Middle East	Jordan	National Climate Change Policy 2013–2020	Sensitive		✓	✓	✓	✓			✓	5
	Tunisia	National Strategy on Climate Change 2012	Sensitive		✓	✓			✓			3
	Palestine	National Adaptation Plan 2016	Sensitive		✓	✓	✓		✓		✓	5
Central Asia	Tajikistan	Tajikistan National Strategy for Adaptation to Climate Change 2030	Sensitive	✓	✓	✓		✓			✓	5

LEGEND Gender Discriminatory — Gender Unaware — Gender Sensitive — Gender Responsive — Gender Transformative

Source Authors (2024). Data compiled from various sources, including national climate change and gender action plans, reports from international organizations (e.g., UN Women, UNDP, ADB), and other relevant literature on gender integration in climate policies

Gender-Responsive Climate Financing Strategy 2023 to align funding with gender equity (UN ESCAP, 2024). In contrast, Sri Lanka and Afghanistan have limited gender integration due to institutional and political challenges.

Among Association of Southeast Asian Nations (ASEAN), Philippines, Cambodia, Thailand, Vietnam, and Indonesia have considerably included genders in climate policies (UN Women, 2022a), while the policies of Laos, Myanmar, Malaysia, Singapore, and Brunei don't yet have any significant inclusion. Philippines integrate gender in Disaster Risk Reduction and Management Act 2010 that mandates gender-sensitive needs assessments during emergencies and encourages inclusive decision-making by women. The National Disaster Risk Reduction and Management Plan 2020–2030 further promotes women's participation across planning and implementation stages. The Climate Change Act 2009 and National Climate Change Action Plan 2011–2028 of Philippines recognize the differentiated impact of climate change on women and appoints women's organization to the Climate Change Commission (ADB, 2021). In Indonesia, the Disaster Management Act 2007 recognizes women as a vulnerable group, although it lacks comprehensive integration mechanisms. Cambodia Climate Change Strategic Plan 2014–2023, Gender and Climate Change Action Plan 2014–2018, and Master Plan on Gender and Climate Change 2018 clearly necessitate identifying gender differentiated needs, but do not institutionalized gender directly. The Guidelines for Mainstreaming Gender in Inclusive Disaster Management 2021 prioritize gender in disaster recovery, while sectoral plans, such as the Climate Change Action Plan for the Energy Sector 2021–2023 and the Strategic Framework in Agriculture 2016–2020, aim to reduce gendered vulnerabilities and empower women (UN Women, 2022a). In Vietnam, Community-based Disaster Risk Management Guidelines 2009 promotes gender mainstreaming, though no specific guidelines exist for meaningful inclusion. Gender Strategy for Agriculture and Rural Development 2001–2010 highlights women's empowerment in agriculture through increased training and leadership roles. Thailand's Revised Climate Change Master Plan 2025–2050 seeks broader gender integration through stakeholder engagement. Laos's Law on Catastrophe Management 2019 provides support for women during emergencies but lacks a broader gender strategy (UN Women, 2022a).

East Asia has made policy advancements in climate sectors but faces challenges related to data and women's participation. The politicization of gender concept in China's climate change plans and strategies is still absent though economic and technological aspects of climate change are well-recognized by the country (Wang, 2023). China's Modern Environmental Governance Guidelines 2020 encourage the participation of social organizations like the Women's Federation in environmental governance. Mongolia sets an example of applying its Law on the Promotion of Gender Equality across environmental policies, such as Environmental Sector Gender Strategy 2014–2030, Mongolian Law on Disaster Protection 2017, Environment Protection Law 2012, and Environmental Impact Assessments Law 1998 (ADB, 2021). Taiwan's Climate Change Response Act 2023 aims for net-zero emissions by 2050, though it lacks explicit gender references. Other supporting frameworks of the country, such as the National Climate Change Adaptation Action Plan, encourage cross-sector collaboration allowing space for gender inclusion. In the Middle East, gender-climate integration is limited despite heightened climate vulnerabilities (Francesca, 2025). Jordan was the first to address gender in its National Climate Change Policy 2013–2020, though implementation remains lacking (UNDP, 2021). Tunisia's National Strategy on Climate Change 2012 acknowledges gender disparities but requires stronger mainstreaming efforts. Countries like United Arab Emirates, Lebanon, Morocco, and Egypt are enhancing women's roles in governance but have not fully embedded gender into climate policy. Palestine's National Adaptation Plan 2016 considers gender a cross-cutting issue, particularly in agriculture and water. Central Asia lacks detailed gender-responsive climate policies, though awareness is increasing (Standal et al., 2023). Women are acknowledged as vulnerable to climate change, necessitating their involvement in decision-making (Kovaleva et al., 2023). Tajikistan's National Strategy for Adaptation to Climate Change 2030 applies a gender lens, particularly in priority sectors. Efforts are ongoing in Kazakhstan, Turkmenistan, Uzbekistan, and Kyrgyzstan to incorporate gender into adaptation strategies.

The conclusion is that gender issues are central to inclusive climate change adaptation in many underdeveloped and developing countries, where climate risks often deepen existing gender inequalities (Pham & Saner, 2021). Some countries currently have extremely good regulations and enforcement structures in place to ensure nondiscrimination, while others lack relevant human rights, gender equality laws, or insufficient integration between these areas. Lack of gender analysis for understanding women's susceptibility as well as their contribution are the main reasons for these absences. The most typical practice for gender inclusion is to ensure services and facilities during catastrophes recognizing women as vulnerable. These strategies do not account for gender inequalities in long-term impacts or preexisting inequity. Effective and meaningful inclusion requires sex-disaggregated data and initiatives by local and national actors. Some countries horizontally integrate gender equality into climate laws through other legislation, i.e., the relationship between the Mongolian Law on Disaster Protection 2017, and the Law on the Promotion of Gender Equality 2011 (ADB, 2021). Similarly, Vietnam Law on Gender Equality 2006, Maldives Gender Equality Act 2016, and Philippine Magna Carta for Women 2008 create provisions for women mainstreaming in climate policy, though these policies do not entirely focus on climate change.

18.2.2 Role Playing Actors for Gender Inclusion in Climate Policies

Among international actors, UN Women, Green Climate Fund, United Nations Development Programme (UNDP), United Nations Framework Convention on Climate Change (UNFCCC), and International Union for Conservation of Nature

(IUCN) advocate for integrating gender equality into climate action, support gender-sensitive climate resilience projects and provide funds. In the long run, national governments play pivotal roles in translating international commitments into localized actionable strategies with ministries remain central to the policy formulation and implementation.

18.2.3 Governments and Policymakers in Integrating Gender Perspectives

In Nepal, the Ministry of Forests and Environment is the lead agency for climate policy, while the Ministry of Women, Children, and Senior Citizens ensures that gender perspectives are incorporated across policies. The National Planning Commission further integrates gender-climate concerns into national development agendas. Bangladesh's Ministry of Environment, Forest, and Climate Change leads climate policy development, notably producing the ccGAP 2024 in collaboration with IUCN and UN Women (Bangladesh Climate Change Trust & UN Women Bangladesh, 2024). Besides, Bangladesh Planning Commission incorporates gender considerations into national plans and budgeting frameworks. Pakistan's ccGAP 2022 was jointly developed by the Ministry of Climate Change and IUCN. Bhutan's National Commission for Women and Children mainstreams gender across sectors in collaboration with National Environment Commission works. In addition, the Ministry of Education is revising curricula to include gender-transformative climate literacy into education system and to empower young women (UN ESCAP, 2024). In India, the Ministry of Environment, Forest, and Climate Change integrates gender within State Action Plans on Climate Change, while the Ministry of Women and Child Development supports women's participation in climate programs. Moreover, the National Disaster Management Authority embeds gender in disaster preparedness and training programs.

The Climate Change Commission of Philippines ensures gender integration in climate governance, supported by the Philippine Commission on Women. In Cambodia, the Ministry of Women's Affairs and the Ministry of Environment jointly lead gender integration through national plans and the Gender and Climate Change Action Plan. Vietnam's Ministry of Natural Resources and Environment incorporates gender into the National Strategy on Climate Change 2021–2030, supported by with the Ministry of Labor, Vietnam Women's Union and UN Women. Indonesia's Ministry of National Development Planning and Ministry of Environment and Forestry lead climate policy development, while the Ministry of Women's Empowerment addresses gender-specific vulnerabilities (UNICEF, 2024). Thailand's Department of Climate Change and Environment is revising its Climate Change Master Plan supported by the Ministry of Social Development. The Office of Women's Affairs and Family Development and the Ministry of Social Development and Human Security support gender-sensitive climate policies, too (UN Women, 2022a). According to

Hung (2024), Thai policymakers must empower women meaningfully by boosting their resilience, resources, and decision-making roles to drive climate action and productivity.

All-China Women's Federation facilitates women's engagement in environmental governance along with the Ministry of Ecology and Environment and UN Women. Japan's Gender Equality Bureau collaborates with the Ministry of the Environment, with support from the Japan International Cooperation Agency for global programs. South Korea's Ministry of Gender Equality and Family and Ministry of Environment lead gender-climate policy integration. Mongolia's Ministry of Environment and Tourism spearheads policy development, working with the National Committee on Gender Equality (ADB, 2021). Taiwan's Environmental Protection Administration takes the leading role for formulating climate strategies, while the Department of Gender Equality promotes gender mainstreaming across sectors. In Central Asia, despite limited engagement in climate policy, governments are responding to growing international attention (Standal et al., 2023). In Tajikistan, the Committee for Environmental Protection and the Committee on Women and Family Affairs play leading roles. Uzbekistan's Ministry of Ecology and Women's Committee and Kazakhstan's Ministry of Ecology and National Commission on Women's Affairs are also active. Middle Eastern countries are slowly aligning gender with climate policy. Jordan's Ministry of Women, Family, and Children conducts gender audits of climate policies and promotes inclusion in rural and urban adaptation programs. Besides, Jordan's National Commission for Women ensures gender analysis in climate plans and supports gender-responsive budgeting (UNDP, 2021). Palestine's Ministry of Women's Affairs and Environmental Quality Authority collaborate with civil society for gender-sensitive climate strategies. Tunisia and Palestine work with UNDP to strengthen community-based gender-climate integration.

18.2.4 Contribution of Civil Society and NGOs

Civil society organizations significantly influence gender-inclusive climate governance. In Nepal, the Local Initiatives for Biodiversity, Research, and Development and Asia Pacific Forum on Women, Law, and Development empower women in climate justice advocacy and policymaking. In Bangladesh, the International Centre for Climate Change and Development and the Red Crescent Society support community-based, gender-responsive adaptation initiatives. India's Swayam Shikshan Prayog promotes women-led climate-resilient farming (UNICEF, 2024), while Bhutan's Gender Expert Group fosters sector-wide integration of gender in climate strategies. Besides, for Integrated Mountain Development (ICIMOD) works across South Asia to promote gender-transformative change (UN Women, 2021a).

In the Philippines, groups like the Asian Peoples' Movement on Debt and Development and Gabriela Women's Party advance gender-responsive climate advocacy. Similarly, Cambodia Development Resource Institute and Plan International integrate gender into local adaptation projects. In Vietnam, institutions like the Institute of Strategy and Policy on Natural Resources and Environment and CARE International develop inclusive policies and grassroots resilience programs. Indonesia's Forest, Nature, and Environment Aceh Foundation empowers women-led conservation teams, while the Thailand Environment Institute fosters regional dialogue on gender-climate policy (UN Women, 2022a).

Social organizations and civil societies in East Asia, i.e., Friends of Nature in China, Women's DRR Network and Kiko Network in Japan, Women Link and Korea Federation for Environmental Movements in Korea, lead awareness campaigns and advocate for equitable disaster response in the respective countries. Mongolia's Women's Fund and the Center for Human Rights and Development promote gender-responsive climate action. In Central Asia, organizations like Zanoni Pokiston in Tajikistan and EcoForum Kazakhstan support women in agriculture and disaster response (UNDP, 2016). In the Middle East, civil societies such as MIFTAH in Palestine, RAJ in Tunisia, and Arab Women Organization in Jordan engage women and youth in climate justice efforts and policy advocacy.

18.3 Gender Inclusion: Successes, Challenges, and Gaps

Current climate crisis policies and institutional approaches across 18 (40%) countries from five different sub-regions of Asia (Table 18.1) demonstrate varying degrees of integration in eight key sectors. While other countries are referenced in various contexts, these policies prominently emphasize critical themes such as gender inclusion level and focusing eight sectors (Table 18.1). A comparative analysis of these regions and countries reveals both shared challenges and emerging opportunities for enhancing gender inclusion in climate action.

18.3.1 Enhancing Women's Leadership in Climate Action: Lessons from Southeast, South and East Asia

Southeast Asia

The "Gender Inclusion Level" in climate policies plays a crucial role in determining the extent to which women are integrated into leadership and decision-making processes. Southeast Asia stands out for its strong commitment to promoting women's leadership, with countries such as Cambodia and Indonesia implementing transformative policies. All ASEAN member states have incorporated gender equality into their Nationally Determined Contributions (NDCs); however, significant gaps remain in their implementation (UN Women, 2022a). Southeast Asia's gender-transformative policies, particularly in Cambodia and Indonesia, demonstrate how institutional commitment can advance women's leadership in climate action. For example, Cambodia's Guidelines for Mainstreaming Gender in Inclusive Disaster

Management (2021) and the Gender Mainstreaming and Strategic Framework in Agriculture (2016–2020) exemplify gender-transformative approaches. These policies prioritize gender equality by ensuring women's leadership in disaster planning and response, addressing gendered vulnerabilities such as unequal access to early warning systems and post-disaster recovery resources (UN Women, 2021b). Additionally, they promote equal opportunities across critical sectors, including agriculture and disaster management. By positioning women in leadership roles within these sectors, the policies aim to reduce gender disparities while enhancing resilience in areas most affected by climate change.

Despite significant progress in East Asia, challenges persist in women's leadership and participation in climate action. For instance, Cambodia's Climate Change Strategic Plan lacks gender-specific monitoring indicators, which limits accountability and progress tracking (UN Women, 2021b). In sectors like energy and agriculture, discriminatory social norms continue to marginalize women's traditional knowledge and contributions (ASEAN, 2022). Furthermore, gender-responsive climate finance remains insufficient in countries like Indonesia, where few projects target women-led enterprises (ASEAN, 2022; UN Women, 2024c). To drive transformative change, several recommendations have been proposed: developing gender-disaggregated indicators for climate policies, as seen in Cambodia's call for enhanced monitoring and evaluation frameworks (UN Women, 2021b); directing climate finance toward women-led initiatives, such as Indonesia's EMPOWER program (UN Women, 2024c); and harmonizing gender-climate policies within ASEAN, utilizing platforms like the ASEAN Committee on Women to share best practices (ASEAN, 2022; UN Women, 2022a). These strategies can help bridge the gaps in women's leadership and participation in climate action.

South Asia

Countries in South Asia have made progress in gender-responsive policies, but these often focus on vulnerability reduction rather than empowering women in leadership roles. Bangladesh's Climate Change and Gender Action Plan (ccGAP), for instance, promotes women's participation in decision-making but does not prioritize leadership equality in sectors like energy or water management (UN Women, 2023, 2024b). While ccGAP emphasizes gender-responsive financing and capacity-building, its actions largely center on committee inclusion rather than advancing women's leadership (UN Women, 2024b). Similarly, India's National Action Plan on Climate Change (NAPCC) incorporates gender perspectives but lacks a strong focus on women's leadership. To overcome these gaps, South Asia can learn from Southeast Asia by integrating leadership quotas and gender-specific indicators, as seen in Cambodia's frameworks (UN Women, 2021b). Additionally, funding women-led enterprises in energy and agriculture, as demonstrated by Indonesia's EMPOWER Program, and strengthening regional models like ASEAN's gender action plans could help harmonize policies and address these leadership gaps (UN Women, 2022a).

East Asia

East Asian countries, such as China and Mongolia, have made progress with gender-sensitive climate policies, but they lag behind Southeast Asia in promoting women's leadership in climate action. While these countries recognize the gendered impacts of climate change, their policies focus more on integrating gender perspectives into environmental governance rather than addressing leadership gaps. For instance, in China and Hong Kong, women-led climate entrepreneurship is growing, but policies remain gender-sensitive instead of transformative. Additionally, the Pacific Women Climate Leaders Program aims to foster East-West collaboration through women's leadership, though national frameworks have yet to reflect this shift (Lee, 2025). These examples highlight the need for policies that not only acknowledge gender disparities but also actively empower women in leadership roles.

18.3.2 Sector-Specific Gender Equality: Southeast Asia Versus South Asia

Southeast Asia and South Asia have both made significant strides in adopting climate-inclusive policies across multiple sectors. Existing policies revealed that countries from both regions, including Cambodia and Thailand from Southeast Asia and Bangladesh and Pakistan from South Asia, have addressed up to seven out of eight sectors in their respective climate policies (Table 18.1). Bangladesh's Climate Change and Gender Action Plan (ccGAP, 2024) and Pakistan's ccGAP (2022) adopt a gender-responsive approach, whereas Cambodia's Gender and Climate Change Action Plan (2014–2018) and Thailand's Climate Change Master Plan (2025–2050) follow a gender-sensitive approach. Notably, Cambodia is the only country in this review with a gender-transformative approach, specifically focusing on sectors such as disaster management and agriculture. Southeast Asian countries, including Cambodia and the Philippines, demonstrate a more targeted approach to gender equality, particularly in agriculture, disaster management, and energy. For example, Cambodia's Master Plan on Gender and Climate Change integrates gender-responsive frameworks in these critical sectors, while the Philippines emphasizes gender mainstreaming in disaster management, recognizing women's vital role during crises. In contrast, South Asian countries like Bangladesh and Pakistan still rely primarily on gender-sensitive or gender-responsive policies, which lack a strong focus on empowering women in leadership roles within sectors like agriculture and disaster management. While East Asia, including China and Taiwan, includes gender considerations in broader environmental governance, their policies remain largely gender-sensitive and do not address sector-specific barriers or promote women's leadership in key areas like agriculture, energy, or water management. Countries from the Middle East and Central Asia remain at the gender-sensitive level, with limited progress in integrating gender equality into climate policies.

Several good practices in gender-specific climate adaptation are particularly evident in Southeast and South Asian countries. For instance, Swayam Shikshan Prayog (SSP), a grassroots initiative in India, has worked with over 65,000 women farmers in Maharashtra to implement climate-resilient farming practices, and is now scaling to other Indian states (Swayam Shikshan Prayog, 2025). This initiative promotes sustainable agriculture, climate-smart technologies, and empowers women farmers to lead community-driven climate resilience efforts. Similarly, in Nepal, Local Initiatives for Biodiversity, Research, and Development (LI-BIRD) employ gender-inclusive strategies for biodiversity conservation and agriculture, enabling women farmers to integrate climate-resilient practices into their livelihoods (weADAPT, 2025). The initiative focuses on developing climate-smart agriculture using women-friendly tools that reduce workloads and enhance food security, while also building women's leadership in climate adaptation through capacity-building and knowledge sharing.

18.3.3 Gender Inclusive Disaster Management: Southeast Asia Leads

Most countries across Southeast, South, and East Asia, along with a very few from the Middle East and Central Asia, have incorporated gender inclusion into their disaster management policies, with many achieving at least a gender-sensitive level of integration. Out of 18 countries in the region, 14 focus on disaster management, making it the sector with the highest gender inclusion. For example, the Philippines has adopted the Philippine Disaster Risk Reduction and Management Act (2010), which includes gender-sensitive needs assessments during emergencies and post-recovery phases, actively promoting women's participation in decision-making and integrating gender perspectives into climate adaptation strategies. The Philippine Climate Change Act of 2009 further allocates climate funds to support rural women, particularly through microcredit schemes addressing the gendered impacts of climate change. In Cambodia, the Gender and Climate Change Action Plan (2014–2018) and the Master Plan on Gender and Climate Change (2018) highlight the importance of recognizing gender-differentiated needs in climate change and emphasize women's leadership in climate action, particularly in agriculture and disaster management. The Cambodia Climate Change Strategic Plan (2014–2023) also integrates gender mainstreaming across sectors like energy and water, ensuring women's involvement in sustainable development practices. In contrast, South Asia has made strides in gender-responsive disaster management, but countries like Bangladesh and Nepal focus more on women's participation rather than leadership. Nepal's Disaster Risk Reduction and Management Act (2017) mandates that at least one-third of disaster management experts be women, ensuring women's representation in governance. In Bangladesh, the Bangladesh Red Crescent Society supports female-led disaster response teams, which are essential in increasing women's participation in climate resilience and disaster management. In East Asia, while countries like Mongolia and Taiwan have gender-sensitive policies recognizing the disproportionate impacts of climate change on women, their focus remains more on addressing vulnerabilities than empowering women in leadership roles. Policies in these countries, such as Mongolia's Environmental Sector Gender Strategy (2014–2030) and Taiwan's Climate Change Response Act (2023), fall short in ensuring women's active roles in disaster preparedness, limiting their potential for gender-transformative change (British Chamber of Commerce Taipei, 2023).

18.3.4 Data, Resources, and Norms: Institutional and Structural Challenges

Initiatives to empower women as agents of change in climate policy have been historically overlooked, despite their potential for significant impact. Despite notable progress in legal reforms and educational equality, deeply rooted social norms and discriminatory practices continue to hinder women's socio-economic contributions across Asia. The institutional and structural challenges to gender inclusion stem from a complex interaction of advancements and persistent barriers, such as inadequate resources, patriarchal norms, and lack of gender-disaggregated data. Specifically, women, who represent 70% of the world's poorest, face barriers in accessing climate finance, limiting their participation in mitigation efforts (ADB, 2024). South Asia, particularly Pakistan and India, faces substantial obstacles, including the absence of gender-disaggregated data and gender audits in climate funding. For example, Pakistan's Nationally Determined Contributions (NDCs) and initiatives like the Climate Change Gender Action Plan (ccGAP) lack provisions for gender-responsive budgeting (Institute of Regional Studies, 2023). In Indonesia, the lack of a shared understanding of gender equality among climate finance implementers undermines the effectiveness of gender-responsive policies (Tomi, 2021).

Entrenched patriarchal norms and gender inequalities restrict women's roles in climate decision-making, particularly in regions like South Asia and parts of East Asia. Social and cultural norms continue to marginalize women's contributions, especially in sectors such as energy, agriculture, and disaster management. In South Asia, women's political representation remains limited despite early suffrage, constraining their influence on climate policies. For instance, in Pakistan, women experience gendered job segregation and significant pay disparities, with a gender wage gap exceeding 30% (Castelo et al., 2024). Similarly, in ASEAN countries, patriarchal norms limit women's access to decision-making roles, resources, and climate finance, exacerbating their vulnerability during disasters (UN Women, 2022a). In East Asia, countries like China and Mongolia also face challenges in overcoming structural barriers within their gender-sensitive policies. For example, Mongolia's National Action Program on Climate Change (NAPCC), while gender-sensitive, lacks the necessary metrics to track women's access to resources or leadership positions, despite including gender equality in its implementation framework (ADB, 2022).

18.3.5 *Gender Inclusion in Asian Climate Policies: Regional Disparities*

While climate vulnerabilities and institutional responses vary across Asia, the degree of gender inclusion in climate policies shows distinct regional patterns. South Asian countries such as Bangladesh, Nepal, and India have integrated gender concerns into national climate plans, often emphasizing women's vulnerability. However, these approaches remain largely gender-responsive, focusing on participation rather than leadership. In contrast, Southeast Asian nations like Cambodia and the Philippines have adopted more gender-transformative frameworks, prioritizing women's leadership, allocating gender-specific resources, and embedding gender considerations across key sectors such as disaster management and agriculture. East Asian countries, including China and Mongolia, have developed gender-sensitive environmental policies, yet lack robust implementation mechanisms for equity. Central Asian countries like Tajikistan and Kyrgyzstan are still in early stages of integration, with limited strategies beyond identifying women as vulnerable. In the Middle East, where gender-inclusive policies remain nascent and fragmented, structural barriers—such as legal and cultural norms—continue to restrict women's roles in climate governance. Compared to other regions, gender inclusion in climate policy here remains significantly underdeveloped (Table 18.2).

Table 18.2 Comparative analysis of various five different sub-regions of Asia

Region	Policy focus	Level of gender integration	Notable countries
South Asia	Vulnerability and participation	Gender-responsive	Bangladesh, Nepal, India
Southeast Asia	Leadership and empowerment	Gender-transformative	Cambodia, Philippines
East Asia	Environmental governance	Gender-sensitive	China, Mongolia
Central Asia	Vulnerability identification	Emerging/minimal	Tajikistan, Kyrgyzstan
Middle East	Basic inclusion in planning	Underdeveloped	Jordan, Palestine

18.4 Concluding Reflections: Gender and Climate Change—Building a More Inclusive and Resilient Future

In conclusion, gender inclusion in climate policies is essential for addressing the compounded impacts of climate change and existing vulnerabilities faced by women and marginalized groups. Regional variation exists in the adoption of gender-inclusive policies, with some regions making progress while others face significant challenges. Despite advancements in certain areas, issues such as patriarchal norms, limited access to leadership roles and resources, and inadequate gender-disaggregated data continue to hinder effective policy implementation.

Southeast Asia stands out for its strong focus on women's leadership and sector-specific gender equality, particularly in countries like Cambodia and the Philippines, with South Asia following closely behind. However, East Asia faces persistent challenges, including structural barriers, a lack of gender-sensitive data, and limited resources. While South Asia has made notable progress in implementing gender-responsive policies, East Asia predominantly emphasizes gender-sensitive approaches, with less attention to transformative change or sector-specific leadership.

In contrast, the Middle East and Central Asia remain in the early stages of integrating gender perspectives into climate strategies. Despite these regional differences, several countries—such as Cambodia, Indonesia, the Philippines, Nepal, Bangladesh, India, and Pakistan—have demonstrated effective practices that contribute to gender integration in climate action. Strengthening gender-inclusive frameworks and reinforcing the need for stronger institutional commitment and multi-sectoral collaboration are crucial for breaking the cycle of gendered vulnerability exacerbated by climate crises, ultimately fostering long-term resilience.

As depicted in Fig. 18.1, gender vulnerabilities are interconnected, creating a cycle that amplifies the challenges women face during climate crises. The framework presented in the figure underscores how gender-inclusive policies address both gender-based inequalities and climate-induced vulnerabilities. These policies empower women by promoting their participation, leadership, and access to resources, thereby mitigating existing vulnerabilities and reducing exposure to climate-related risks. To ensure progress in both climate resilience and gender equality, it is essential that gender considerations be systematically integrated into climate policies at all levels.

The challenges to gender-inclusive climate policies vary significantly across regions, though some common issues persist. Existing literature provides several key recommendations to address these challenges:

1. **Gender-Responsive Climate Governance**: Governance frameworks must integrate gender considerations to ensure women's participation in decision-making and leadership roles within climate-related sectors.
2. **Targeted Financing and Performance-Based Budgeting**: Research shows that gender-responsive climate budgeting can mitigate the adverse effects of climate change on women by reallocating resources effectively. For example,

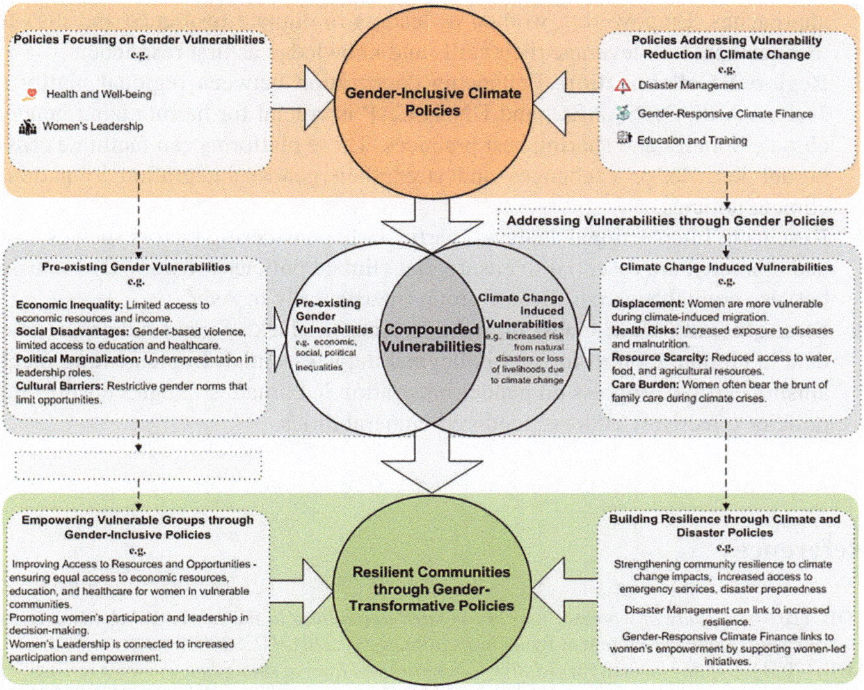

Fig. 18.1 A framework for gender-inclusive climate policies to build resilient communities

allocating quota-based funds for women-led climate enterprises is crucial, as seen in Australia's Gender Equality Fund. Additionally, Performance-Based Budgeting (PBB) can promote gender equality and poverty reduction, provided that government ministries and agencies recognize the importance of gender equality, acknowledge the role of women and marginalized groups, and leverage past experiences.

3. **Multi-sectoral Collaboration for Climate Resilience and Gender Justice**: A collaborative approach across sectors is essential for advancing both climate resilience and gender justice, creating a more inclusive response to climate challenges.
4. **Improved Data Collection and Assessment**: Developing gender-disaggregated data systems are necessary to track the effectiveness of gender-inclusive policies and ensure equitable climate action. Mandating gender impact assessments for all climate-related projects will help evaluate the differentiated impacts on men and women, ensuring that climate strategies do not exacerbate existing gender inequalities.
5. **Gender-Transformative Disaster Risk Reduction (DRR) Initiatives**: There is a need to move beyond gender-responsive to gender-transformative DRR

approaches. Empowering women as leaders in climate resilience and disaster management will leverage their skills and knowledge as first responders.
6. **Regional Collaboration**: Enhancing cooperation between regional platforms such as ASEAN, SAARC, and UN ESCAP is crucial for harmonizing gender-climate policies and sharing best practices. These platforms can facilitate cross-border knowledge exchanges and strengthen gender integration in national climate policies.
7. **Research**: Intersectional analyses, particularly considering factors such as caste and ethnicity, are essential to ensure that climate policies are not only inclusive but also equitable across diverse groups, particularly in Asia.
8. **Strengthening Data Collection**: Addressing the lack of gender-disaggregated data in climate monitoring and policymaking is essential. Implementing mechanisms to track progress on gender integration in climate strategies ensures that policies effectively address gendered vulnerabilities.

References

ADB. (2020). *A study of women's role in irrigated agriculture in the Lower Vaksh River Basin, Tajikistan.* Asian Development Bank. https://doi.org/10.22617/TCS200384-2

ADB. (2021). *Gender-inclusive legislative framework and laws to strengthen women's resilience to climate change and disaster.* Asian Development Bank. https://doi.org/10.22617/TCS210482-2

ADB. (2022). *Women's resilience in Mongolia how laws and policies promote gender equality in climate change and disaster risk management.* Asian Development Bank. file:///C:/Users/s5243987/Downloads/Temp/women-resilience-mongolia.pdf

ADB. (2024). *Sustainable finance benefits from a gender lens.* Asian Development Bank. https://doi.org/10.22617/ARM240017-2

Ancha, S. (2025). *Mainstreaming climate change concerns in development: Issues and challenges for Asia* (pp. 76–97). Institute for Global Environmental Strategies. https://www.iges.or.jp/en/publication_documents/pub/reportchapter/en/123/CP-05-016.pdf

Anjum, G., & Aziz, M. (2025). Climate change and gendered vulnerability: A systematic review of women's health. *Women's Health, 21*, 17455057251323644.

ASEAN. (2022). *The state of gender and climate change in ASEAN.* https://asean.org/wp-content/uploads/2022/08/ASEAN_Policy-Brief_Digital_FINAL-1.pdf

Bangladesh Climate Change Trust & UN Women Bangladesh. (2024). *Bangladesh climate change and gender action plan.* Ministry of Environment of Forest, Government of the People's Republic.

Benjamin, L., & Thomas, A. (2020). Gender and climate justice-implications for policy formation in the Caribbean region. *Loyola Law Review, 66*, 329.

British Chamber of Commerce Taipei. (2023). *Women in business climate action Taiwan.* https://www.bcctaipei.com/sites/default/files/2023-03/Women%20in%20Business%20Climate%20Action%20Taiwan%20(2023)%20Report_2.pdf

Castelo, S., Antunes, L., & Ashrafuzzaman, Md. (2024). The impact of the climate crisis on gender inequality. Looking to the frontlines in search of priorities for policy. *Frontiers in Sustainable Cities,6*, 1304535. https://doi.org/10.3389/frsc.2024.1304535

Chidakwa, P., Mabhena, C., Mucherera, B., Chikuni, J., & Mudavanhu, C. (2020). Women's vulnerability to climate change: Gender-skewed implications on agro-based livelihoods in Rural Zvishavane, Zimbabwe. *Indian Journal of Gender Studies, 27*(2), 259–281. https://doi.org/10.1177/0971521520910969

Dando, W. A. (1987). Asia, climate of. In *Climatology. Encyclopedia of Earth Science* (pp. 116–126). Kluwer Academic Publishers. https://doi.org/10.1007/0-387-30749-4_18

Dando, W. A. (1998). Asia: Climate. In *Encyclopedia of hydrology and lakes. Encyclopedia of Earth Science* (pp. 89–95). Springer Netherlands. https://doi.org/10.1007/1-4020-4497-6_23

Desai, Z., & Zhang, Y. (2021). Climate change and women's health: A scoping review. *Geohealth,5*(9), e2021GH000386.

Deshar, R., & Koirala, M. (2020). Climate change and gender policy. In V. Venkatramanan, S. Shah, & R. Prasad (Eds.), *Global climate change and environmental policy* (pp. 411–422). Springer Singapore. https://doi.org/10.1007/978-981-13-9570-3_14

Eastin, J. (2018). Climate change and gender equality in developing states. *World Development, 107*, 289–305.

Fatema, S. R., East, L., Islam, S., & Usher, K. (2023). Gender-based vulnerabilities for women during natural disasters in Bangladesh. *Frontiers in Communication, 8*, 1180406.

Francesca, E. (2025). Introduction: Gender and climate action in the Middle East and North Africa—Understanding the intersections. *Studi Magrebini, 22*(2), 143–150. https://doi.org/10.1163/2590034x-20240105

Hung, J. (2024). Policy implications for the nexus of climate change and gender-based issues in Thailand: Case study of waste management. *Environmental Reviews, 32*(4), 726–733.

Institute of Regional Studies. (2023). *Pakistan's climate change policies: A gender perspective (Spotlight No. 100).* https://irs.org.pk/Spotlight/SP10012023.pdf

IPCC. (2014). *Climate change 2014: Impacts, adaptation, and vulnerability: Working Group II contribution to the fifth assessment report of the Intergovernmental Panel on Climate Change.* In C. B. Field, V. R. Barros, & Intergovernmental Panel on Climate Change (Eds.). Cambridge University Press.

Jafry, T. (2016). Making the case for gender sensitive climate policy—Lessons from South Asia/IGP. *International Journal of Climate Change Strategies and Management, 8*(4), 559–577. https://doi.org/10.1108/IJCCSM-04-2015-0049

Kovaleva, M., Leal Filho, W., Borgemeister, C., & Komagaeva, J. (2023). Central Asia: Exploring insights on gender considerations in climate change. *Sustainability, 15*(16), 12667. https://doi.org/10.3390/su151612667

Lee, J. (2025, March 27). *Women-centered climate action coalition launches in Hong Kong*. Earth.Org. https://earth.org/new-women-centered-climate-action-coalition-launches-in-hong-kong/

Memon, F. S. (2020). Climate change and violence against women: Study of a flood-affected population in the rural area of Sindh, Pakistan. *Pakistan Journal of Women's Studies = Alam-e-Niswan = Alam-i Nisvan,27*(1), 65–85.

Oliver, J. E. (2008). *Encyclopedia of world climatology*. Springer Science & Business Media.

Pham, H., & Saner, M. (2021). A systematic literature review of inclusive climate change adaption. *Sustainability, 13*(19), 10617. https://doi.org/10.3390/su131910617

Prakash, A., Ley, D., & Thamari, M. (2024). How gender-sensitive are environmental institutions, climate adaptation, and mitigation actions? A narrative from the Global South. *Annual Review of Environment and Resources, 49*(1), 449–474. https://doi.org/10.1146/annurev-environ-121322-073202

Qayum, M., Mohmand, S., & Arooj, H. (2012). Frequency and physical factors associated with gender-based violence in the internally displaced people of Pakistan. *Journal of the College of Physicians and Surgeons Pakistan, 22*(1), 63–65.

Roy, C. (2022). *Climate finance in Asia: Assessing the state of climate finance in one of the world's most climate vulnerable regions.* Oxfam International. https://doi.org/10.21201/2022.9837

Sanderson, J., & Islam, S. M. N. (2000). Climate change in Asia: Issues and policy options. *Natural Resources Forum, 24*(1), 39–48. https://doi.org/10.1111/j.1477-8947.2000.tb00928.x

Sidun, N. M., & Gibbons, J. L. (2024). Women, girls, and climate change: Human rights, vulnerabilities, and opportunities. *International Journal of Psychology, 59*(2), 257–266.

Standal, K., Daloz, A. S., & Kim, E. (2023). A gendered approach to understanding climate change impacts in rural Kyrgyzstan. In *Climate change in Central Asia: Decarbonization, energy transition and climate policy* (pp. 123–134). Springer Nature Switzerland.

Swayam Shikshan Prayog. (2025). *Women-Led Climate Resilient Farming (WCRF)*. https://swayamshikshanprayog.org/what-we-do-sectors-climate-resilient-agriculture/

Tanou, M., Kishida, T., & Kamiya, Y. (2022). The effects of precipitation and temperature on birth weight: A cross-sectional study from the Republic of Benin. *Global Journal of Health Science, 14*(7), 19.

Tomi, S. (2021). *Gender-responsive climate finance needed to lift more women out of poverty*. UNDP. https://www.undp.org/indonesia/press-releases/gender-responsive-climate-finance-needed-lift-more-women-out-poverty

UN ESCAP. (2023a). *Seizing the moment: Targeting transformative disaster risk resilience*. UN ESCAP.

UN ESCAP. (2023b). *The race to net zero: Accelerating climate action in Asia and the Pacific*. United Nations. https://doi.org/10.18356/9789210025706

UN ESCAP. (2024). *Gender-responsive financing solutions for sustainable and resilient energy and food systems in Asia-Pacific*. https://hdl.handle.net/20.500.12870/6921

UN Women. (2021a). *Basics of mainstreaming gender into climate change and disaster risk reduction policies*. UN Women.

UN Women. (2021b). *State of gender equality and climate change in Cambodia*. https://www.uncclearn.org/wp-content/uploads/library/final-digital_cambodia_report.pdf

UN Women. (2022a). *State of gender equality and climate change in ASEAN*. https://asean.org/wp-content/uploads/2022/08/State-of-Gender-Equality-and-Climate-Change-in-ASEAN_FINAL-1.pdf

UN Women. (2022b). *State of gender equality and climate change in South Asia and Hindu Kush Himalaya report*. https://asiapacific.unwomen.org/sites/default/files/2023-05/state-of-ge-and-cc-in-sa-and-hkh-2d933.pdf

UN Women. (2023). *Review of implementation: Bangladesh's climate change gender action plan (ccGAP)*. https://asiapacific.unwomen.org/sites/default/files/2023-04/final_ccgap_rapid%20appraisal_bco_9-dec-clean-copy.pdf

UN Women. (2024a). *ASEAN gender outlook achieving the SDGs for all and leaving no woman or girl behind*. https://data.unwomen.org/sites/default/files/documents/Publications/2024/ASEAN-gender-outlook_2024.pdf

UN Women. (2024b). *Bangladesh climate change and gender action plan (ccGAP)*. https://asiapacific.unwomen.org/sites/default/files/2024-07/bd-ccgap-final-2024.pdf

UN Women. (2024c). *Empower: Women for climate-resilient societies in Indonesia*. UN Women. https://asiapacific.unwomen.org/sites/default/files/2024-05/empower-in-indonesia-brief_may-2024.pdf

UNDP. (2016). *Gender equality in national climate action: Planning for gender responsive nationally determined contributions*. https://www.undp.org/sites/g/files/zskgke326/files/publications/Gender_Equality_in_National_Climate_Action.pdf

UNDP. (2021). *Gender equality and climate change in Jordan—Exploratory gender analysis*. https://www.undp.org/sites/g/files/zskgke326/files/migration/jo/GenderEqualityonline.pdf

UNDP. (2024). *Human development report 2023/24. Breaking the gridlock: Reimagining cooperation in a polarized world*. https://www.ssrn.com/abstract=4934148

UNDRR. (2021). *Asia-Pacific action plan 2021–2024 for implementation of the Sendai framework for disaster risk reduction 2015–2030*. https://www.undrr.org/media/76306/download?startDownload=20250420

UNFCCC. (2025). *United Nations climate change. A new decade for advancing gender-responsive climate policies that benefit all.* https://unfccc.int/news/a-new-decade-for-advancing-gender-responsive-climate-policies-that-benefit-all

UNICEF. (2024). *Gender transformative climate action: Best practice case studies.* UNICEF. https://www.unicef.org/eap/media/17336/file/Case%20studies_Gender%20transformative%20Climate%20programming.pdf.pdf

UN-IOM. (2024). *How disasters have gendered impacts on the climate migration and what youth can do about it: The case of Pakistan.* https://environmentalmigration.iom.int/blogs/how-disasters-have-gendered-impacts-climate-migration-and-what-youth-can-do-about-it-case-pakistan

United Nations Environment Programme. (2016). *Global gender and environment outlook 2016.* UN. https://doi.org/10.18356/0b979453-en

Wang, T. (2023). The absence of gendered management of climate change in China: An analysis of adaptation policies. In M. K. De Santo & S. E. Domptail (Eds.), *Degrowth decolonization and development* (pp. 173–201). Springer International Publishing. https://doi.org/10.1007/978-3-031-25945-6_8

weADAPT. (2025). *LI-BIRD.* weADAPT. https://weadapt.org/organisation/li-bird/

WHO. (2014). *Gender, climate change and health.* World Health Organization.

WMO. (2024). *State of the climate in Asia 2023.* World Meteorological Organization.

Wu, N., Wang, C., Ausseil, A. G., Alhafedh, Y., Broadhurst, L., Lin, H. J., Axmacher, J., Okubo, S., Turney, C., & Onuma, A. (2018). Direct and indirect drivers of change in biodiversity and nature's contributions to people. In M. Karki, S. S. Sellamuttu, S. Okayasu, & W. Suzuki (Eds.), *The IPBES regional assessment report on biodiversity and ecosystem services for Asia and the Pacific.* Secretariat of the Intergovernmental Science-Policy Platform on Biodiversity and Ecosystem Services, Bonn.

MIX
Papier aus verantwortungsvollen Quellen
Paper from responsible sources
FSC® C105338

If you have any concerns about our products,
you can contact us on
ProductSafety@springernature.com

In case Publisher is established outside the EU,
the EU authorized representative is:
Springer Nature Customer Service Center GmbH
Europaplatz 3, 69115 Heidelberg, Germany

Printed by Libri Plureos GmbH
in Hamburg, Germany